Fiber Optic
Communications

Other McGraw-Hill Telecommunication Books of Interest

To order or receive additional information on these or any other McGraw-Hill titles, in the United States please call 1-800-722-4726. Or visit us at www.ee.mcgraw-hill.com In other countries, contact your local McGraw-Hill representative.

Fiber Optic Communications

Systems, Analysis, and Enhancements

Gerard Lachs

Department of Electrical Engineering
University of South Florida
Tampa, Florida

McGraw-Hill

New York   San Francisco   Washington, D.C.   Auckland   Bogotá
Caracas   Lisbon   London   Madrid   Mexico City   Milan
Montreal   New Delhi   San Juan   Singapore
Sydney   Tokyo   Toronto

Library of Congress Cataloging-in-Publication Data

Lachs, Gerard.
    Fiber-optic communications / Gerard Lachs.
        p.   cm.
    Includes index.
    ISBN 0-07-038279-4
    1. Optical communications. 2. Fiber optics. I. Title.
TK5103.59.L27 1998
621.382'75—dc21   97-440321
                                                        CIP

McGraw-Hill

A Division of The McGraw-Hill Companies

1 2 3 4 5 6 7 8 9 0 FGR/FGR 9 0 3 2 1 0 9 8

ISBN 0-07-038279-4

The sponsoring editor for this book was Stephen S. Chapman, the editing supervisor was Paul R. Sobel, and the production supervisor was Tina Cameron. It was set in Century Schoolbook by Priscilla Beer of McGraw-Hill's Professional Book Group composition unit.

Printed and bound by Quebecor Fairfield.

McGraw-Hill books are available at special quantity discounts to use as premiums and sales promotions, or for use in corporate training programs. For more information, please write to the Director of Special Sales, McGraw-Hill, 11 West 19th Street, New York, NY 10011. Or contact your local bookstore.

This book is printed on recycled, acid-free paper containing a minimum of 50% recycled, de-inked fiber.

Contents

Preface

Many textbooks on fiber optic communication systems (FOCS) have been published in the last two decades. Almost all of them emphasize the physical aspects of FOCS. This text has been written specifically for a reader already familiar with basic communication systems terminology such as modulation, time-division and frequency-division multiplexing, digital error rates and channel bandwidth. It is not unusual for readers in this category to have lost touch with many of the basic concepts of the theory of electromagnetic fields that they learned as undergraduates. There is a far greater emphasis on detection statistics for FOCS in this text and the role they play in system design.

Chapters 2 and 3 provide the basics for propagation of light in optical fibers. In Chap. 2, we focus on a simple non-physical model known as a slab waveguide. The electromagnetic field concepts and mathematics required to understand propagation through a slab waveguide are relatively simple and provide an excellent introduction to the graphical techniques employed in Chap. 3 to explain propagation cylindrical fibers.

In Chap. 4 we describe lasers and the functional aspects of optical amplifiers.

Chap. 5 describes optical detectors. The most complicated portion of optical detection is the description of the effects of noise on the detection process, a subject that is very mathematical in nature. This and related aspects of optical detection are covered in Chaps. 6, 7, and 8.

Chapter 6 focuses on the mathematical examination of the fundamental noise sources in FOCS. Some of the material in this chapter has not as yet appeared in any FOCS text.

In Chap. 7 we present some practical aspects of FOCS as a whole, including modulation and multiplexing, concepts, as well as system performance criteria such as error rate.

Chapter 8 is concerned with optical heterodyne (also known as coherent detection) systems. Here we examine such topics as system performance, the effects of phase noise and some very basic methods for dealing with this problem. The material in Chap. 8 draws heavily on Chaps. 6 and 7.

There has been an ongoing debate in the fiber-optics community about the question of coherent detection versus direct detection and much of Chap. 9 is concerned with this question. Some published papers are utilized to compare arguments on both sides. In Chap. 10 we examine this question from a fresh perspective by applying the Bohr Correspondence Principle to the design of high-speed and high sensitivity coherent detection to FOCS. This material has not as yet appeared in any textbook.

This text evolved from a set of notes for a Graduate level fiberoptic communications course taught at the University of South Florida. I would like to acknowledge the assistance received from the feedback from all of those students. In particular I would like to acknowledge the enormous help that I received from Carlos Montemayor, and John Obara. As can be seen from the references at the end of the chapters this text owes much to my graduate students Y. K. Choi, S. M. Zaidi, A. Singh and H. G. Kim.

On a personal note, I want to thank Lee Stefanakos for his encouragement and cooperation, and Dave Snider and Paul Flikkema for their valuable feedback. Thanks to the secretarial support of the E.E. Department, namely Gayla Montgomery and Becky Brenner. A note of appreciation to my wife, Sandy, and son, Greg, for their grammatical and editorial input.

Gerard Lachs

Acronyms

AM	Amplitude Modulation
AMI	Alternate Mark Inversion
APD	Avalanche PhotoDiode
AR	AntiReflection
ASE	Amplified Spontaneous Emmission
BCP	Bohr Correspondence Principle
BPF	BandPass Filter
BPSK	Binary PSK
BW	BandWidth
C^3	Cleaved Coupled Cavity
CATV	Cable TV
CD*	Coherent Detection
CDM	Code Division Muliiplexing
CDMA	Code Divisionltiple Access
CPFSK	Continuous Phase FSK
CWG	Cylindrical Wave Guide
DBR	Distributed Bragg Reflection
DD	Direct Detection
DFB	Distributed Fedback
DMD	Delay and Muliply Discriminator
DPSK	Differential PSK
DR	Depletion Region
DSB/SC	Double Sideband Supressed Carrier
DSP	Digital Signal Processing
DSPP	Doubly Stochastic Poisson Process
E&M	Electromagnetic Fields
EDF	Erbium-Doped Fiber
EDFA	Erbium-Doped fiber Amplifier
EMI	ElectroMagnetic Interference
FDM	Frequency Division Multiplexing
FM	Frequency Modulation
FMFB	FM Feedback
FOCS	Fiberoptic Communication Systems

---

*Occasionally means Compact Disk.

FSK	Frequency Shift Keying
FWHM	Full Width at Half Maximum
G&R	Green and Ramaswami
Gb/s	Gigabit per second
GRV	Gaussian Random Variables
HD	Heterodyne Detection
HJL	HeteroJunction Laser
IF	Intermediate Frequency
IMDD	Intensity Modulation Direct Detection
KLE	Karhunen-Loeve Equation
LAN	Local Area Network
LCP	Left Circularly Polarized
LED	Light Emitting Diodes
LHS	Left Hand Side
LIDAR	Optical Radar
LP	Linear Polarized
Mb/s	MegaBits per Second
NA	Numerical Aperture
NFU	Normalized Frequency Unit
OA	Optical Amplifier
OEIC	OptoElectronic Integrated Circuits
OHD	Optical Heterodyne Dection
OOK	On-Off Keyed
OP	Optical Preamplifier
PCE	Power Conversion Efficiency
PCM	Pulse Code Modulation
PI	Population Inversion
PLL	Phase Locked Loop
PM	Phase Modulation
PMT	PhotoMultier Tube
POM	Photon OPtics Model
ppb	photons per bit
PSD	Power Spectral Density
PSK	Pase Shift Keying
PTE	Periodic Table of Elements
QAM	Quadrature Amplitude Modulation
QED	Quantum ElectroDynamics
QPSK	Quadrature PSK
QWL	Quantum Well Laser
RCP	Right Circularly Polarized
RF	Radio Frequency
RHS	Right Hand Side
RIN	Relative Intensity Noise
S/N	Signal-to-Noise
SLD	Semiconductor Laser Diode
SNR	Signal-to-Noise Ratio
SOA	Semiconductor Optical Amplifier
SSB	Single SideBand
SWG	Slab Wave Guide

TDM	Time Division Multiplexing
TW	Time-Bandwidth product
TWOA	Travelling Wave Optical Amplifier
VCO	Voltage Controlled Oscillator
VSB	Vestigal SideBand
W&L	Wagner and Linke
WDM	Wave Division Multiiplexing
WSC	Wavelength Selective Coupler

Fiber Optic
Communications

1

Introduction

1.1 General Introduction to Text

This text has been written for readers already familiar with basic communication systems terminology such as modulation, time-division and frequency-division multiplexing, digital error rates, and channel bandwidth. It is not unusual for readers in this category to have lost touch with many of the basic concepts of the theory of electromagnetic fields that they learned as undergraduates. However, it turns out that fiber-optic communications is an area where the systems engineer must be acutely aware of developments in the physical devices associated with fiber-optic communication systems. Therefore, the principal goal of this text is to present the physics of optical fibers, sources, and detectors as simply as possible, while still maintaining the essence of those phenomena that affect the communication systems aspects of fiber-optic channels.

Two main thrusts existed in the commercial development of communication systems at the time this text was written. One such thrust involves subscriber services for "untethered" communications employing a system of satellites; these systems will have far greater capabilities than the "cellular" communication systems in vogue today. Such systems have the distinct advantage over fiber-optic systems of not requiring a hard-wire (fiber) physical connection between transmitter and receiver. For example, vehicular communications would not be feasible using optical fiber interconnects. Conversely, these free-space channels need to share both regional and local frequency band allocations with other users and are often severely affected by environmental conditions. This alone will considerably limit the data rates available to individual users. Because these systems involve completely different communication environments, they really aren't in competition with each other. At present, there is no technology that has the data rate potential for point-to-point communications as fiber-optic communication systems. We introduce optical fiber communications by describing why there is a growing need for data rates in excess of 1 gigabit per second (Gbit/s).

1.2 Why Fiber Optics?

1.2.1 Need for ultra-high-speed communications

The massive increase in the availability of inexpensive, high-speed personal computers has led to the deployment of large-scale distributed computing networks with the capability of serving a large number of users. Such systems permit the users to share software and databases to accomplish tasks that were once the sole province of mainframe timeshare computers; they are characterized by centrally located file servers which store the main programs and back up the PC systems, thus avoiding disk crash calamities. Direct communication between computers is now necessary in many working environments, resulting in greatly expanded requirements for high-speed digital communication channels. There are many applications that require extremely high-speed interconnections:

1. Rapid access to very large databases. This application is one of the primary reasons for the presence of high-speed PCs in many offices today. Processor speeds of 100 MHz are certainly not required for word-processing applications.
2. High-definition image transmission which includes
   a. Medical images such as X rays, magnetic resonant images (MRIs), and computerized axial tomography (CAT) scans for intercity medical teleconferencing
   b. Video cables transmitting high-definition TV
   c. Three-dimensional (3D) images for robotics and next generation surveillance and tracking systems
3. Computer-computer communications and internal communication within supercomputers. In particular, there has been interest in employing optical communications within massively parallel machines.
4. "Information Superhighway" communications.
5. TV cable with a massive number of channels.

At the present time fiber-optic communication systems (FOCS) is the strongest candidate for achieving the communication speeds required for the above-mentioned tasks. Data rates in excess of 10 Gbits/s have been achieved in the laboratory, and practical systems with data rates in excess of 1 Gbits/s are being deployed by major communication carriers throughout the world. In the remainder of the text we shall see why FOCS has the potential to achieve the aforementioned tasks and thereby fulfill foreseeable communication requirements well into the twenty-first century. Although the only potential competitor for FOCS at these data rates is superconducting cables, that technology is barely in its infancy.

1.2.2 History

The most important physical process that makes a laser possible is *stimulated emission*—a phenomenon first proposed by Einstein in 1917. This concept was first harnessed at microwave frequencies in a device known as a *maser* (microwave amplification by stimulated emission of radiation). Maiman achieved the first optical frequency version of the maser (originally called an *optical maser*). In 1960 the advent of what is now known as the laser generated a furious burst of activity in optical communications. This new device seemed to be just what was needed for transmitting almost unlimited data over long distances without cables of any sort. However, it was discovered that random fluctuations in the index of refraction caused by atmospheric turbulence created such severe propagation distortion that ultra-high-speed data transmission through the Earth's atmosphere was almost impossible. For example, a laser beam traveling though air at surface level loses its coherence over a few hundred meters; this negated many of the advantages of optical communication systems. Therefore, the concept of direct communication at Gbit/s data rates over "line-of-sight paths" between cities was temporarily shelved. However, there are still many uses for free-space optical communications such as satellite-to-satellite communications as well as *lidar* (laser infrared radar; optical radar).

While the laser was responsible for the initial burst of activity, it was the totally separate development of suitable, low-loss optical fibers in the 1970s and 1980s that is responsible for the great interest in optical communications that exists today. The optical fibers in existence in the 1960s were far too lossy ($\sim$100 dB/km and greater) for long-range data communications since they would require an enormous number of repeaters to attain a reasonable signal-to-noise ratio. Although bundles of the original fibers had some applications in observing events in difficult-to-reach locations (say, inside a nuclear reactor), they were not suitable for long-range communications.

It was the development of low-loss fibers with loss coefficients of less than 1 dB/km (at appropriate wavelengths) that has enabled the deployment of high-speed FOCS. These low-loss optical fibers are so important that the characteristics of these fibers dominate the design process of the overall system. Such basic design parameters as optical carrier frequency, modulation format, and repeater spacing are determined almost completely by the physical properties of these low-loss fibers. In general, the design of FOCS is driven primarily by advances in the manufacture and improvements in performance of optical fibers, sources, and detectors. As an example of the dominance of device development in the FOCS design process, we note the fairly recent development of practical optical amplifiers. These devices have completely revised both the near-term and future system plans of the major communication carriers (such as the telephone companies) for the deploying FOCS. Interestingly, the most popular optical amplifiers in use today are designed to operate at wavelengths near the 1.55 μm absorption loss minimum of modern fibers, and there is great interest in designing practical optical amplifiers that operate near the 1.3-μm loss minimum.

1.2.3 Channel losses

One of the most important characteristics of any digital communication system is its reliability as expressed by the error rate (probability of error). In virtually all digital communication systems the probability of error is primarily a function of the energy per bit and noise level at the receiver. Both of these quantities will be described at length in later chapters of the text, but for this introduction we concentrate on the status of the energy level of the received signal. In particular, sufficient energy per pulse at the receiving terminus is necessary to achieve an acceptable error rate. In most radio-frequency systems this parameter is often determined by the receiver signal-to-noise ratio. However, in the case of FOCS, the key parameter employed for determining system sensitivity is the average number of photons per bit as measured at the input to the receiver. *Probably the single most important reason to use fiber optics for communication channels is their ability to convey optical energy a long distance with very small loss coefficients.*

Before we consider optical fibers, however, let us first consider the loss characteristics of a copper cable. Loss coefficient for a typical coax cable is shown in Fig. 1.1. It is clear from the data plotted in this figure that losses in

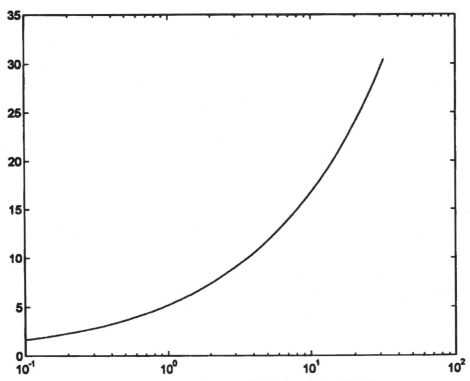

Figure 1.1 Some typical loss coefficients for coax cable. (*Adapted from Ref. 4.*)

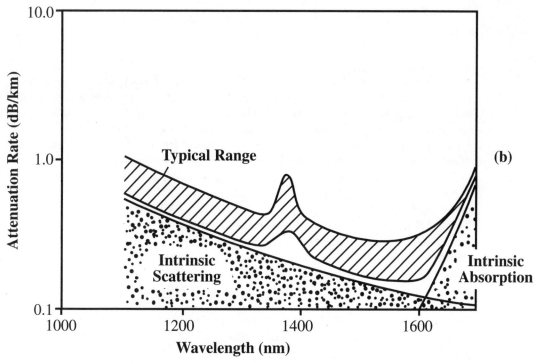

Figure 1.2 Loss Coefficient for a set of optical fibers. (*Reproduced with permission from Ref. 1, Fig. 1*b.)

copper cables are in the range of approximate tens of decibels per kilometer or more for high-frequency usage. The gigahertz range isn't even on the scale here. On the other hand, consider the loss curve for high-quality fiber-optic cable that is shown in Fig. 1.2. The data plotted here is for some actual production runs on optical fibers manufactured by the Corning company [1]. While there is some variation in the responses of individual fibers, all have their loss minima at about the same wavelengths. The two most important minima are at 1.55 and 1.3 μm. On the average the losses are ~0.2 dB/km in the 1.55-μm wavelength range and $b \sim 0.7$ for the 1.3-μm region. This represents an enormous improvement over copper because the effects of these absorption coefficients are manifested in the exponent. For example, the loss coefficient for fiber-optic propagation at ~1.55 μm will have an improvement over typical copper cable by a factor of more than 50 *in the exponent:*

$$\frac{P_{out}}{P_{in}} = 10^{-(\alpha L/10)}$$

where α = dB/km loss coefficient and L = length in kilometers. Thus for L = 100 km

$$\frac{P_{\text{out}}}{P_{\text{in}}} \sim \begin{cases} 10^{-100} & \text{for coax (coaxial cable)} \\ 10^{-2} & \text{for fiber} \end{cases}$$

Other power considerations play a role in comparing communication channels. For example, much higher power insertion levels are possible on coaxial cable as opposed to optical fibers where high insertion powers may lead to severe nonlinear distortion of signals. Nevertheless, the exponential effect of loss coefficient dominates these comparisons; fiber-optic channels permit much larger (albeit more expensive) spacing between repeaters.

To further demonstrate the importance of the physical characteristics of the properties of the optical fiber itself, we note that almost every long-distance high-speed fiber-optic system being deployed today operates at either 1.3 or 1.55 μm, and that these particular wavelengths coincide with the two principal minima in the loss coefficient curves. It is apparent that the fiber characteristics were not designed to fit existing lasers. Instead, it was the other way around; when these excellent fibers first appeared on the market, there was furious activity to develop suitable semiconductor diode lasers to match these wavelengths. This relatively new class of laser diodes, known as *heterojunction laser diodes,* also turns out to have the desirable property of lower lasing thresholds than their predecessors. In Chap. 4, we'll discuss some of these terms, as well as the reasons why low threshold current is a distinct advantage.

1.2.4 Usable bandwidth considerations

Another important parameter that affects data rates is the channel bandwidth, for it is well known that achievable data rates increase as the usable bandwidth increases. For example, Shannon's formula for channel capacity in bits per second, $C = W \log_2 [1 + (S/N)]$, states that theoretically achievable data rates are far more sensitive to bandwidth than to signal-to-noise ratio. Furthermore, in principle, if not in practice, it is possible to achieve arbitrarily low error rates as long as we do not exceed the channel capacity C.

We note that the frequency spread of the two primary minima in the loss coefficient curve are $\sim 5 \times 10^{12}$ Hz near 1.3 μm and $\sim 12 \times 10^{12}$ near 1.55 μm. Just these two bandwidths alone are more than three orders of magnitude larger than the usable bandwidths of coax cables. Because optical communication can take place in other spectral regions over modest distances such as those encountered within small cities, college campuses, and research institutes, this is only part of the story. Fiber-optic systems in operation today make use of only a small fraction of the available bandwidth. There is the potential for massive increases in data rates over bit rates in existing systems. Limited electronic processing capability is one major obstacle to achieving the potential of data rates in the range of terabits per second (Tbits/s).

One way to ease the frequency response requirements on the electronics is to transmit information over many parallel, slower-speed, nonoverlapping channels—thus filling up the available frequency space with data and yet processing that data with present-day electronics. However, this approach will require a large number of lasers at both transmitters and receivers and may be cost-prohibitive. We shall discuss such wavelength-division-multiplexed systems in Chaps. 8 and 9.

Most nonoptical communication systems are designed with firm bandwidth constraints imposed by the need to share resources among many users. One design parameter that is often used as a figure of merit is bits per Hertz (bits/Hz). The larger this parameter, the better the system. Because far more bandwidth is available than is actually used in fiber-optic systems, bits/Hz is not as important a consideration in the design of FOCS today as it is in wireless systems.

1.2.5 Confinement of radiation

The low level of material absorption for modern optical fiber materials, however, is only part of the story. A light beam traveling in a uniform, homogeneous material will, as a consequence of the laws of physics (Maxwell's equations), diverge as it propagates along a path [2]. For example, consider a cylindrically symmetric single-mode laser beam such as is generated by many lasers. These laser beams also are such that

$$|E(r)| \propto \exp\left[\frac{-r^2}{w^2}\right]$$

where w is a parameter called the *spot size*. The spot size is equal to the radial distance from the axis of symmetry to a point where the field amplitude has decreased by a factor of $1/e$. Let us now consider such a circularly symmetric *gaussian beam* with the spot size $= w_0$ and having an equiphase contour that lies in a plane. It can be shown that the optical field can propagate in either or both directions from that plane in a manner such as that shown in Fig. 1.3 below.

It is clear that the beam spreads out spatially as it propagates in either direction, and that w_0 represents the *minimum spot size* known as the "waist" for such a beam in a uniform lossless unbounded space. The light contained within the spot size contains approximately 86 percent of the total energy of the beam. The parameter z_R, known as the *Rayleigh range,* is the distance from the waist at which the energy density drops off to half of its value at the waist. As such, it is often used as a measure of the collimation length of an optical beam. The Rayleigh length is given by

$$Z_R = \frac{\pi w_0^2}{\lambda} \tag{1.1}$$

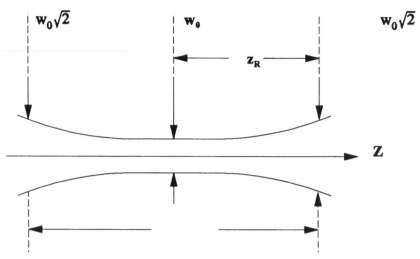

Figure 1.3 Propagation of a cylindrically symmetric gaussian beam.

and the spot size at a distance z from the waist is given by

$$w(z) = w_0 \sqrt{1 + \left(\frac{z}{z_R}\right)^2} \tag{1.2}$$

which, for $z \gg z_R$, simplifies to

$$w(z) \approx w_0 \frac{z}{z_R} = \frac{\lambda z}{\pi w_0} \tag{1.3}$$

Clearly, the spot size grows linearly with distance for $z \gg z_R$ and the cross section of the optical beam grows quadratically with distance. Thus, the energy density decreases quadratically as the distance of the waist increases, and some way must be found to confine the beam. We also note that $z_R \propto w_0^2$, and one way to extend the collimation length is to increase w_0. Free-space optical data links can be employed for short-range linkups such as adjacent computer consoles or in outer space; however, it is impractical to use enlarged beams for confining these beams. This is the case for even moderate distances in the Earth's atmosphere. On the other hand, the ability to collimate an optical fiber is proportional to the area of the transmitter measured in square wavelengths. When compared to microwaves, the much smaller wavelengths of light make free-space optical communications a candidate for interplanetary communications.

We next indicate how very thin modern optical fibers can achieve nearly total optical confinement without the beam spreading according to Eq. (1.2). In a coaxial cable, the electromagnetic field is confined to the region

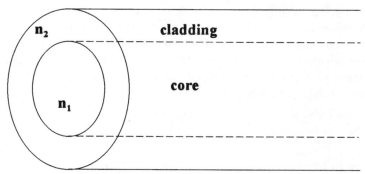

Figure 1.4 Typical cylindrical optical fiber structure without soft opaque outer cover

between the conductors; the transmitted signal is a traveling wave that propagates through the insulator between the inner and outer conductors. The outer conducting shell also serves to shield the inner signal from outside electrical disturbances. However, powerful low-frequency fields such as those found in many heavy manufacturing facilities can significantly interfere with signals inside the coax. A typical modern optical fiber structure is shown in Fig. 1.4.

The physical mechanism responsible for trapping the optical beam within the core is called *total internal reflection,* a phenomenon that was discovered by Tyndall [3]. We shall discuss this phenomenon in detail in Chap. 2, but we note here that total internal reflection is a direct consequence of Snell's law. When a ray (or beam of light) encounters the core cladding interface at a sufficiently shallow angle on the core side of the interface, and the index of refraction in the core is larger than in the index of refraction in the cladding, total internal reflection will occur. Some salient features of this phenomenon that we shall encounter in conjunction with high-speed data transmission over optical fibers are as follows:

1. There is practically no loss whatsoever on reflection from the core side of the core–cladding interface. This type of reflector is far more efficient than silvered mirrors.

2. This property, along with the low loss property of optical fibers, forms the principal energy retention mechanism for fields propagating down fiber-optic systems.

3. While the condition that $n_1 > n_2$ is required to trap the beam, it turns out also that $n_1 \approx n_2$ is also desirable. In addition to attenuation, it is important to maintain the integrity of optical pulses in digital fiber-optic communication systems. For high-speed data transmission, there is a need to prevent the signal pulses from spreading in time (dispersing) and causing overlapping pulses at the receiving terminal.

4. For example, consider a simpler optical fiber structure consisting of merely a high-quality glass core surrounded by air. Such a fiber would certainly trap the light, but it would also result (as we shall demonstrate in Chap. 3) in massive pulse spreading and intersymbol interference. Therefore the dispersion problem eliminates this simple fiber from consideration: for high speed communication systems.

5. One consequence of the solution to the dispersion problem is that the radius of the core will turn out to be very small (often on the order of a few micrometers).

6. We shall also see in Chap. 3 that in spite of the total internal reflection, there is an electromagnetic field in the cladding region. Also, the cladding must be constructed of high-quality glass because some of the light travels down the cladding. We shall see there that the term *total internal reflection* is somewhat of an overstatement when taken out of context.

Thus, in summary, a good optical fiber consists of high-quality glass in both the core and the cladding with $n_1 > n_2$. The interface between these two regions must be very smooth and dimensionally consistent.

In addition, there is also a soft opaque casing around the cladding (not shown in Fig. 1.4) which provides both complete optical isolation from other fibers in a bundle as well as a degree of mechanical isolation. The development of manufacturing techniques whereby multi-kilometer lengths of high-quality, small-core-radius, optical fibers with extremely smooth and regular core-cladding interfaces was a great achievement and is largely responsible for the rapid progress that has been made in the deployment of these systems.

1.3 Summary of Fiber-Optics Advantages

1. Low loss coefficient.
   a. An improvement of at least three orders of magnitude in the dB/km loss coefficient over coax.
   b. ~0.2 dB/km at a wavelength ~ 1.55 μm.
2. Extremely large usable bandwidth.
   a. ~5×10^{12} Hz at 1.3 μm and 12×10^{12} at 1.55 μm.
   b. Leads to very flexible multiplexing and multiple-access systems, as we shall see in later chapters.
   c. Also yields potential data rates in the Tbit/s range and below; next-generation *Superhighway*.
3. Energy confinement.
   a. The energy is trapped so effectively within the optical fiber that the fiber channel is virtually immune to the effects of external fields.
      (1) Thus, optical fibers may be bundled in groups without any crosstalk between fibers.

(2) One area for utilization of optical fibers is in an environment with heavy electrical machinery. *No electromagnetic interference* (EMI) *problem.*

(3) It is also difficult to eavesdrop on an optical fiber.

b. As described in Sec. 1.2.1, the optical fiber confines the optical signal to the dimensions of the optical fiber, thereby avoiding the $1/r^2$ energy density reduction experienced by electromagnetic fields propagating through free space.

c. The key to the confinement properties of the optical fiber is the consequence of Snell's law called *total internal reflection.*

d. The difference between the indices of refraction in the core and the cladding must be very slight to avoid excessive pulse spreading.

In the remainder of this section we shall list some salient features of FOCS, including some of the differences and similarities between FOCS and radio-frequency (rf) communication systems (defined here to include microwave frequency signals as well as rf signals). All the topics mentioned in this subsection are described in detail within the text.

While the signals used in both optical frequency and rf signals are transmitted as electromagnetic radiation, the vast difference in wavelength generates some huge differences between these types of systems. In particular, detection of rf radiation is basically a linear process, whereas detection at optical frequencies is more or less a quadratic process with the sum frequency terms being discarded. One consequence of this difference is that heterodyne detection at optical frequencies is achieved by addition of the local laser to the received signal instead of multiplication as used in rf systems. Also, the noise characteristics and signal amplification processes are quite different for rf versus FOCS.

Nevertheless, almost every type of modulation, multiplexing, digital signaling technique that has been employed at rf has also been generated in FOCS. These include the following: (See List of Acronyms for explanation of notation.)

1. AM, FM, and PM (amplitude, frequency, and phase modulation) analog modulation systems

2. ASK, FSK, PSK (amplitude, frequency, and phase shift keying), and QAM digital signaling techniques on both direct-carrier and subcarrier structures

3. Time-division, frequency-division, and code-division multiplexing

4. Time-division, frequency-division, and code-division multiple access networks

There are some additional modulation and multiplexing schemes that can be used only in conjunction with FOCS. Consider, for example, Fig. 9.6, which displays the full range of multiplexing possibilities available in FOCS. In particular, the use of multiple fibers in a cable results in a spatial form of multi-plexing that is made feasible by the small fiber dimensions and the complete isolation of fields within each fiber of the bundle. An opaque covering on each individual fiber provides almost perfect isolation for each fiber; this is in stark contrast to the electromagnetic interference problem that would be encountered in a similarly constructed conductor-based cable. In essence, each fiber is its own communication universe with data rate potential that exceed any communication trunk line in existence today.

The wave-division multiplexing system (WDM, indicated in Fig. 9.6) is also unique to FOCS. Mathematically, WDM is a version of frequency-division multiplexing. When applied to FOCS, it results in an enormous distinct hier-archy of communication channels which greatly increases the data rate capa-bility of each individual optical fiber. In contrast to the situation in rf commu-nication systems, the broad bandwidth possibilities created by wave-division multiplexing leads to traveling wave-optical amplifiers assuming a greater role in repeaters than regenerative repeaters. Further discussion of this topic will arise in ensuing chapters.

1.4   Overview of Text

As noted earlier the single most important component in fiber-optic communi-cation systems (FOCS) is the optical fiber itself. We therefore begin the tech-nical portion of the textbook with an introduction to optical fibers, which is presented in Chaps. 2 and 3. We use Chap. 2 to introduce (or review) mode concepts and determine the conditions for single-mode propagation (a neces-sary condition for long-range high-speed data transmission). To simplify the mathematical description of this material, we begin our analysis of optical fiber propagation by examining a simplified rectangular structure known as a "slab" waveguide. Although the slab waveguide as described in Chap. 2 is not practical,[1] it has the virtue of yielding relatively simple mode solutions. In this chapter we also introduce graphical procedures for obtaining numerical values for key parameters that describe the behavior of propagating modes. Then when we study propagation in the cylindrically shaped optical fibers described in Chap. 3, we employ similar graphical methods to specify modal solutions. Some other effects that we describe in Chap. 3 are dispersion (pulse spreading), scattering, and absorption losses, as well as a variety of other phenomena that affect propagation in cylindrical fibers.

Another important component of a modern high-speed FOCS is the *semi-conductor laser diode* (SLD), which is modulated[2] to produce the input signal

---

[1]We note here that some variations of the slab waveguide can be applied to the analysis of optoelectronic integrated chips.

[2]This modulation may be accomplished by direct control of the SLD driving current or indirect-ly by an external modulator. (See Chap. 7 for details.)

to the optical fiber. This device is also used to pump optical amplifiers and as a local laser in optical heterodyne detection receivers. In Chap. 4, we describe the basic processes that lead to optical amplification, which is a primary requisite for laser action. Other important aspects of SLDs such as single-frequency operation and tunability are examined in this chapter. Chapter 4 also contains material on the operational characteristics of optical amplifiers which are also important components in a modern FOCS. Since our primary interest in this textbook is in high-speed data transmission, we will omit light-emitting diodes.

Almost all detectors in high-speed FOCS are one form or another of photodiodes. In Chap. 5, we describe the operation of several types of optical photodiodes, including avalanche and $p-i-n$ photodiodes. A number of important parameters that determine receiver sensitivity such as quantum efficiency, cutoff wavelength, and responsivity will be described in this chapter. Some of the effects of these parameters on system performance are discussed in Chaps. 5 and 7.

To properly analyze the reliability of a communication system as represented by the probability of error, it is necessary to accurately model the noise aspects of the system. Although this topic will appear in all the remaining chapters of the text, most of the fundamentals will be covered in Chaps. 6 to 8. In Chap. 6 we focus on optical detection statistics as represented by photon-count distributions and how they relate to system performance. We first find the photon-count distribution for some basic types of optical fields and then determine the probability of error for detection based on photon-count statistics. In Chap. 6 we also examine the connection between photon-count statistics and shot noise, including a discussion of photon bunching noise. The chapter also contains a description of the noise aspects of optical amplifiers, including such basic amplifier characteristics as noise figure and signal-to-noise ratio. In Chap. 7, we examine a variety of FOCS and describe some performance criteria for several of these systems. In this chapter, however, we shall employ more realistic detection models based on integrated current as the detection variable as opposed to the nonpractical photon-counting-based detection.

The analysis of optical heterodyne detection (OHD) systems and how they are affected by phase noise is described in Chap. 8. Phase noise is a manifestation of random phase fluctuations generated by lasers in the system. They are particularly troublesome in optical heterodyne systems, and this explains why this noise process is as described in Chap. 8. In this chapter we shall investigate the performance of the OHD versions of OOK (ON/OFF keying), FSK, PSK, and DPSK. Also we shall examine in detail the design and performance of an actual high-speed FSK system using OHD.

Virtually all rf and microwave frequency receivers are heterodyne receivers. In Chap. 9 we consider four recently published papers comparing the performance of optical heterodyne receivers and direct-detection (non-heterodyne-detection) receivers. The first two papers discuss the pros and cons.

The third paper, mentioned in Chap. 9, describes some aspects for a proposed large-scale optical fiber network that may employ coherent optical detection. The final paper discussed in this chapter presents one solution to the image channel problem for the optical heterodyne detection of densely packed optical channels (transmitted over a single optical fiber).

Finally, in Chap. 10, we describe a proposed optical heterodyne detection system that has much greater detector sensitivity, as expressed by a photons/bit criterion, than systems being deployed today.

1.5   Some Textbooks on Fiber-Optic Communications

1. Textbooks that have been used in the past by the author:
   a. G. Keiser, *Optical Communications,* McGraw-Hill, New York, 1982. (There is now a second edition of this text published in 1991 by the same publisher.)
   b. J. Senior, *Optical Fiber Communications: Principles and Practice,* Prentice-Hall, Englewood Cliffs, N.J., 1985.
   c. B. E. A. Saleh, and M. C. Teich, *Fundamentals of Photonics,* Wiley, New York, 1991. (*Note:* This not a text devoted to fiber optics per se, but it contains in-depth descriptions of all the relevant phenomena scattered throughout the text. If there had been more systems concepts included in their text, I never would have been moved to write this text.)
2. Some recently published texts:
   a. G. P. Agrawal, *Fiber-Optic Transmission Systems,* Wiley, New York, 1992.
   b. P. E. Green, Jr., *Fiber-Optic Networks,* Prentice-Hall, Englewood Cliffs, N.J., 1993.
   c. J. C. Palais, *Fiber Optic Communications,* 3d ed., Prentice-Hall, Englewood Cliffs, N.J., 1992.

References

1. D. B. Keck, "Fundamentals of optical waveguides, fibers," *IEEE Commun. Mag.* **23,** 18 (May 1985).
2. A. E. Siegman, *An Introduction to Lasers and Masers,* McGraw-Hill, New York, 1971, Chap. 8.
3. J. Tyndall, *Proc. Roy. Inst.* **1,** 446 (1954).
4. D. R. Smith, *Digital Transmission Systems,* Van Nostrand Reinhold, New York, 1985 (see pp. 317–318).

2

Basic Electromagnetic Field Concepts

2.1 Maxwell's Equations

2.1.1 Introduction

To design a complete radio-frequency communication system such as a satellite relay, we usually require the collaboration of three types of electrical engineers. One group consists of communication systems engineers, another group consists of electronics engineers, and finally we include a group of electrical engineers with expertise in antennas and propagation. Except for a sharing of overall goals and specifications, these groups often perform their design tasks separately with relatively little interaction. This separation of tasks is not possible in the design of fiber-optic communication systems (FOCS). As discussed in Chap. 1, the area of fiber-optic communications has been driven by developments in materials and devices. It is imperative that an FOCS designer be aware, quantitatively, of how physical properties of the channel and various devices affect system performance. In this chapter we endeavor to present the principal effects that optical fiber characteristics have on FOCS design. The material presented here will concentrate primarily on those aspects that affect system design, with no attempt to either present the detailed derivation or cover all the subtle nuances of the principal results.

In this section we first derive the wave equation for homogeneous, isotropic materials. We then examine a particular set of solutions to these equations called *uniform plane waves.* Using such waves and well-known boundary conditions for electromagnetic waves, we shall derive Snell's law. Finally we shall utilize Snell's law to arrive at *total internal reflection,* and we shall see there why this term is somewhat of an exaggeration. In the next chapter we shall apply these results to propagation in an optical waveguide.

2.1.2 Equations for a source-free region

An excellent reference for the propagation of electromagnetic fields in dielectric waveguides as described in both Chaps. 2 and 3 is Cherin [1].

The sources of electromagnetic fields are free electric charge and electric currents. A source-free region is such that both

$$\rho = 0 \quad \text{and} \quad \mathbf{J} = 0$$

where ρ = volume charge density and $\mathbf{J}$ = current density. We note that ρ is a scalar quantity while $\mathbf{J}$ is a vector quantity. In such a region Maxwell's equations are

$$\nabla \times \mathbf{E} = -\frac{\partial \mathbf{B}}{\partial t} \tag{2.1}$$

$$\nabla \times \mathbf{H} = \frac{\partial \mathbf{D}}{\partial t} \tag{2.2}$$

$$\nabla \cdot \mathbf{D} = \rho = 0 \tag{2.3}$$

$$\nabla \cdot \mathbf{B} = 0 \tag{2.4}$$

In isotropic media we also have $\mathbf{D} = \varepsilon\mathbf{E}$ and $\mathbf{B} = \mu\mathbf{H}$.

At the interface of two distinguishable homogeneous, isotropic materials there exist constraints on the values that both $\mathbf{E}$ and $\mathbf{H}$ can take on either side of the interface. These are simply called *boundary conditions,* and for source-free regions these are

1. $E_{t,1} = E_{t,2}$ and $H_{t,1} = H_{t,2}$
2. $D_{n,1} = D_{n,2}$ and $B_{n,1} = B_{n,2}$

Here the subscripts 1 and 2 identify the side of the interface that we are referring to, while the subscripts t and n refer to the tangential and normal components (referenced to the boundary) of the fields, respectively. Note that these fields apply only at the interface (or an infinitesimal distance within the appropriate side) and the fields may vary considerably from the preceding relations as the distance from the interface increases.

Now for the types of nonmagnetic materials utilized in FOCS, we find that $\mu_1 = \mu_2 = \mu_0$. This is, of course, not a general relation for all materials, but in this case it leads to the following simplification on the boundary conditions for the normal components of the magnetic field: $H_{n,1} = H_{n,2}$.

2.1.3 Wave equation

It can be seen from Eq. (2.1) that the existence of a time-varying magnetic field at some point in space gives rise to an electric field at that point; while at the time a time varying electric field gives rise to a magnetic field [as shown by Eq. (2.2)]. *This regeneration process is such that once a time-varying electromagnetic field is created somewhere in a uniform medium, it will continue to propagate indefinitely through that medium in a manner governed by Maxwell's equations.* These equations can be combined to arrive at a single-vector equation to describe this propagation; this is called the *wave equation.* To obtain this equation, take the curl of both sides of Eq. (2.2) and employ a well-known vector identity as shown below.

$$\nabla \times \nabla \times \mathbf{E} = -\nabla \times \left(\mu \frac{\partial \mathbf{H}}{\partial t}\right) = -\mu \frac{\partial}{\partial t}(\nabla \times \mathbf{H}) = -\mu\varepsilon \frac{\partial^2 \mathbf{E}}{\partial t^2}$$

$$\nabla(\nabla \cdot \mathbf{E}) - \nabla^2 \mathbf{E} = -\mu\varepsilon \frac{\partial^2 \mathbf{E}}{\partial t^2}$$

using Eq. (2.3), we obtain

$$\nabla^2 \mathbf{E} = \varepsilon\mu \frac{\partial^2 \mathbf{E}}{\partial t^2} \tag{2.5}$$

Similarly, we begin with the curl of both sides of Eq. (2.1) to obtain a similar value for the magnetic field.

$$\nabla^2 \mathbf{H} = \varepsilon\mu \frac{\partial^2 \mathbf{H}}{\partial t^2} \tag{2.6}$$

Equations (2.5) and (2.6) are both wave equations; furthermore, each of these two equations is really three equations, one for each component in three-dimensional space. The symmetry of the physical configuration usually determines the logical choice of coordinate system. Thus, for example, cylindrical coordinates are a very suitable choice to describe electromagnetic propagation within optical fibers.

2.1.4 Uniform plane wave traveling in z direction

Let there exist an electromagnetic field propagating in the z direction within a homogeneous material. Furthermore, let both the $\mathbf{E}$ and $\mathbf{H}$ fields be uniform in any plane perpendicular to the z axis. Thus, if we use cartesian coordinates, $\mathbf{E}$ and $\mathbf{H}$ are not functions of either x or y. Such a wave is said to be a *uniform plane wave traveling in the* z direction. It can be readily shown that any continuous function of the following form will satisfy the wave equation:

$$E_i = f_1(z - vt) + f_2(z + vt) \tag{2.7}$$

where $i = x$, y, or z. The functions $f_1(\cdot)$ and $f_2(\cdot)$ are arbitrary, twice differentiable functions.

We note that

$$\frac{\partial^2 \mathbf{E}}{\partial t^2} = v^2\, \mathbf{E}'' \qquad v^2\, \mathbf{E} = \frac{\partial^2 E_z}{\partial^2 z} = E''$$

Thus the wave equation yields

$$E' = \mu\varepsilon\, v^2 E'' \Rightarrow 1 = \varepsilon\mu v^2 \Rightarrow v = \frac{1}{\sqrt{\mu\,\varepsilon}}$$

We next consider separately the situations in which the argument of $f_1(\cdot)$ is a constant,

$$z - vt = \text{constant} \Rightarrow \frac{dz}{dt} = v \Rightarrow \text{wave traveling in the } +z \text{ direction}$$

and where the argument of $f_2(\cdot)$ is a constant to arrive at

$$z + vt = \text{constant} \Rightarrow \frac{dz}{dt} = -v \Rightarrow \text{wave traveling in the } -z \text{ direction}$$

We may then conclude that the uniform plane wave described here will travel at a velocity v in the $\pm z$ direction, and that the velocity is determined by the physical parameters ε and μ. Specifically, we have

$$v = \frac{1}{\sqrt{\varepsilon\mu}} \tag{2.8}$$

It is instructive to examine the properties of a single-frequency, sinusoidal, uniform plane wave. The propagation properties of more complex functions can be obtained by from these waves by employing Fourier analysis techniques. It is especially important to note here that the wave equation is a linear differential equation, provided ε and μ are not dependent on the magnitude of the electromagnetic field. This is the case for almost all materials, including glass at moderate power levels, but nonlinear effects can occur for intense fields. We shall postpone discussion of these nonlinear effects until later in the course, and assume that linearity is valid until indicated otherwise. Consider

$$\cos(\omega t - \beta z) = \cos[-\beta(z - vt)]$$

where $v = \omega/\beta$. Note how this latter form is in the style of Eq. (2.7) for a plane wave traveling in the $+z$ direction. Now let us imagine that at some instant, say, $t = t_0$, we freeze in time this wave. Now select the smallest λ with the property that

$$\cos[-\beta(z + \lambda - vt_0)] = \cos[-\beta(z - vt_0)] \qquad \forall\, z$$

This spatial period of repetition λ is called the *wavelength*.

Next let us imagine that we are sitting at the point $z = z_0$; we shall examine how this field varies with time. We note the following about this scenario:

1. Cosine oscillation at frequency $f = \omega/2\pi$ is observed at that point.

2. In one second, f complete cycles are observed to occur. This implies that the number of wavelengths that pass the point in one second $= f$. Thus

$$v = f\lambda \tag{2.9}$$

A fundamental property of a linear system is that when it is stimulated by a steady-state sinusoid, the response will be a sinusoid at the same frequency. We shall soon be examining what happens when a sinusoidal plane wave is incident on an interface between two dielectrics with different ε and different propagation velocities. It will be helpful to bear in mind that the frequency will be the same on both sides of the interface, but the wavelengths will be different. Thus, if $v_2 = av_1$, then $\lambda_2 = a\lambda_1$.

As long as we are discussing wavelength, we note that almost always laser sources are identified by their wavelength in vacuum. It is too awkward to identify them by frequency because it is awkwardly large: $\sim 10^{15}$. As a demonstration of this, consider the loss coefficient–wavelength curve that we encountered in Chap. 1. Even though these fibers are made of glass with propagation velocities less than that of vacuum, the abscissa notation is in terms of the vacuum wavelength. In this text we shall endeavor to use λ_0 to represent wavelength in vacuum. We shall use other subscripts to indicate wavelength in specific nonvacuum materials as needed, and omit the subscript when referring to wavelength in the generic sense.

Another aspect of uniform plane waves is that they are transverse waves. This means that neither the **E** field nor the **H** field points in the direction of propagation. For the specific geometry we have here, this means that $E_z = 0$, and $H_z = 0$. In order to demonstrate this we note that since both E_x and E_y are functions of z and t only, it follows that

$$\nabla \cdot \mathbf{E} = \frac{\partial E_x}{\partial_x} + \frac{\partial E_y}{\partial y} + \frac{\partial E_z}{\partial z} = 0 + 0 + \frac{\partial E_z}{\partial z} = 0$$

then

$$\frac{\partial E_z}{\partial z} = 0 \Rightarrow E_z = \text{constant} = 0$$

Where we arbitrarily assume that the constant is zero since we have no interest here in DC fields here. Similarly, by using $\nabla \cdot \mathbf{H} = 0$, one can show that $H_z = 0$. We shall see later on that this transverse property is not generally valid for electromagnetic fields in enclosed regions such as those that we shall encounter when we study guided waves.

2.2 Plane Waves, Reflection, and Transmission at Interfaces

2.2.1 Plane wave traveling in an arbitrary direction

Let $\hat{\mathbf{n}}$ be a unit vector in the direction of propagation. Let us now define

$$\mathbf{k} = \frac{2\pi}{\lambda} \hat{\mathbf{n}} = k\hat{\mathbf{n}}$$

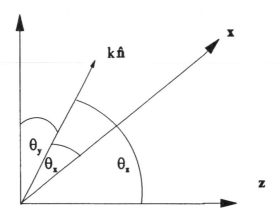

Figure 2.1 Plane wave travelling in an arbitrary direction.

where $\hat{\mathbf{n}} = \cos\theta_x \hat{\mathbf{n}}\mathbf{x} + \cos\theta_y \hat{\mathbf{y}} + \cos\theta_z \hat{\mathbf{z}}$ and where the angles θ_x, θ_y, and θ_z are the angles between the propagation vector and the coordinate axes as is shown in Fig. 2.1. The electric field observed at a particular point in space generated by such a field is given by the following expression:

$$\mathbf{E}(x,y,z,t) = \text{Re}\{\mathbf{E}_0(x,y,z)\exp[j(\omega t - \mathbf{k}\cdot\mathbf{r})]\} \tag{2.10}$$

where $\mathbf{r} = x\hat{\mathbf{x}} + y\hat{\mathbf{y}} + z\hat{\mathbf{z}}$.

Equation (2.10) can be written in the following more explicit form:

$$\mathbf{E}(x,y,z,t) = \text{Re}\left\{\mathbf{E}_0(x,y,z)e^{j\omega t}\exp\left[-j\frac{2\pi}{\lambda}\left\{x\cos\theta_x + y\cos\theta_y + z\cos\theta_z\right\}\right]\right\} \tag{2.11}$$

which can, in turn, be condensed to

$$\mathbf{E}(x,y,z,t) = \text{Re}\left\{\mathbf{E}_0(x,y,z)\,e^{j\omega t}\,\exp\left[-j\{x\beta_x + y\beta_y + z\beta_z\}\right]\right\}$$

where

$$\beta_i = \frac{2\pi}{\lambda}\cos\theta_i \tag{2.12}$$

Equation (2.12) can lead to some interesting interpretations of phase velocity. We already noted that the phase velocity $v = \omega/\beta$. Now consider Eq. (2.12) for β_i and note that when $\theta_i \sim 90°$ the phase constant β_i will be very small indeed and the phase velocity can be larger than c (the velocity of light in vacuum). In order to explain this phenomenon, consider Fig. 2.2. We shall rectify this apparent violation of special relativity when we discuss the propagation of light within optical fibers in Chap. 3. Note, however, that the preceding solutions are steady-state sinusoidal solutions that do not include transients such as propagating digital pulses. For now, however, the task in hand is to

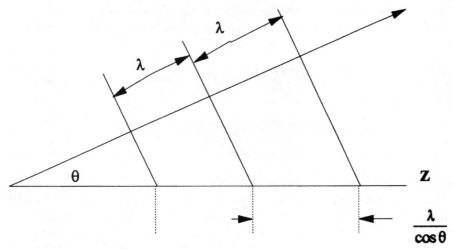

Figure 2.2 Propagation velocity projection effect.

derive Snell's law, and for that task we shall employ the forms represented by Eqs. (2.10) and (2.11). However, we need to review a few more properties of these waves before we attack Snell's law.

We shall first verify that the preceding general form agrees with forms encountered for plane waves traveling in the $+z$ direction. Such a wave will make the following angles with coordinate axes:

$$\theta_x = \theta_y = \frac{\pi}{2} \quad \text{and} \quad \theta_z = 0$$

Then

$$\mathbf{E}(x,y,z,t) = \mathbf{E}_0(x,y) \cos(\omega t - kz)$$

where $\mathbf{E}_0(x,y)$ is perpendicular to z axis and where $k = 2\pi/\lambda$ was used. To verify that this agrees with the wave equation, we note

$$\nabla^2 \mathbf{E} = \frac{\partial^2 E}{\partial^2 z} = -k^2 \cos(\omega t - kz) = \mu\varepsilon \frac{\partial^2 E}{\partial^2 t} = -\mu\varepsilon v^2 \cos(\omega t - kz)$$

and we have

$$k = \sqrt{\mu\varepsilon}\,\omega = \frac{2\pi f}{v} = \frac{2\pi f}{f\lambda} = \frac{2\pi}{\lambda}$$

Thus we see that this particular form for $\mathbf{E}(x,y,z,t)$ and this particular value for k are consistent with the wave equation. Note also that since λ is a function of the index of refraction of the material, k depends on n.

An important quantity that is very useful when determining the energy storage and power flow in electromagnetic fields is the *Poynting vector*. This may be expressed in terms of the following phasor relation for a sinusoidal field:

$$\tilde{\mathbf{S}} = \tfrac{1}{2} [\mathbf{E} \times \mathbf{H^*}] = \mathbf{S}_{av} + j\mathbf{S}_{react}$$

where

$$\mathbf{S}_{av} = \tfrac{1}{2} \operatorname{Re} [\mathbf{E} \times \mathbf{H^*}] \qquad \text{(actual power flow)}$$

$$\mathbf{S}_{react} = \tfrac{1}{2} \operatorname{Im} [\mathbf{E} \times \mathbf{H^*}] \qquad \text{(stored power)}$$

At this point it is useful to observe that when the Poynting vector is a pure imaginary quantity in a particular region then in that region the field is not propagating energy. An example of this is the field inside an ideal capacitor. There is energy flowing in and out, but no energy is dissipated inside the device. We shall soon see that the Poynting vector in the cladding of an ideal optical fiber is pure imaginary; therefore, there is a nontrivial field there, but no power flow.

2.2.2 Polarization

To understand what happens to light at an interface, we first need to introduce the concept of polarization. We therefore digress for a moment to introduce this subject.

Let us once again consider a uniform plane wave traveling in the z direction, specifically, $\hat{\mathbf{n}} = \hat{\mathbf{z}}$. Then, since we have previously shown $E_z = 0$, for this situation, we have

$$\mathbf{E} = \hat{\mathbf{x}} E_x + \hat{\mathbf{y}} E_y$$

If $E_y = 0$ while $E_x \neq 0$, then the field is said to be *linearly polarized in x direction*.

If $E_x = 0$ while $E_y \neq 0$, then the field is said to be *linearly polarized in y direction*.

Thus a uniform plane wave traveling in the z direction can be expressed as a linear combination of two orthogonal linear polarizations. If the components of the field are such that

$$E_x = E_{x,0} \cos(\omega t - kz) \qquad \text{and} \qquad E_y = E_{y,0} \cos(\omega t - kz)$$

then the field is also linearly polarized at an angle θ to the X axis as shown in Fig. 2.3. Note that the key point here is that in order for resultant of the x polarization and the y polarization to be linearly polarized, it is necessary that both of these constituent components be in phase. The orientation angle that the net polarization makes with respect to the X and Y axes does, howev-

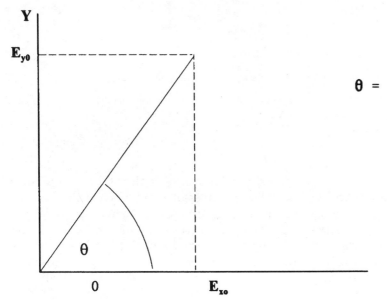

Figure 2.3 Linear polarization at an angle.

er, depend on the amplitudes $E_{x,0}$ and E_{y0}. If those components are not in phase (except for 180°), then the net polarization is not linear regardless of the amplitudes.

We next consider the specific case in which both E_x and E_y have the same amplitude but different phases:

$$E_x = E_0 \cos(\omega t - kz) \qquad E_y = E_0 \cos(\omega t - kz + \theta)$$

We further limit our interest here to the cases in which $\theta = \pm (\pi/2)$. Then $E_y = - \pm E_0 \sin(\omega t - kz)$.

At a fixed point in space the resultant **E** traces a circle in the x–y plane, and the field is said to be circularly polarized. Let $t = 0$ and $z = 0$. Now imagine that you observe the field in the same direction that it is propagating. Then when $\theta = -(\pi/2)$, the wave will progress, as t increases, in the same direction as a right-hand screw would progress if it rotated in the same direction as the field resultant rotates. Thus this is called right circular polarization (RCP), and for similar reasons $\theta = +(\pi/2)$ results in a left circularly polarized (LCP) field. (See Fig. 2.4.)

2.2.3 Birefringence

Birefringence is a property of some crystals whereby different polarizations travel at different velocities through the crystal. This is a consequence of interactions between the propagating fields and the internal electric fields

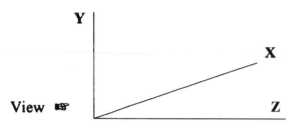

Figure 2.4 Circular polarization reference.

that hold the crystal structure together. Furthermore the effect is dependent on the alignment of the field with specific axes of symmetry within the crystal. In particular, some crystals exhibit electro-optic properties in which the difference in velocity between the two polarizations is affected by an applied electric field. This effect can be employed to modulate a laser beam, a topic we will consider in more detail in a later chapter. At this point we shall examine the birefringence phenomenon in order to develop some facility for manipulating polarized plane waves before we tackle Snell's law.

Consider the following phasor expression for circularly polarized fields:

$$\text{LCP} \Rightarrow (\hat{\mathbf{x}} + j\,\hat{\mathbf{y}})\tilde{\mathbf{E}}_0\,e^{j(\omega t - kz)}$$

$$\text{RCP} \Rightarrow (\hat{\mathbf{x}} - j\,\hat{\mathbf{y}})\tilde{\mathbf{E}}_0\,e^{j(\omega t - kz)}$$

where $\tilde{\mathbf{E}}_0$ is a phasor representing the amplitude and phase at $z = 0$ and $t = 0$. The $+j$ in LCP and the $-j$ in RCP represent the appropriate $90°$ phase shifts of the y component from the x component. Now

$$\text{LCP} + \text{RCP} = 2\,\hat{\mathbf{x}}\tilde{\mathbf{E}}_0\,e^{j(\omega t - kz)} = \text{LP in x direction}$$

$$\text{LCP} - \text{RCP} = 2j\,\hat{\mathbf{y}}\tilde{\mathbf{E}}_0\,e^{j(\omega t - kz)} = \text{LP in y direction}$$

Thus linear polarization (LP) may be decomposed into equal and opposite circular polarizations. Now suppose that we illuminate the crystal with linear polarization aligned with the appropriate crystal axes of a birefringent crystal. In this case two circularly polarized beams will travel at different velocities through the crystal. An important consequence of this is that the LCP and RCP are no longer in phase at the far end of the crystal. At the far end we then obtain

$$\text{LCP} \Rightarrow (\hat{\mathbf{x}} + j\,\hat{\mathbf{y}})\tilde{\mathbf{E}}_0\,e^{j(\omega t - kz)}$$

and

$$\text{RCP} \Rightarrow e^{j\phi}(\mathbf{x} - j\,\hat{\mathbf{y}})\,\mathbf{x}_0\,e^{j(\omega t - kz)}$$

the field which propagates inside the crystal is the superposition of these two circularly polarized fields. We now take the real part of the resultant and separate the E_x and E_y components to obtain the following:

$$E_x = \left[2\tilde{\mathbf{E}}_0 \cos\left(\frac{\phi}{2}\right)\right]\cos\left\{\omega t - kz + \frac{\phi}{2}\right\} = \tilde{\mathbf{E}}_{x,0} \cos\left\{\omega t - kz + \frac{\phi}{2}\right\}$$

$$E_y = \left[2\tilde{\mathbf{E}}_0 \sin\left(\frac{\phi}{2}\right)\right]\cos\left\{\omega t - kz + \frac{\phi}{2}\right\} = \tilde{\mathbf{E}}_{y,0} \cos\left\{\omega t - kz + \frac{\phi}{2}\right\}$$

where

$$\tilde{\mathbf{E}}_{x,0} = 2\tilde{\mathbf{E}}_0 \cos\left(\frac{\phi}{2}\right) \qquad \text{and} \qquad \tilde{\mathbf{E}}_{y,0} = 2\tilde{\mathbf{E}}_0 \sin\left(\frac{\phi}{2}\right)$$

Clearly the components E_x and E_y are in phase but have different amplitudes. This constitutes linear polarization at an angle θ with respect to the x axis (see Fig. 2.3), which is given by

$$\theta = \tan^{-1}\left(\frac{E_{y,0}}{E_{x,0}}\right) = \tan^{-1}\left\{\frac{\sin(\phi/2)}{\cos(\phi/2)}\right\} = \frac{\phi}{2}$$

For some crystals ϕ can be controlled by a voltage applied to appropriately situated electrodes. This property, along with linear polarizers, can be used to intensity-modulate laser beams.

We have so far only considered linear and circular polarizations, and this is all we shall need in this text. However, when the two orthogonal components have different amplitudes and phases (other than multiples of 90°), the polarizations will be such that the resultant traces out an ellipse. The most general form of polarization is elliptical, and they come in independent pairs. We shall not pursue this matter in this text beyond the brief discussion in the next paragraph.

Optical fibers have a slight birefringent effect which can be become very important in some systems because of the length of many fiber channels. It is a pressure-dependent effect which can result in random fluctuations in polarization observed from the output of the fiber channel at receiving terminals. Furthermore these polarizations fluctuate in polarization type as well as orientation. Thus a simple electrooptic device such as the one we described above cannot compensate for these generalized polarized variations. Some detection systems are not affected by this phenomenon, but those that are affected require more elaborate detection systems, as we shall describe in a later chapter. Finally we note here that there exist polarization preserving fibers, but these are significantly more expensive.

2.2.4 Snell's law

Consider a uniform plane wave obliquely incident to a flat dielectric interface as shown in Fig. 2.5. The subscripts i, R, and T refer to the incident wave, the reflected wave, and the transmitted (through the interface) wave, respective-

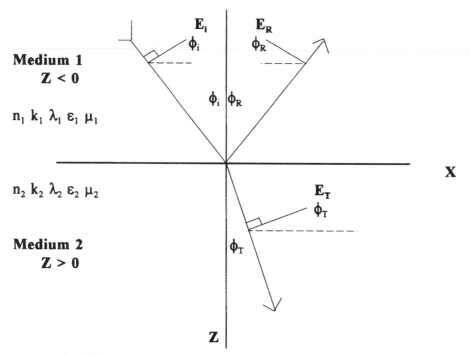

Figure 2.5 Snell's Law.

ly. Each field in this particular example is linearly polarized in the plane of the page, and perpendicular to the direction of propagation as indicated in Fig. 2.5.

This particular polarization, called *parallel incidence,* is said to be in the plane of incidence. Snell's law can also be derived using polarization perpendicular to the plane of incidence. The consequences of Snell's law which concern relations between the angles ϕ_i, ϕ_R, and ϕ_T are independent of the polarization of the incident radiation. There are some other aspects of the reflected and transmitted waves, however, that are polarization-dependent. We will have more to say about this subject later on.

Let

$$\xi_i = \hat{\mathbf{n}} \cdot \mathbf{r} \text{ (incident wave)} = z \cos \phi_i + x \sin \phi_i$$

$$\xi_R = \hat{\mathbf{n}} \cdot \mathbf{r} \text{ (reflected wave)} = -z \cos \phi_R + x \sin \phi_R$$

$$\xi_T = \hat{\mathbf{n}} \cdot \mathbf{r} \text{ (transmitted wave)} = z \cos \phi_T + x \sin \phi_T$$

The electric field then may be written in phasor format where we suppress the $e^{j\omega t}$ part as

For $z < 0$ (above the interface in region 1):

$$\tilde{\mathbf{E}} = \mathrm{E}_0(\hat{\mathbf{x}} \cos \phi_i - \hat{\mathbf{z}} \sin \phi_i)e^{-jk_1\xi_i} + E_{0,\mathrm{R}} (-\hat{\mathbf{x}} \cos \phi_\mathrm{R} - \hat{\mathbf{z}} \sin \phi_\mathrm{R})e^{-jk_1\xi_\mathrm{R}} \quad (2.13a)$$

For $z > 0$ (below the interface in region 2):

$$\tilde{\mathbf{E}} = \mathrm{E}_{0,\mathrm{T}} (\hat{\mathbf{x}} \cos \phi_\mathrm{T} - \hat{\mathbf{z}} \sin \phi_\mathrm{T})e^{-jk_1\xi_\mathrm{R}} \quad (2.13b)$$

At $z = 0$ the boundary condition $E_{t,1} = E_{t,2} \Rightarrow E_{x1} = E_{x2}$. Now by applying Eqs. (2.13a) and (2.13b) to these boundary conditions, we obtain

$$E_0 \cos \phi_i \, e^{-jk_1 x \sin \phi_i} - E_{0,\mathrm{R}} \cos \phi_\mathrm{R} \, e^{-jk_1 x \sin \phi_\mathrm{R}} = E_{0,\mathrm{T}} \cos \phi_\mathrm{T} \, e^{-jk_2 x \sin \phi_\mathrm{T}} \quad (2.14)$$

First note that every term in Eq. (2.14) has an exponential factor containing the variable x. Then we note that for any given situation ϕ_i, ϕ_R, and ϕ_T are constants. The only way that this equation can be satisfied for each and every value of x at the interface (*recall that the uniform plane-wave assumption spreads field over all space and x*) is for the multipliers of x in each exponential term of Eq. (2.14) to be identical. Thus we have

$$k_1 \sin \phi_i = k_1 \sin \phi_\mathrm{R} = k_2 \sin\phi_\mathrm{T}$$

The first two terms yield the well-known result $\phi_\mathrm{R} = \phi_i$, while the first and last yield

$$\frac{\sin \phi_\mathrm{T}}{\sin \phi_i} = \frac{k_1}{k_2} = \frac{2\pi/\lambda_1}{2\pi/\lambda_2} = \frac{\lambda_2}{\lambda_1} = \frac{\lambda_2 f}{\lambda_1 f} = \frac{v_2}{v_1} \quad \text{(Recall same } f \text{ in both media)}$$

The index of refraction is defined by

$$n\ell = \frac{c}{v_\ell} \qquad \text{(where } \ell = 1, 2\text{). Thus}$$

$$\frac{k_1}{k_2} = \frac{v_2}{v_1} = \frac{n_1}{n_2}$$

This leads to Snell's law:

$$n_1 \sin \phi_1 = n_2 \sin \phi_2 \quad (2.15)$$

We now adopt the usual notation for Snell's law where subscript 1 refers to the input and reflected waves wave and subscript 2 refers to the wave which propagates through medium 2. In the remainder of the text we shall use ϕ_1 for both ϕ_i and ϕ_R, and we shall use ϕ_2 in place of ϕ_T.

2.2.5 Total internal reflection (critical angle)

Snell's law is usually covered in an engineering curriculum (without being derived) in undergraduate physics courses. At that level the main interest is in the angles of propagation of the various optical rays and perhaps a short discussion of total internal reflection. This phenomenon is the principal mechanism for confining the optical field within the fiber, and we need to understand exactly what that term really means. Another reason for including this detailed derivation of Snell's law is that it provides the reader with a good example of how the electric and magnetic (E&M) boundary conditions affect the propagation properties of waves at interfaces. Only one boundary condition was required to obtain Snell's law, and the result is independent of polarization. We did not, however, determine the amplitudes and phases of the reflected and transmitted fields. To obtain numerical values for these quantities, we require most or all the of the E&M boundary conditions. *Those results will turn out to be polarization-dependent.* We shall not pursue this matter further at this time. Instead we shall now examine the "total reflection" phenomenon.

From Snell's law we have

$$\sin \phi_2 = \frac{n_1}{n_2} \sin \phi_1$$

Now let $n_1 > n_2$ and let us define the "critical angle" ϕ_c as the ϕ_1 at which

$$\phi_2 = \frac{\pi}{2} \qquad \text{or} \qquad 1 = \frac{n_1}{n_2} \sin \phi_c$$

Then

$$\phi_c = \sin^{-1} \left[\frac{n_1}{n_2} \right] \tag{2.16}$$

Thus when $\phi_1 = \phi_c$ the T wave moves parallel to (not through) the interface.

We next consider what happens when $\phi_1 > \phi_c$. From Snell's law we have

$$\sin \phi_2 = \frac{n_1}{n_2} \sin \phi_1 \tag{2.17}$$

$$\cos \phi_2 = \sqrt{1 - \sin^2 \phi_2} = \sqrt{1 - \left[\frac{n_1}{n_2} \right]^2 \sin^2 (\phi_1)} \tag{2.18}$$

Now from Eq. (2.11) the field in medium 2 propagates according to

$$\tilde{\mathbf{E}}_T = \tilde{\mathbf{E}}_{T,0} \, e^{-jk_2 \xi_T} = \tilde{\mathbf{E}}_{T,0} \, \exp[-jk_2 (z \cos \phi_2 + x \sin \phi_2)]$$

where $\tilde{\mathbf{E}}_{T,0} = E_{T,0} (\hat{\mathbf{x}} \cos \phi_T - \hat{\mathbf{z}} \sin \phi_T)$. Then

$$\tilde{\mathbf{E}}_T = \tilde{\mathbf{E}}_{T,0} \, \exp\left[-jk_2 \left\{ z \sqrt{1 - \left(\frac{n_1}{n_2} \right)^2 \sin^2 (\phi_1)} + x \left[\frac{n_1}{n_2} \sin \phi_1 \right] \right\} \right]$$

When $\phi_1 = \phi_c$

$$\left(\frac{n_1}{n_2} \right) \sin \phi_1 = 1 \qquad \text{so} \qquad 1 - \left(\frac{n_1}{n_2} \right)^2 \sin^2 \phi_1 = 0$$

Then

$$\tilde{\mathbf{E}}_T = \tilde{\mathbf{E}}_{T,0} \, e^{-jk_2 x}$$

This represents propagation in the x direction only. There is no propagation in the z direction (through the interface).

Next consider the field in medium 2 when $\phi_1 > \phi_c$, which implies that

$$\sin \phi_1 > \frac{n_2}{n_1}$$

Note that the sine function is monotonically increasing in the first quadrant. We next demonstrate that under these conditions a nontrivial field exists on the other side of the interface. This field, however, does not represent energy being transported through the interface. Note that

$$\cos \phi_2 = \sqrt{1 - \left[\frac{n_1}{n_2}\right]^2 \sin^2 \phi_1} = -j \sqrt{\left[\frac{n_1}{n_2}\right]^2 \sin^2 \phi_1 - 1}$$

where we retain only the $-j$ form because $+j$ leads to a nonphysical solution which grows exponentially as the wave propagates in the $+z$ direction. Recall that

$$\tilde{\mathbf{E}}_T = \mathbf{E}_{T,0} \exp[-jk_2 (z \cos \phi_2 + x \sin \phi_2)]$$

$$= \mathbf{E}_{T,0} \exp\left\{-zk_2 \sqrt{\left[\frac{n_1}{n_2}\right]^2 \sin^2 \phi_1 - 1}\right\} \exp\left\{-jxk_2 \left\{\left(\frac{n_1}{n_2}\right) \sin \phi_1\right\}\right\}$$

where

$$\tilde{\mathbf{E}}_t = \mathbf{E}_{T,0} \, e^{-\alpha z} \, e^{-j\beta_{x,2} x} \tag{2.19a}$$

$$\alpha = k_2 \sqrt{\left[\frac{n_1}{n_2}\right]^2 \sin^2 \phi_1 - 1} \tag{2.19b}$$

$$\beta_{x,2} = k_2 \left\{\left(\frac{n_1}{n_2}\right) \sin \phi_1\right\} \tag{2.19c}$$

Thus there exists a field on the other side of the interface. *The first exponential factor on the right-hand side of Eq. (2.19a) represents exponential decay, but not propagation in the z direction. The second exponential term shows that any propagation is in the x direction which is parallel to the interface* between regions 1 and 2, but not across it. A field of the type of energy storage field that exists in region 2 when $\phi_1 > \phi_c$ is called an *evanescent* field and plays an important role in the design of optical couplers.

2.2.6 Poynting vector considerations for Snell's law

From Eq. (2.1) we have

$$-\frac{\partial \mathbf{B}}{\partial't} = \begin{vmatrix} \hat{\mathbf{x}} & \hat{\mathbf{y}} & \hat{\mathbf{z}} \\ \dfrac{\partial}{\partial x} & \dfrac{\partial}{\partial y} & \dfrac{\partial}{\partial z} \\ E_x & E_y & E_z \end{vmatrix}$$

while Eq. (2.13b) yields

$$E_x = E_0 \cos \phi_2 \, e^{j(\omega t - k_2 \xi_T)}$$

$$E_y = 0$$

$$E_Z = -E_0 \sin \phi_2 \, e^{j(\omega t - k_2 \xi_T)}$$

$$\xi_T = z \cos \phi_T + x \sin \phi_T$$

The term ξ_T is the only part of these field components that depends on the coordinates, and further *we note that all of these components are independent of y*. We then have

$$-\frac{\partial \mathbf{B}}{\partial't} = \begin{vmatrix} \hat{\mathbf{x}} & \hat{\mathbf{y}} & \hat{\mathbf{z}} \\ \dfrac{\partial}{\partial x} & 0 & \dfrac{\partial}{\partial z} \\ E_x & 0 & E_z \end{vmatrix} \qquad \therefore \qquad -\mu \frac{\partial \mathbf{H}}{\partial t} = -\hat{\mathbf{y}} \left[\frac{\partial E_z}{\partial x} - \frac{\partial E_x}{\partial z} \right]$$

Now by using jT and then differentiating the fields given above, we obtain

$$\frac{\partial E_z}{\partial x} = -E_0 \sin \phi_2 \, \{-jk_2\} \sin \phi_2 \, e^{j(\omega t - k_2 \xi_T)} = -jk_2 \, E_0 \sin^2 \phi_2 \, e^{j(\omega t - k_2 \xi_T)}$$

$$\frac{\partial E_x}{\partial z} = E_0 \cos \phi_2 \, \{-jk_2\} \cos \phi_2 \, e^{j(\omega t - k_2 \xi_T)} = -jk_2 \, E_0 \cos^2 \phi_2 \, e^{j(\omega t - k_2 \xi_T)}$$

Then

$$\mu \frac{\partial \mathbf{H}}{\partial t} = \hat{\mathbf{y}} \, jk_2 \, E_0 \, e^{j(\omega t - k_2 \xi_T)}$$

The $e^{j\omega t}$ time dependence of the fields results in $\partial \mathbf{H}/\partial t = j\omega \mathbf{H}$ and

$$H_y = \frac{k_2}{\mu\omega} E_0 \, e^{j(\omega t - k_2 \xi_T)} = \sqrt{\frac{\varepsilon}{\mu}} \, E_0 \, e^{j(\omega t - k_2 \xi_T)}$$

These calculations lead to the following complex Poynting vector for the field on the transmitted side of the interface when $\phi > \phi_c$:

$$\tilde{\mathbf{S}} = \frac{1}{2}\,[\mathbf{E} \times \mathbf{H}^*] = S_{av} + jS_{react}$$

where

1. The complex conjugate of $\mathbf{H}$ removes the $e^{j(\omega t\,-\,k_2 \xi_T)}$ propagation term from the Poynting vector.

2. $S_x = -E_z H_y^* = +\sqrt{\varepsilon/\mu}\, E_0^{\,2} \sin \phi_2$ = real power flow in the x direction.

3. $S_z = E_x H_y^* = \sqrt{\varepsilon/\mu}\, E_0^{\,2} \cos \phi_2$

In this case, however, the $\cos \phi_2$ factor is pure imaginary, and therefore this represents reactive power flow in the z direction.

2.3 Dielectric Slab Waveguide

2.3.1 General introduction

In Chap. 1 we described how the optical confinement capability of fibers is one of the primary reasons for employing them for high-speed data communication systems. We also mentioned there that a physical phenomenon called *dispersion* results in pulse spreading in time. To understand how these characteristics affect system design, we need to evaluate optical propagation within the bounded space of a fiber-optic cable. Specifically, we shall need to find the solutions (use of the plural here is not a misprint) of the wave equation in a framework that involves multiple boundary conditions. This text has been written with communication engineers in mind. To obtain what we need to know to design fiber-optic communication systems with a minimum of discomfort, we shall cover optical waveguides in the following sequence:

1. We first examine propagation within the rectangular geometry of the slab waveguide (SWG) shown in Fig. 2.6. Solutions of the wave equation in cartesian coordinates are easier to obtain than solutions for cylindrical waveguides (CWG). Nevertheless, our analysis of the SWG will incorporate most of the pertinent terms and computation processes that we will encounter when we analyze CWG. We shall encounter specific solutions called *modes*, utilize graphical techniques to obtain their propagation properties, and determine the conditions for single-mode propagation.

2. In Chap. 3 we present an abbreviated study of the modal solutions for CWG. Here we shall get to graphical solution techniques as quickly as possible. We remind the reader here that the design of fiber-optic communication systems is affected by the physical properties of the channel to a far greater degree than in most rf communication systems.

3. Then we shall discuss the loss and dispersion properties of cylindrical waveguides only (also in Chap. 3).

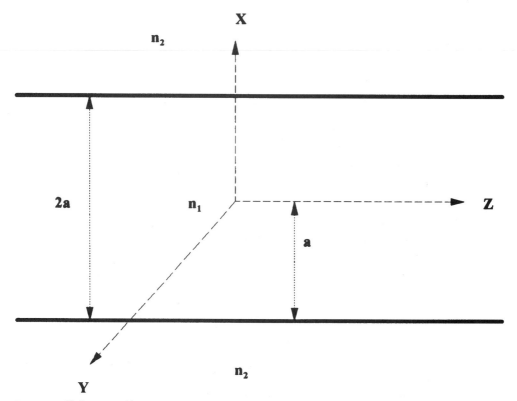

Figure 2.6 Slab waveguide.

2.3.2   Introduction to slab waveguides

The geometry of the slab waveguide is depicted in Fig. 2.6. It consists of three layers, all extending from $-\infty$ to $+\infty$ in both the y and z directions. As preparation for terminology that we will employ when we describe cylindrical fibers, we shall call the central portion the *core,* and the remaining regions, the *cladding.* The core is characterized by an index of refraction n_1 while the cladding regions have an index of refraction of n_2 where $n_1 > n_2$. We have chosen to denote the thickness of the slab as $2a$. This will simplify our eventual transition to cylindrical coordinates where similarity will be found between a and the radius of the cylindrical fiber core.

We shall assume that the field is propagating in the $+z$ direction and, because of the symmetry of the slab, that all fields are independent of the y coordinate. Therefore all $\partial\psi/\partial y = 0$, where ψ is any component of $\mathbf{E}$ or $\mathbf{H}$. This last assumption is brought about by the absence of propagation and/or boundaries in the y direction. Thus we arrive at the following form for the $\mathbf{E}$ field:

$$\mathbf{E} = E_0(x)\, e^{-j\beta z}\, e^{j\omega t}$$

Now we consider Eqs. (4.a–4.d) of App. A, Sec. A2, and set all $\partial/\partial y$ terms = 0. Then utilizing $k = \omega^2 \mu\varepsilon = 2\pi/\lambda$, we obtain the following equations for the non-zero-field components.

$$E_x = \frac{-j\beta}{\kappa^2} \frac{\partial E_z}{\partial x} \tag{2.20a}$$

$$E_y = \frac{-j\omega\mu}{\kappa^2} \frac{\partial H_z}{\partial x} \tag{2.20b}$$

$$H_x = \frac{-j\beta}{\kappa^2} \frac{\partial H_z}{\partial x} \tag{2.20c}$$

$$H_y = \frac{-j\omega\varepsilon}{\kappa^2} \frac{\partial E_z}{\partial x} \tag{2.20d}$$

where $\kappa^2 = k^2 + \beta^2$.

The geometry of the slab waveguide leads to two sets of solutions: transverse electric (TE) modes, where $E_z = 0$, and transverse magnetic (TM) modes, where $H_z = 0$. By using Eqs. (2.20) it is easily seen that these modes are characterized by

1. For TE modes $E_z = 0 \Rightarrow E_x, H_y = 0$ and $E_y, H_x, H_z \neq 0$
2. For TM modes $H_z = 0 \Rightarrow E_y, H_x = 0$ and $E_x, E_z, H_y \neq 0$

2.3.3   TE Modes (General)

For TE modes the only nonzero component of **E** is E_y. The remaining nonzero magnetic field components may be obtained from

$$H_x = \frac{-\beta}{\omega\mu} E_y \qquad \text{(App. 2A, Sec. 2A.2, Eq. 3.d)} \tag{2.21a}$$

$$H_z = \frac{-1}{j\omega\mu} \frac{\partial E_y}{\partial x} \qquad \text{(App. 2A, Sec. 2A.2, Eq. 3.f)} \tag{2.21b}$$

We now assume that $E_y = E_{0,y}(x)\, e^{-j\beta z}\, e^{j\omega t}$. Thus

$$\frac{\partial^2 E_y}{\partial t^2} = (j\omega)^2 E_y = -\omega^2 E_y \qquad \text{and similarly} \qquad \frac{\partial^2 E_y}{\partial z^2} = -\beta^2 E_y$$

The wave equation then becomes

$$\nabla^2 E_y = -\omega^2\, \varepsilon\mu\, E_y = -k^2 E_y$$

or

$$\nabla^2 E_y + k^2 E_y = 0 \qquad \text{(the Helmholtz equation)}$$

We next find an explicit differential equation for E_y by combining these results with the geometry of the slab waveguide to get

$$\nabla^2 E_y = \frac{\partial^2 E_y}{\partial x^2} + 0 - \beta^2 E_y = -\omega^2 \varepsilon\mu\, E_y = -k^2 E_y$$

or

$$\frac{\partial^2 E_y}{\partial x^2} + \kappa^2 E_y = 0 \Rightarrow \frac{\partial^2 E_y}{\partial x^2} = -\kappa^2 E_y$$

where

$$\kappa^2 = k^2 - \beta^2$$

First we solve the preceding single-variable Helmholtz equation for $E_y(x)$. Once E_y has been determined, we find the two remaining nonzero components H_x and H_z from Eqs. (2.21a) and (2.21b).

There are, however, two classes of solutions to this equation. One will apply only to the cladding region, and the other will apply only to the core region. We shall subdivide our solutions between those where $\kappa^2 < 0$ and those where $\kappa^2 > 0$.

Case 1. $\kappa^2 < 0$. Now let $\gamma^2 = -\kappa^2$. Then

$$\frac{\partial^2 E_y}{\partial x^2} = \gamma^2 E$$

and

$$E_y = C_1 e^{-\gamma x} + C_2 e^{\gamma x}$$

The first term may be recognized from our derivation of Snell's law as the field on the n_2 side of a dielectric when $n_1 > n_2$, and when $\phi > \phi_c$ (i.e., total internal reflection). Clearly this solution applies to the cladding regions. We shall discard the second term in the preceding solution as being nonphysical because of its exponential growth with increasing x.

In the cladding region the solution takes the form

$$E_y = C \exp\{-\gamma(|x| - a)\}\, e^{j(\omega t - \beta z)} \qquad \text{for} \qquad |x| > a$$

where this single form can be used for cladding regions both above and below the central core of the slab. Also we note that

$$\gamma^2 = \beta^2 - k_2^2 = \beta^2 - n_2^2 k_0^2 > 0 \qquad \text{for} \qquad |x| > a$$

Case 2. Next we consider the case where $\kappa^2 > 0$ *or* $n_1^2 k_0^2 - \beta^2 > 0$. This leads to

$$\frac{\partial^2 E_y}{\partial x^2} = -\kappa^2 E_y$$

which, in turn, yields the following form for the solution:

$$E_y = A \cos(\kappa x) + B \sin(\kappa x) \quad \text{for} \quad |x| < a$$

Other equivalent forms of this solution are also possible here, but this particular form best fits our purposes here.

The condition that a mode propagate in the core and not leak energy into the cladding is that β satisfy both

$$\kappa^2 = n_1^2 k_0^2 - \beta^2 > 0 \quad \text{and} \quad \gamma^2 = \beta^2 - n_2^2 k_0^2 > 0$$

or

$$n_2^2 k_0^2 \leq \beta^2 \leq n_1^2 k_0^2 \tag{2.22}$$

We note that the condition $n_1 > n_2$ makes the two inequalities represented by Eq. (2.22) possible. This restricts β to the continuous range of values between $n_2 k_0$ and $n_1 k_0$. We have not as yet considered the constraints imposed by the E&M boundary conditions at the core–cladding interface. These boundary conditions, along with the constraint represented by Eq. (2.22), will lead to a discrete set of possible values for β, each of which characterizes specific solutions to the wave equation that we call *modes*.

2.3.4 TE modes (specific solutions)

We shall separate our solutions into those with even symmetry and those with odd symmetry. This is a second-order subdivision when compared with the TE and TM subdivision. First we now consider modes with *even symmetry* in the x direction.

Core region, $|x| < a$: Let $E_{y1} = A \cos(\kappa x) e^{j(\omega t - \beta z)}$

Then

$$H_{x,1} = \frac{-\beta}{\omega\mu} A \cos(\kappa x) e^{j(\omega t - \beta z)} \quad \text{(App. 2A, Sec. 2A.2, Eq. 3.}d\text{)}$$

$$H_{z,1} = \frac{-j\kappa}{\omega\mu} A \sin(\kappa x) e^{j(\omega t - \beta z)} \quad \text{(App. 2A, Sec. 2A.2, Eq. 3.}f\text{)}$$

Cladding region, $|x| > a$: Let $E_{y,2} = C \exp\{-\gamma(|x| - a)\} e^{j(\omega t - \beta z)}$

$$H_{x,2} = \frac{-\beta}{\omega\mu} C \exp\{-\gamma(|x| - a)\} e^{j(\omega t - \beta z)}$$

$$H_{z,2} = \frac{-j\gamma}{\omega\mu} \{\frac{-x}{|x|}\} C \exp\{-\gamma(|x| - a)\} e^{j(\omega t - \beta z)}$$

where $\{-x/|x|\} = -1$ for $x > a$ and where $\{-x/|x|\} = +1$ for $x < -a$, which allows us to use a single equation for both sides of the core.

The next step is to apply the boundary conditions at $x = a$. We also note that the $e^{j(\omega t - \beta z)}$ term is common to both sides of the boundary condition equations, and we shall cancel this term out before we write the pertinent equations. The $E_{\tan}$ boundary condition then yields

$$E_{t,1} = E_{t,2} \Rightarrow E_{y,1} = E_{y,2}$$

Thus $A \cos(\kappa a) = C \, e^0 = C$, and we can replace C by $A \cos(\kappa a)$. Also

$$H_{t,1} = H_{t,2} \Rightarrow H_{z,1} = H_{z,2}$$

$$\frac{-j\kappa}{\omega\mu} A \sin(\kappa a) = C \frac{j\gamma}{\omega\mu} \frac{-a}{|a|} = A \cos(\kappa a) \frac{-j\gamma}{\omega\mu}$$

Thus

$$\kappa \sin(\kappa a) = \gamma \cos(\kappa a)$$

which leads to

$$\tan(\kappa a) = \frac{\gamma}{\kappa} = \frac{\gamma a}{\kappa a} \tag{2.23a}$$

For TE modes with *odd symmetry,* one can show that by choosing

$$E_{y,1} = B \sin(\kappa x) \, e^{j(\omega t - \beta z)} \quad \text{(in core region)}$$

$$E_{y,2} = C \exp\{-\gamma(|x| - a)\} e^{j(\omega t - \beta z)} \quad \text{(in cladding region)}$$

and

$$\tan(\kappa a) = \frac{-\kappa}{\gamma} = \frac{\kappa a}{\gamma a} \tag{2.23b}$$

2.3.5   Graphical solution for TE mode parameters

Consider again the range restraints on β:

$$\kappa^2 = n_1^2 k_0^2 - \beta^2 > 0 \tag{2.24a}$$

$$\gamma^2 = \beta^2 - n_2^2 k_0^2 > 0 \tag{2.24b}$$

Next we multiply both sides of both Eqs. (2.24a) and (2.24b) by a^2, and then add them to arrive at

$$(\kappa a)^2 + (\gamma a)^2 = k_0^2 a^2 (n_1^2 - n_2^2)$$

Now let $X = \kappa a$ and $Y = \gamma a$; then the preceding equation becomes

$$X^2 + Y^2 = V^2 \tag{2.25a}$$

where

$$V = k_0 a \sqrt{n_1^2 - n_2^2} \tag{2.25b}$$

We shall see in Chap. 3 that practical optical fiber communication systems have the additional condition that $\Delta \ll 1$ where

Now

$$\Delta \doteq \frac{n_1 - n_2}{n_1} \ll 1$$

and

$$n_1 \Delta = n_1 - n_2 \quad \rightarrow \quad n_2 = n_1 (1 - \Delta)$$

which leads to

$$n_2^2 = n_1^2 (1 - 2\Delta + \Delta^2) \approx n_1^2 - 2\Delta \, n_1^2$$

$$V \approx n_1 k_0 a \sqrt{2 \, \Delta} \tag{2.25c}$$

Clearly Eq. (2.25a) is the equation of a circle with radius V. Here, too, we select this particular notation in anticipation of our analysis of cylindrical fibers in the next chapter. Equations (2.23a) and (2.23b) may also be recast with this notation in the following forms:

$$Y = X \tan(X) \qquad \text{(for even-symmetry TE modes)}$$

$$Y = -X \cot(X) \qquad \text{(for odd-symmetry TE modes)}$$

These nonlinear equations can be solved simultaneously using graphical techniques. First we evaluate the parameter V numerically using the fiber system specifications, namely, λ_0, n_1, n_2 (or Δ), and the slab halfwidth a. Then we draw a quarter of a circle of radius V on the graph shown in Fig. 2.7. Each intersection between this arc and the $X \tan X$ and $-X \cot X$ curves represents a TE mode that can propagate through that particular slab waveguide. Then the phase propagation constant for each mode may be obtained by finding either the X or Y coordinates for each intersection and employing whichever of Eqs. (2.24) is convenient.

Example 2.1   Let $a = 4.0$ μm, $n_1 = 1.463$, $\Delta = 0.01$, $\lambda_0 = 1.3$ μm, and

$$V = \frac{2\pi}{1.3 \times 10^{-6}} \, (1.463) \sqrt{0.02} \, (4 \times 10^{-6}) \approx 4.00$$

An arc of radius 4 has been drawn on the slab waveguide curves shown in Fig. 2.7. The intersections of the curves now determine κ from the X coordinates of the intersections and/or determine γ from the Y coordinates of the intersections. Only one set is necessary to find the β for each TE mode. In this example we shall use the Y coordinates for each of the TE modes that propagate

$$Y^2 = \gamma^2 a^2 = a^2(\beta^2 - n_2^2 k_0^2)$$

and

$$\beta^2 = \frac{Y^2}{a^2} + n_2^2 k_0^2$$

Also note that

$$v_z = \frac{\omega}{\beta} = \frac{2\pi f}{\beta} = \frac{2\pi c}{\beta \lambda_0}$$

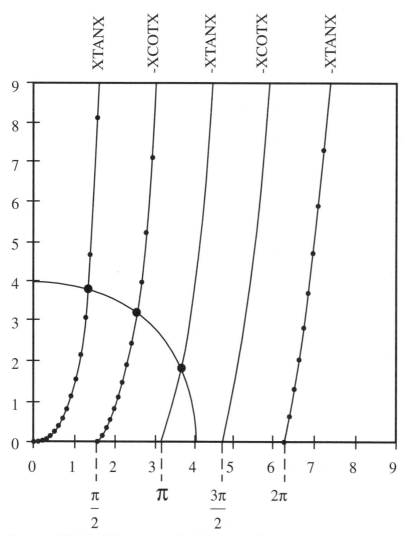

Figure 2.7   Modal solution example for slab waveguide.

TABLE 2.1 **Solutions to Example 2.1***

TE mode	Y	β_z^2	β_z	v_z	θ_z, degrees
0	3.8	4.9863×10^{13}	7.0614×10^6	2.0534×10^8	2.99
1	3.2	4.9613×10^{13}	7.0437×10^6	2.0582×10^8	5.04
2	1.8	4.9197×10^{13}	7.0140×10^6	2.0672×10^8	7.28

Note: Only certain discrete values of β that satisfy all the E&M boundary conditions and the Helmholtz eqautions in both regions (core and cladding) are given in this table.

Numerical values for β_z, v_z and the angle with the Z axis are given in Table 2.1.

The angle θ_z was obtained by using Eq. (2.12). Thus $\theta_z = \cos^{-1}\{\beta\lambda_0/2\pi n_1\}$. Note also how the higher phase velocity coincides with the largest θ_z, which is in agreement with our discussion concerning Fig. 2.2.

The results of Table 2.1 can be misleading. The quantity v_z is a phase velocity, not a group velocity, and therefore these results do not imply that a real signal will propagate more rapidly for the TE_2 mode than the TE_0 mode. In fact, the opposite is true since it is known that

$$v_{phase}\, v_{group} = \frac{1}{\mu\,\epsilon} \tag{2.25d}$$

Therefore the group velocity (which more properly represents the velocity of propagation of information-bearing signals) is actually larger for the TE_0 mode.

It turns out, however, that the identity of the fastest traveling mode is irrelevant. It is a practical impossibility to transmit only one of these modes at a time for a reasonable distance down the slab fiber. Because of imperfections at the core–ladding boundary, all modes that can propagate will propagate, and the fact that each mode has a different velocity can result in serious signal distortion regardless of which mode arrives first. This type of distortion is called *intermodal dispersion*. We note that although the variation of β is very slight, the variation in phase velocity is relatively small. The total phase variation, however, may be very large because the lengths of many fibers are measured in kilometers.

The TE modes for Example 2.1 are displayed in Fig. 2.8.

The specific values for the κ_{kh} and γ_k employed in Fig. 2.8 were obtained for Example 2.1 from the Y intercepts listed in Table 2.1 and are listed in Table 2.2.

These values of κ_k and γ_k lead to TE_k modes that are continuous (within graphical accuracy) at $x = |a|$ (see Fig. 2.8).

2.3.6 Conditions for single-mode propagation

One way to avoid the intermodal dispersion problem is to design the slab fiber so that only a single mode can propagate. As we shall see in Chap. 3, this does not entirely eliminate dispersion distortion. We note from the slab waveguide graphs that we have employed to determine the modal solutions, that when $V < \pi/2$, only the TE0 mode will propagate (*and there is no cutoff for this mode*). This forms the foundation for single-mode propagation condi-

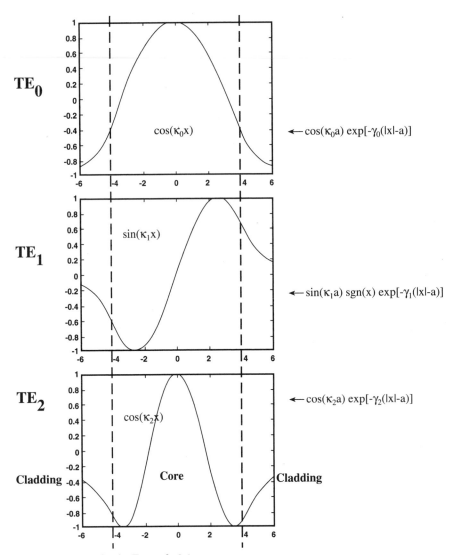

Figure 2.8   TE modes for Example 2.1.

TABLE 2.2 κ_k and γ_k used in Fig. 2.8

k	Y	$X = V^2 - Y^2$	$\kappa_k = \dfrac{X}{a}$	$\gamma_k = \dfrac{Y}{a}$
0	3.8	1.25	0.3125	0.95
1	3.2	2.4	0.6	0.8
2	1.8	3.572	0.893	0.45

tions. With a view toward our study of practical cylindrical fibers in Chap. 3, we shall concentrate on slab fibers that are such that $\Delta \ll 1$. This simplifies the single-mode design task. Then

$$V = k_0 a \sqrt{n_1^2 - n_2^2} \approx n_1 k_0 a \sqrt{2\Delta} < \frac{\pi}{2}$$

Since $k_0 = 2\pi/\lambda_0$, where λ_0 = wavelength in vacuum, the single-mode propagation condition is either

$$\lambda_0 > 4n_1 a \sqrt{2\Delta}$$

or

$$a < \frac{\lambda_0}{4n_1 \sqrt{2\Delta}}$$

For cylindrical fibers, we shall arrive at formulas very similar to the two bounds on a and λ_0 given above. In Chap. 3 the slab halfwidth a will be replaced by the radius of the core. In the cylindrical fiber the $\pi/2$ cutoff condition on V turns out to be 2.405. In practice the absorption characteristics of the fiber usually determine the λ_0 for the channel, so the radius of the core is the primary single-mode fiber design parameter.

The TE_1 mode won't be stimulated unless $n_1 k_0 a \sqrt{2\Delta} > \pi/2$, and the TE_m mode cannot propagate a reasonable distance unless $n_1 k_0 a \sqrt{2\Delta} > m\pi/2$.

2.3.7 TM modes

For TM modes $H_z = 0$ and H_y, E_x, and $E_z \neq 0$. Here we assume the following form for the only nonzero H component:

$$H_y = H_{0,y}(x) e^{-j\beta z} e^{j\omega t}$$

Then E_x and E_z are obtained from Maxwell's equations to arrive at

$$E_x = \frac{\beta}{\omega\epsilon} H_y \qquad \text{(App. 2A, Sec. 2A.2, Eq. 3.}a\text{)} \qquad (2.26a)$$

$$E_z = \frac{-j}{\omega\epsilon} \frac{\partial H_y}{\partial x} \qquad \text{(App. 2A, Sec. 2A.2, Eq. 3.}c\text{)} \qquad 2.26b$$

The solutions have forms similar to those obtained for TE modes and by matching the tangential boundary conditions the following equations are obtained:

For even TM modes

$$\tan(\kappa a) = \frac{n_1^2}{n_2^2} \frac{\gamma}{\kappa} = \frac{n_1^2}{n_2^2} \frac{\gamma a}{\kappa a} \qquad (2.27a)$$

For odd TM modes

$$\tan{(\kappa a)} = \frac{n_2^{\,2}}{n_1^{\,2}} \frac{-\kappa}{\gamma} = \frac{n_2^{\,2}}{n_1^{\,2}} \frac{-\kappa a}{\gamma a} \qquad (2.27b)$$

These equations are very similar to Eqs. (2.23) that we obtained for the TE modes. The only difference between the TM and TE equations is a factor of $n_1^{\,2}/n_2^{\,2}$ on the right-hand side of Eq. (2.27a) and its inverse on the right-hand side of Eq. (2.7b).

Let us now assume that $\Delta \ll 1$. Then this particular factor is so close to unity that it will not noticeably alter the values of β as obtained graphically. Equation (2.25a) is the same for both TE and TM modes. We can then pair up TE and TM modes since they have approximately the same propagation constant β. The electric field for the TM modes is perpendicular to the electric field of the corresponding TE mode. Thus one could combine a TE_ℓ and TM_ℓ into a single mode with two orthogonal degrees of polarization. Furthermore the single-mode cutoff condition is virtually the same for both types of modes.

2.3.8 Ray optics description of modes

The solutions for the slab waveguide were obtained using the wave equation and the boundary conditions for Maxwell's equations. It is also possible to use ray optics to determine the modes that propagate. This does not present any results that we have not already obtained, but it will give the student some useful insight into the concept of modes in waveguides. The ray optics approach to modes requires that we know the phase shift incurred by the reaction of the process at the core–cladding interface, and this is where we shall begin our coverage of this subject.

When we derived Snell's law, we employed only a single boundary condition and the results were independent of polarization. To obtain the magnitude and phase of both the reflected wave and the transmitted wave, we require all the boundary conditions. Furthermore the results will be dependent on the polarization of the incident beam. Our interest here is only in the reflected rays since they will constitute the modes propagating within the core. We use perpendicular and parallel polarization to obtain the following equations:

$$\frac{E_{0R}^{\perp}}{E_0^{\perp}} = \frac{n_1 \cos \phi_1 - n_2 \, coa \, \phi_2}{n_1 \cos \phi_1 + n_2 \cos \phi_2} \qquad (2.28a)$$

$$\frac{E_{0R}^{\parallel}}{E_0^{\parallel}} = \frac{n_1 \cos \phi_2 - n_2 \cos \phi_1}{n_1 \cos \phi_2 + n_2 \cos \phi_1} \qquad (2.28b)$$

We are interested in these equations only for the total internal reflection condition, specifically, when $\phi_1 > \phi_c$. We recall [see Eq. (2.18)] that under these conditions

$$\cos \phi_2 = -jB \qquad \text{where} \qquad B = \sqrt{\left[\frac{n_1}{n_2}\right]^2 \sin^2 \phi_1 - 1} > 0$$

Next we divide the numerator and the denominator of Eq. (2.28a) by n_2 and arrive at

$$\frac{E_{0,R}^{\perp}}{E_0^{\perp}} = \frac{A + jB}{A - jB} = 1\angle 2\theta_{\perp}$$

where

$$A = \frac{n_1}{n_2} \cos \phi_1 \qquad \text{and} \qquad \theta_{\perp} = \tan^{-1}\left[\frac{B}{A}\right] \qquad (2.29a)$$

Thus the phase shift $\theta_{\perp}$ is a function of ϕ_1. It is important to note that this form is valid only for $\phi_1 > \phi_c$. Otherwise $\cos \phi_2$ is real and B is imaginary, thereby altering the form of the preceding equations. We note the following extremes for $\theta_{\perp}$ for entire internal reflection range:

As $\theta_i \to \theta_c, B \to 0 \Rightarrow \theta_{\perp} \to 0$.

As $\theta_i \to \pi/2, A \to 0 \Rightarrow \theta_{\perp} \to \pi$.

Similarly, by using Eq. (2.28b), one can show that the phase shift for ∥ polarized incidence is given by a phase shift $= 2 \theta_{\|}$, where

$$\theta_{\|} = \frac{n_1^2}{n_2^2} \tan^{-1}\left[\frac{B}{A}\right] \qquad (2.29b)$$

We now turn our attention to the ray optics view of modes. Consider Fig. 2.9.

In Fig 2.9 the arrows depict the direction of the rays. The key to this ray forming a mode is that the net phase shift along path ABC must be an integer multiple of 2π, a phenomenon known as the *self-consistency condition*. This path includes the two phase shifts that occur on reflection from the boundaries of the core. The situation depicted in Fig. 2.9, however, is not representative of the situation in a practical fibers because the $\theta < 45°$ condition would permit too many modes to propagate. We included Fig. 2.9 here because it presents a clearer picture of the self-consistency condition than the more realistic situation depicted in Fig. 2.10.

In Fig. 2.10 $\phi = \overline{\theta}$ is the complement of the θ depicted in Fig. 2.9 and in this figure (Fig. 2.10). The extension of the ray outside the confines of the core represented by the dashed line between A and B is there to help explain the self-consistency condition satisfied by ray modes for angle $\theta > 45°$. In Fig. 2.9 we have $45° < \theta < 90°$ (in fact, $\theta \approx 90°$) which corresponds to a representation of a ray mode for weakly guided modes ($\Delta \ll 1$). As we shall see in the next chapter, weakly guided modes are the primary transmission modes employed in fiber-optic communication systems.

Now the dotted line that runs through points B and C in Fig. 2.10 is an equiphase plane which is perpendicular to the rays at these points. The self-

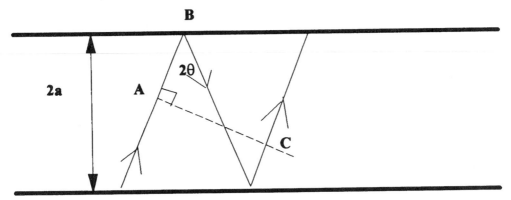

Figure 2.9   Ray version of mode with $\Theta < 45°$.

consistency condition applied to Fig. 2.10 means that the path difference measured in radians between the path lengths AB and AC must be such that

$$\frac{2\pi}{\lambda}(AC - AB) - 2\theta_r = 2\pi\ell \qquad (2.30a)$$

where ℓ = integer, and where θ_r = reflection phase shift for either perpendicular or parallel polarization of the ray, which, in turn, corresponds to TE and TM modes, respectively.

We now reformulate Eq. (2.30a) so that it may be expressed in terms of the angle ϕ shown in Fig. 2.10. We note that

$$\sin \phi = \frac{2a}{AC} \qquad \text{thus} \qquad AC = \frac{2a}{\sin \phi}$$

$$\cos(2\phi) = \frac{AB}{AC} \qquad \text{thus} \qquad AB = AC \cos(2\phi)$$

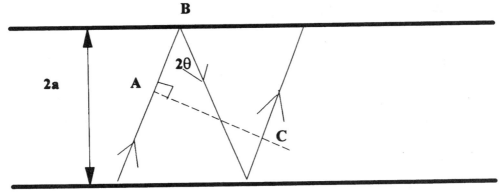

Figure 2.10   Ray Version of Modes $\Theta > 45°$.

Thus

$$AC - AB = AC\,[1 - \cos(2\phi)] = 2\,AC\,\sin^2\phi = 4a\,\sin\,\phi$$

where we substituted for AC. Then the self-consistency condition equation becomes

$$\frac{2\pi}{\lambda}\,4a\,\sin\,\phi_\ell - 2\theta_r = 2\pi\ell \qquad (2.30b)$$

In general, there are a finite number of discrete angles ϕ_l that satisfy Eq. (2.30b). This is because only isolated angles θ_ℓ will satisfy both the total internal reflection condition and simultaneously match the phase shift at the reflection condition expressed by Eq. (2.30b).

It is interesting to note here that under the $\Delta \ll 1$ condition the reflection phase shifts θ_r are almost identical for perpendicular polarization [Eq. (2.29a)] and parallel polarization [Eq. (2.29b)]. We found that for the slab waveguide, the TE modes have only a nonzero E_y component, which means that the TE modes are perpendicularly polarized, while in a similar manner the TM modes are parallel polarized. Once again we see that TE and TM modes of the same index will have the same ray trajectories, and that they can be considered as two orthogonal polarizations of a single mode.

We shall encounter a similar simplification of mode structures when we study cylindrical fibers. There, too, it will be brought about by the condition that $\Delta \ll 1$. This condition is almost universally the case for practical cylindrical fibers because, as we shall see in the next chapter, this condition helps minimize dispersion. The rectangular geometry of the slab waveguide itself is not of great physical interest, but some other variations of this enclosed finite geometry have application to optical channels embedded in integrated optical devices. There is usually at least one boundary with air often from a large-index-of-refraction material such as GaAs or InP, and we can no longer assume that $\Delta \ll 1$ for such situations. We cannot, therefore, use $\Delta \ll 1$ to simplify the problem of finding the modes that propagate in such channels. For example, if the index of refraction in the upper cladding region differs from that of the lower cladding region, it is possible to obtain a total cutoff condition where even the fundamental mode has a cutoff and will not propagate [2,3].

The ray mode picture provides an alternative viewpoint that shows why the mode with the fastest group velocity coincides with the mode with the slowest phase velocity. This may be obtained by examining Figs. 2.2 and 2.9 and noting that the phase velocity concept depicted in Fig. 2.2 applies only to a steady-state solution. A pulse traveling down the fiber will follow the ray trajectory indicated in Fig. 2.9, and the fastest mode will correspond to the ray with the shortest trajectory down the fiber. This will correspond to the smallest angle of incidence with the core–cladding interface (θ in Fig. 2.9), and this coincides with the smallest phase velocity as depicted by Fig. 2.2. This viewpoint by itself, however, does not generate Eq. (2.25d).

2.3.9 Power loss for normal incidence

The fraction of power reflected at an interface between two dielectrics can be found from Eqs. (2.28a) and (2.28b). In particular, at normal incidence when $\phi_1 = \phi_2 = 0$, these equations yield the same result for all polarizations. Let r = the fraction of power reflected and note that this term depends or the reflection coefficient squared to arrive at

$$r = \left\{ \frac{n_1 - n_2}{n_1 + n_2} \right\}^2 \tag{2.31}$$

This simple result is valid only for normal incidence. It is clear from Eqs. (2.28) that in general the reflection coefficient is a function of the angle of incidence.

Problems

1   A chunk of gallium arsenide (GaAs) with $n = 3.6$ lies at the bottom of a tank of water ($n = 1.33$). The tank is illuminated from above the water by a laser beam of $\lambda_0 = 1.3$ μm in such a way that the it falls on the top surface of the GaAs block. Let the angle of incidence at the air–water interface be 30° measured from the normal. Answer the following:
   (a) Find the wavelength and frequency in each medium.
   (b) Determine the angle with respect to the normal in each material.
   (c) Now let the input laser beam be normal to the air–water interface. What fraction of the incident power would enter the GaAs? Also, what is the decibel loss in this process?

2   Consider Fig. 2.5, which we used to derive Snell's law, and let the x axis and z axis be as indicated in the figure. Now consider a uniform plane wave that is incident on this page from above at angle of 45° with the x axis and 55° with the y axis. Also let $\lambda_0 = 1.3$ μm, $n_1 = 3.6$ (GaAs), and $n_2 = 1.45$.
   (a) Find θ_z. From the information given above
   (b) What is the critical angle for this interface?
   (c) What is the direction of the y axis in Fig. 2.5, and why?
   (d) Which angle is the incident angle that applies to Snell's law?
   (e) What is the decay constant on the n_2 side of the interface?

3   Let a 1.55-μm laser beam illuminate the core of a slab waveguide with the following parameters: $a = 3$ μm, $n_1 = 1.45$, and $\Delta = 0.01$.
   (a) Determine which TE modes are excited.
   (b) For each excited TE mode, determine the actual propagation constant β.
   (c) For each excited TE mode, determine the propagation velocity.
   (d) Determine the largest slab halfwidth a that will support only one TE mode.
   (e) Which TM modes will propagate?

Appendix 2A. Some Basic Electromagnetic Equations

2A.1 Maxwell's equations

$$1.\ \nabla \times \mathbf{E} = \frac{-\partial \mathbf{B}}{\partial t}$$

2. $\nabla \times \mathbf{H} = \dfrac{\partial \mathbf{D}}{\partial t}$

3. $\nabla \cdot \mathbf{D} = \rho = 0$ (fo free charges)
4. $\nabla \cdot \mathbf{B} = 0$ (no free magnetic poles)
5. In isotropic media $\mathbf{D} = \varepsilon \mathbf{E}$ and $\mathbf{B} = \mu \mathbf{H}$
6. Boundary conditions
   a. $E_{\text{tan},1} = E_{\text{tan},2}$; $H_{\text{tan},1} = H_{\text{tan},2}$
   b. $D_{\text{norm},1} = D_{\text{norm},2}$; $B_{\text{norm},1} = B_{\text{norm},2}$
7. Wave equations

   a. $\nabla^2 \mathbf{E} = \mu \varepsilon \dfrac{\partial^2 E}{\partial t^2}$ (three equations)

   b. $\nabla^2 \mathbf{H} = \mu \epsilon \dfrac{\partial^2 H}{\partial t^2}$ (three equations)

   c. $\nabla^2 \psi = \mu \epsilon \dfrac{\partial^2 \psi}{\partial t^2}$ (generic version: ψ may be any component of $\mathbf{E}$ or $\mathbf{H}$)

2A.2 Plane waves (cartesian-coordinated)

1. $\mathbf{E}(r,t) = \text{Re}\{E_{)}(r) \exp[j(\omega t - \mathbf{k} \cdot \mathbf{r})]\}$
   where $\mathbf{k} = \beta \{\hat{\mathbf{x}} \cos \theta_x + \hat{\mathbf{y}} \cos \theta_y = \hat{\mathbf{z}} \cos \theta_z\}$
   and $\mathbf{r} = \hat{\mathbf{x}} + \hat{\mathbf{y}} + \hat{\mathbf{z}}$
2. Plane waves traveling in the z direction. *This constraint will apply to the remainder of this section (2A.2).*
   a. $\mathbf{E} = \text{Re}\{\mathbf{E}_0(x,y) \exp[j(\omega t - \beta z)]\}$
   b. $\mathbf{H} = \text{Re}\{\mathbf{H}_0(x,y) \exp[j(\omega t - \beta z)]\}$
   In the following two equations ψ represents any component of either $\mathbf{E}$ or $\mathbf{H}$:

   c. $\dfrac{\partial \psi}{\partial z} = -j\beta \psi$

   d. $\dfrac{\partial \psi}{\partial t} = j\omega \psi$

3. We now substitute Eqs. 2.*c* and 2.*d* in this section into Eqs. 1 and 2 in Sec. 2A.1 to obtain

   a. $\dfrac{\partial H_z}{\partial y} + j\beta H_y = j\omega \epsilon E_x$

   b. $-j\beta H_x - \dfrac{\partial H_z}{\partial x} = j\omega \varepsilon E_y$

c. $\dfrac{\partial H_y}{\partial x} - \dfrac{\partial H_x}{\partial y} = j\omega\epsilon E_z$

d. $\dfrac{\partial E_z}{\partial y} + j\beta E_y = -j\omega\mu H_x$

e. $-j\beta E_x - \dfrac{\partial E_z}{\partial x} = -j\omega\mu H_y$

f. $\dfrac{\partial E_y}{\partial x} - \dfrac{\partial E_x}{\partial y} = -j\omega\mu H_z$

4. By further manipulation, we obtain the following equations which give the x and y components of the fields in terms of E_z and H_z:

a. $E_x = \dfrac{-j}{\kappa^2}\left\{\omega\mu\,\dfrac{\partial H_z}{\partial y} + \beta\,\dfrac{\partial E_z}{\partial x}\right\}$

b. $E_y = \dfrac{-j}{\kappa^2}\left\{\beta\,\dfrac{\partial E_z}{\partial_y} - \omega\mu\,\dfrac{\partial H_z}{\partial x}\right\}$

c. $H_x = \dfrac{-j}{\kappa^2}\left\{\beta\,\dfrac{\partial H_z}{\partial x} - \omega\epsilon\,\dfrac{\partial E_z}{\partial y}\right\}$

d. $H_y = \dfrac{-j}{\kappa^2}\left\{\beta\,\dfrac{\partial H_z}{\partial y} + \omega\epsilon\,\dfrac{\partial E_z}{\partial x}\right\}$

where $\kappa^2 = k_0^2 - \beta^2$ and $k_0 = \omega\sqrt{\mu_0\epsilon_0}\ k_0 = \dfrac{2\pi}{\lambda}$

5. Using Eqs. 4.c and 4.d in Eq. 3.c and multiplying through by $\dfrac{j\kappa^2}{\omega\epsilon}$, we obtain

a. $\dfrac{\partial^2 E_z}{\partial x^2} + \dfrac{\partial^2 E_z}{\partial y^2} + \kappa^2 E_z = 0$

Similarly, using Eqs. 4.a, 4.b, and 3.f, one obtains

b. $\dfrac{\partial^2 H_z}{\partial x^2} + \dfrac{\partial^2 H_z}{\partial y^2} + \kappa^2 H_z = 0$

2A.3 Plane waves (cylindrical coordinates)

1. Curl and laplacian in cylindrical coordinates

a. $\nabla \times \mathbf{A} = \hat{\mathbf{r}}\left\{\dfrac{1}{r}\dfrac{\partial H_z}{\partial\phi} - \dfrac{\partial A_\phi}{\partial z}\right\} + \hat{\boldsymbol{\phi}}\left\{\dfrac{\partial A_r}{\partial z} - \dfrac{\partial A_z}{\partial r}\right\} + \hat{\mathbf{z}}\,\dfrac{1}{r}\left\{\dfrac{\partial}{\partial r}\,rA_\phi - \dfrac{\partial A_r}{\partial\phi}\right\}$

  b. $\nabla \psi = \dfrac{\partial^2 \psi}{\partial r^2} + \dfrac{1}{r}\dfrac{\partial \psi}{\partial r} + \dfrac{1}{r^2}\dfrac{\partial^2 \psi}{\partial \phi^2} + \dfrac{\partial^2 \phi}{\partial z^2}$

2. By manipulating the curl equations in a manner similar to that employed to obtain Eqs. 4.*a* to 4.*d* in Sec. 2A.2, one can derive

  a. $E_r = \dfrac{-j}{\kappa^2}\left\{\beta\,\dfrac{\partial E_z}{\partial r} + \omega\mu\,\dfrac{1}{r}\dfrac{\partial H_z}{\partial \phi}\right\}$

  b. $E_\phi = \dfrac{-j}{\kappa^2}\left\{\beta\,\dfrac{1}{r}\dfrac{\partial E_z}{\partial \phi} - \omega\mu\,\dfrac{\partial H_z}{\partial r}\right\}$

  c. $H_r = \dfrac{-j}{\kappa^2}\left\{\beta\,\dfrac{\partial H_z}{\partial r} - \omega\epsilon\,\dfrac{1}{r}\dfrac{\partial E_z}{\partial \phi}\right\}$

  d. $H_\phi = \dfrac{-j}{\kappa^2}\left\{\beta\,\dfrac{1}{r}\dfrac{\partial H_z}{\partial \phi} - \omega\epsilon\,\dfrac{\partial E_z}{\partial r}\right\}$

References

1. A. Cherin, *An Introduction to Optical Fibers,* McGraw-Hill, New York, 1983.
2. B. E. A. Saleh and M. C. Teich, *Fundamentals of Photonics,* Wiley, New York, 1991.
3. P. Bhattacharya, *Semiconductor Optoelectronic Devices,* Prentice-Hall, Englewood Cliffs, N.J., 1994, p. 242.

3

Cylindrical Waveguide

3.1 Modes for Cylindrical Fibers

3.1.1 Introduction

In the previous chapter we saw how a light beam can be confined within a slab waveguide, and how the propagating radiation could be expressed in terms of discrete modes. Graphical techniques were presented there for determining which modes can propagate in a particular slab waveguide structure. We also were able to determine the phase velocity for each mode. Our coverage of optical fibers will rely heavily on the physical intuition we obtained in Chap. 2, as we briefly (*relatively speaking*) examine propagation in a cylindrical waveguide. We shall encounter a somewhat similar graphical technique for determining the corresponding information about the identity and physical properties of the modes that propagate through cylindrical optical fibers. As we shall see, the fact that cylindrical fibers are enclosed in two dimensions as opposed to one for the slab waveguide will lead to a more complicated mathematical description of a diverse set of modes.

The typical structure for a step index cylindrical fiber is shown in Fig. 3.1. Although there are variations of this structure that we shall discuss later in this chapter, we shall simply refer to this as the *fiber* or the *optical fiber*. It is, after all, by far the most common fiber structure in use today.

As was the case with the slab waveguide, the optical fiber has a core with an index of refraction $n_1 > n_2$, the index of refraction of the cladding. There is also a soft opaque outer covering which provides complete optical isolation as well as a degree of mechanical isolation from external forces. The cladding completely encases the core in a concentric structure, and the fiber is open only at the ends. The cylindrical symmetry of these fibers will result in a more complex mode structure than we experienced with the relatively open structure of the slab waveguide. For example, it will be necessary to use two subscripts to designate modes for this structure.

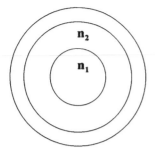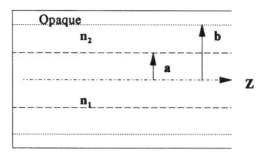

Figure 3.1   Basic cylindrical fiber structure.

3.1.2   Solution to wave equation in cylindrical coordinates

The laplacian (∇^2) in the wave equation will be expressed in cylindrical coordinates for the geometry of the fiber depicted in Fig. 3.1. That form may be found in App. 2A, Sec. 2A.3, Eq. 1.*b* and leads to the following version of the wave equation:

$$\frac{\partial^2 U}{\partial r^2} + \frac{1}{r}\frac{\partial U}{\partial r} + \frac{1}{r^2}\frac{\partial^2 U}{\partial \phi^2} + \frac{\partial^2 U}{\partial z^2} = \epsilon\mu\frac{\partial^2 U}{\partial t^2} \tag{3.1}$$

where U may be any cylindrical coordinate component of either **E** or **H.** The primary propagation direction of interest in FOCS will be directed down the fiber; hence it will be generally in the z direction. However, as was the case in the slab fiber, rays for the propagating modes generally travel at angles to the z axis. We therefore assume that traveling in the z direction has a solution in the following form:

$$U(r,\phi,z) = AF_1(r)F_2(\phi)\,e^{j(\omega t - \beta z)} \tag{3.2}$$

The last $e^{j(\omega t - \beta z)}$ factor represents this propagation term, while the product $F_1(r)\,F_2(\phi)$ represents the usual separation of variables technique often employed to solve this type of partial differential equation.

We now utilize the following equations, most of which were previously encountered in Chap. 2:

$$\frac{\partial^2 U}{\partial t^2} = -\omega^2 U \quad\text{and}\quad \frac{\partial^2 U}{\partial z^2} = -\beta^2 U$$

$$k_1 = n_1 k_0 = \omega\sqrt{\epsilon_1\mu_0}$$

$$k_2 = n_2 k_0 = \omega\sqrt{\epsilon_2\mu_0} \quad\text{where}\quad k_0 = \frac{2\pi}{\lambda_0}$$

As in Chap. 2, let $\kappa^2 = n_1^2 k_0^2 - \beta^2$ and $\gamma^2 = \beta^2 - n_2^2 k_0^2$.

Substituting Eq. (3.2) and the preceding equations into Eq. (3.1) and dividing through by the product $F_1(r) F_2(\phi)$, one obtains the following equations:

$$\left\{ \frac{1}{F_2} \left[\frac{d^2F_1}{dr^2} + \frac{1}{r} \frac{dF_1}{dr} \right] + r^2\kappa^2 \right\} + \left\{ \frac{1}{F_2} \frac{d^2F_2}{d\phi^2} \right\} = 0 \qquad (3.3a)$$

$$\left\{ \frac{r^2}{F_1} \left[\frac{d^2F_1}{dr^2} + \frac{1}{r} \frac{dF_1}{dr} \right] - r^2\gamma^2 \right\} + \left\{ \frac{1}{F_2} \frac{d^2F_2}{d\phi^2} \right\} = 0 \qquad (3.3b)$$

function of r only   function of ϕ only

Equation (3.3a) applies to the core region where n_1 is the index of refraction and κ applies, while Eq. (3.3b) applies to the cladding region. We note that each of the two major brackets in Eqs. (3.3a) and (3.3b) depends on only one variable. To ensure that the left-hand side adds up to zero for all values of r and ϕ, each bracket separately must equal constants of opposite sign. With a view toward the symmetry of the fiber, we equate the two brackets to the following constants that we select somewhat arbitrarily. These constants must sum to zero.

Function of r only $= \ell^2$ and   function of ϕ only $= -\ell^2$

3.1.3 Solution of $F_2(\phi)$ equation

The reason for the preceding particular choice of ℓ^2 as the constant will now be explained. Specifically, the second bracket in both Eqs. (3.3a) and (3.3b) may be expressed as

$$\frac{d^2F_2}{d\phi^2} = -\ell^2 F_2 \qquad (3.4)$$

The solutions of this familiar equation may be expressed as follows

$F_2(\phi) = A_1 e^{j\ell\phi} + A_2 e^{-j\ell\phi}$ or $B \cos \ell\phi + C \sin \ell\phi$

Now, because of the cylindrical symmetry of the fiber, we expect that $F_2(\phi)$ must be periodic in ϕ with period 2π. Thus

$F_2(\phi + \ell 2\pi) = F_2(\phi)$ $\forall$ ϕ and $\forall$ integer ℓ

The periodicity implied by fiber symmetry requires $\ell =$ integer. In addition, the parameter ℓ will be one of the two basic indices used to denote modes for optical fibers. Note that the preceding periodic solution for $F_2(\phi)$ applies to both core and cladding regions. *It is the $F_1(r)$ solutions that will be different for these two regions.*

3.1.4   Solutions for $F_1(r)$

The equation for $F_1(r)$ will be obtained from the first bracket of Eqs. (3.3), but here we shall encounter different forms for the solutions in the core and the cladding regions. Eventually we shall match boundary conditions at the interface of these regions. The parameter κ applies in the core, while the parameter γ applies in the cladding region. The equations for the two regions are

$$\frac{d^2F_1}{dr^2} + \frac{1}{r}\frac{dF_1}{dr} + \left[\kappa^2 - \frac{\ell^2}{r^2}\right]F_1 = 0 \qquad \text{(for } r < a) \qquad (3.5)$$

$$\frac{d^2F_1}{dr^2} + \frac{1}{r}\frac{dF_1}{dr} - \left[\gamma^2 + \frac{\ell^2}{r^2}\right]F_1 = 0 \qquad \text{(for } r > a) \qquad (3.6)$$

The differential Eqs. (3.5) and (3.6) frequently are found in physical situations with cylindrical symmetry that occur in other branches of science and engineering, and the solutions to these equations are therefore well known. Furthermore since these are second-order differential equations, there will be, as was the case in the slab waveguide, two mathematical solutions for each region. Here, however, in contradistinction to our slab waveguide analysis, one of the solutions will be discarded out of hand on physical realizability grounds in *both* the core and cladding regions. The general solution of Eq. (3.5) is

$$F_1(r) = A\,J_\ell(\kappa r) + B\,Y_\ell(\kappa r) \qquad \text{(for } r < a)$$

The $Y_\ell(\kappa r)$ solution is nonphysical in the core region because $|Y_\ell(\kappa r)| \to \infty$, as $r \to \infty$. The other solution $J_\ell(\kappa r)$ is often simply called the *Bessel function* in engineering publications. It is the same function that many students of communication theory encounter when they compute the bandwidth of analog frequency modulation systems. $J_\ell(\kappa r)$ remains finite for all real arguments, and therefore it is indeed a physically realizable solution. Now let us define the following variables that we shall use extensively in this section.

$$R = \frac{r}{a} \qquad \text{and} \qquad X = \kappa a = a\sqrt{n_1^2 k_0^2 - \beta^2} \qquad (3.7a)$$

Then

$$F_1(R) = A\,J_\ell(XR) \qquad \text{(for } R < 1) \qquad (3.7b)$$

Similarly, for the cladding region, we obtain the following solution to Eq. (3.6):

$$F_1(R) = C\,K_l(YR) + D\,I_l(YR) \qquad \text{(for } R > 1)$$

where

$$Y = a\gamma = a\sqrt{\beta^2 - n_2^2 k_0^2} \qquad (3.7c)$$

The primary characteristic for physically realizable solutions for the cladding region is the asymptotic behavior of K_ℓ (YR) and I_ℓ (YR) as R approaches infinity. These turn out to be

$$K_\ell(YR) \approx \sqrt{\frac{\pi}{2YR}}\, e^{-YR} \qquad (\text{as } R \to \infty)$$

and

$$I_\ell(YR) \approx \frac{e^{YR}}{\sqrt{\pi YR}} \qquad (\text{as } R \to \infty)$$

General trends of these functions may be seen in Fig. 3.2. Reasons for adopting the preceding notation for X, Y, and R will soon be made apparent.

Clearly the $I_\ell(YR)$ grows exponentially with increasing R while the $K_\ell(YR)$ term approaches zero. Therefore $I_\ell(YR)$ is a nonphysical solution for the cladding region, and we may express the cladding solution in the following form:

$$F_1(R) = CK_\ell(YR) \qquad (\text{for } R > 1) \qquad (3.8)$$

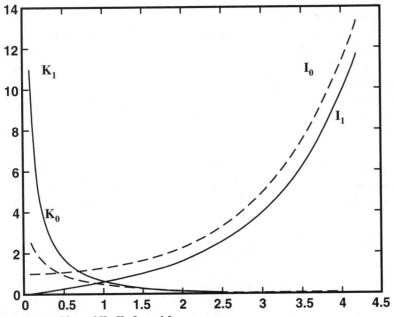

Figure 3.2 Plots of K_0, K_1, I_0, and I_1.

The asymptotic form for $K_\ell(YR)$ is in good agreement with our slab waveguide results. As R increases, the surface area of a cylindrical shell increases in direct proportion to R. Thus the energy density is diluted in direct proportion to $1/R$. This, in turn, causes $|\mathbf{E}|$ to decrease as $R^{-1/2}$, which is in good agreement with the asymptotic form for $K_\ell(YR)$.

It is useful at this point to again define a parameter V (the "normalized frequency") in the same form as we did in Chap. 2:

$$V \doteq \sqrt{X^2 + Y^2} = k_0 \, a \sqrt{n_1^2 - n_2^2}$$

$$V \approx \frac{2\pi}{\lambda_0} n_1 a \sqrt{2\Delta} \qquad \text{(for } \Delta \ll 1) \tag{3.9}$$

We note here that V is virtually identical to the radius parameter that we employed in the graphical analysis of the slab waveguide. This similarity is one reason for introducing the X and Y while the normalized radius R has been introduced for convenience. The term V will also be used to obtain modal solutions graphically for cylindrical waveguides (although somewhat differently than it was used in the slab waveguide analysis).

Another useful parameter that is usually introduced at this point in the analysis is the *normalized propagation* constant b, which is defined as

$$b \doteq 1 - \frac{X^2}{V^2} = \frac{Y^2}{V^2} = \frac{(\beta/k_0)^2 - n_2^2}{n_1^2 - n_2^2} \tag{3.10}$$

where $0 < b < 1$ since $n_2 < \beta/k_0 < n_1$.

The normalized frequency and propagation constants often are employed in the literature to study modes. Specifically, they will be used to determine which modes can propagate, and for determining the value of β for each mode. We shall also find the cutoff condition for single-mode propagation from graphs plotted in terms of these two parameters.

3.1.5 Boundary conditions

In order to begin the solution process, we first select forms for E_z and H_z on the basis of our preceding analysis. Thus from Eqs. (3.7) and (3.8), along with the periodic structure for $F_2(\phi)$, we obtain the following for the core region:

$$E_z = AJ_\ell(XR)e^{j\ell\phi} \tag{3.11a}$$

$$H_z = BJ_\ell(XR)e^{j\ell\phi} \qquad \text{(for } R < 1) \tag{3.11b}$$

while the nonzero fields that occur in the cladding region are given by

$$E_z = CK_\ell(YR)e^{j\ell\phi} \tag{3.12a}$$

$$H_z = DK_\ell(YR)e^{j\ell\phi} \qquad \text{(for } R > 1\text{)} \tag{3.12b}$$

The tangential boundary conditions at the core–cladding interface for cylindrical fibers imply that we must match the E_z, H_z, E_ϕ, and H_ϕ at $R = 1$. We already have forms for E_z and H_z, and we obtain the remaining two forms (which are derived from the curl equations for **E** and **H**):

$$E_\phi = \frac{-j}{\kappa^2}\left\{\beta\,\frac{1}{r}\,\frac{\partial E_z}{\partial\phi} - \omega\mu\,\frac{\partial H_z}{\partial r}\right\} \tag{3.13a}$$

$$H_\phi = \frac{-j}{\kappa^2}\left\{\beta\,\frac{1}{r}\,\frac{\partial H_z}{\partial\phi} + \omega\epsilon\,\frac{\partial E_z}{\partial r}\right\} \tag{3.13b}$$

Core region solutions for E_ϕ and H_ϕ are obtained by substituting Eqs. (3.11) into Eqs. (3.13), while cladding region solutions are obtained from Eqs. (3.12) and (3.13). By matching these four pairs of equations at $R = 1$, we can obtain four simultaneous equations for the unspecified constants A, B, C, and D.

Since we have four equations in four unknowns, it would seem at first glance that it is possible to obtain specific answers for each of the four unknowns. This, however, is a physical impossibility since we would be specifying the electromagnetic field completely without any knowledge of the input optical signal power. Somehow the four parameters A, B, C, and D must be, in some way, proportional to the input signal strength. Therefore these equations cannot be independent, and this realization leads to

$$\mathbb{N}\begin{bmatrix} A \\ B \\ C \\ D \end{bmatrix} = 0 \tag{3.14}$$

where $\mathbb{N}$ is a 4×4 whose matrix elements consist of Bessel functions, first derivatives of Bessel functions, and some constants. We shall not use the specific elements here, but students interested in more detail can find it in the references listed at the end of the chapter.

The elements of the matrix $\mathbb{N}$ are in general functions of β explicitly, as well as implicitly through their dependence on the quantities X and Y that appear within the arguments of the Bessel functions (and their first derivatives where they occur). Now, in order for there to exist nontrivial solutions to Eq. (3.1), specifically, other than $A = B = C = D = 0$, it is necessary that det $\mathbb{N} = 0$. This results in the following equation, which specifies both the modes and their associated value of β:

$$\{\mathcal{I}_\ell + \mathcal{L}_\ell\}\{k_1{}^2 J_\ell + k_2{}^2 \mathcal{K}_\ell\} = \left(\frac{\beta\ell}{a}\right)^2 \left[\frac{1}{X^2} + \frac{1}{Y^2}\right]^2 \tag{3.15}$$

where

$$\mathcal{I}_\ell = \frac{J'_\ell(Xa)}{XJ_\ell(Xa)} \quad \text{and} \quad \mathcal{K}_\ell = \frac{K'_\ell(Ya)}{YK_\ell(Ya)}$$

Clearly, Eq. (3.15) is a very complex equation. Note that the $J_\ell(\bullet)$ form of the Bessel functions and its derivatives are oscillatory with nonperiodic zero crossings, This leads to multiple solutions to Eq. (3.15), which results in multiple solutions for the propagation constant β. Each of these individual solutions for the propagation constant β corresponds to a distinct mode. Furthermore, each of these modes is really equivalent to two modes because of the two degrees of freedom regarding polarization. Numerical techniques and computers are usually employed to obtain solutions to these equations, and the results of such calculations are summarized in graphs that we shall soon be using to obtain specific solutions. To interpret those solutions, we first need to discuss some general properties and mode designation notation of fiber modes in cylindrical coordinates.

3.1.6 Mode designations

As a consequence of the two-dimensional constraints imposed on the cylindrical waveguide structure, we shall require two indices to denote the modes. The two major types of modes found in cylindrical optical fibers are

1. TE modes ($E_z = 0$)
2. TM modes ($H_z = 0$)

Cylindrical fibers also support modes in which neither E_z nor $H_z = 0$; these are called *hybrid modes*. For hybrid modes, the first letter designates which component makes the largest contribution to the transverse field. The ray trajectory for these modes forms a nearly complete spiral generated by straight-line sections between reflections at the core–cladding boundary as it progresses down the fiber without passing through the principal axis of the fiber. For this reason these hybrid modes are sometimes called "corkscrew" modes. On the other hand, the TE modes and TM modes are called *meridional rays* because their ray trajectory does pass through the principal axis of the fiber.

This potpourri of modes can be simplified greatly when $\Delta \ll 1$, a situation which is applicable to virtually all fibers used for high-speed FOCS. We noted in Chap. 2 how the combined conditions $n_1 > n_2$ but also $n_1 \approx n_2$ led to a pairing up of TE and corresponding TM modes in the slab waveguide. An important consequence of this approximation was that it led to a single set of criteria for single-mode propagation for both TE and TM modes. This simplification also occurs in cylindrical waveguides. We call these "weakly

**TABLE 3.1 Examples of Weakly
Guided Modes**

$LP_{\ell m}$	Constituent Modes
LP_{01}	HE_{11}
LP_{02}	HE_{12}
LP_{03}	HE_{13}
LP_{11}	TE_{01}, TM_{01}, HE_{21}
LP_{12}	TE_{02}, TM_{02}, HE_{22}
LP_{13}	TE_{03}, TM_{03}, HE_{23}
LP_{21}	HE_{31}, EH_{11}
LP_{22}	HE_{32}, EH_{12}
LP_{23}	HE_{33}, EH_{13}

guided" [2,3] modes *linearly polarized modes,* designated $LP_{\ell m}$. Each $LP_{\ell m}$ mode consists of a group of modes containing some hybrid modes and/or TE and TM modes. Table 3.1 lists the constituent modes of a few weakly guided modes.

We begin by noting that Eq. (3.15) for the weakly guided case (LP modes) simplifies to Eq. (3.16) [12]:

$$X \frac{J_{\ell \pm 1}(X)}{J_\ell(X)} = \pm \frac{Y K_{\ell \pm 1}(Y)}{K_\ell(Y)} \tag{3.16}$$

The solutions to this equation are depicted by the graph displayed in Fig. 3.4. The general patterns for the constituents of the $LP_{\ell m}$ are

1. $\ell = 0 \Rightarrow HE_{1m}$
2. $\ell = 1 \Rightarrow HE_{2m}$, TE_{0m}, and TM_{0m}
3. $\ell > 1 \Rightarrow HE_{\ell+1,m}$ and $EH_{\ell-1,m}$

All members of a particular $LP_{\ell m}$ group have nearly identical propagation constants β. Furthermore, with the exception of rotation, each member of the LP group has nearly identical field patterns. The mode patterns for a few LP modes are presented in Fig. 3.3.

Each $LP_{\ell m}$ mode behaves like a single mode with two degrees of polarization. Some specific characteristics which can be seen to hold for the three modes are displayed in Fig. 3.3. Some of these trends and an explanation of $LP_{\ell m}$ mode notation follow:

1. There are 2ℓ maxima around the perimeter of the fiber core, and in addition there are m field minima along the radius vector, including the outside of the pattern.

LP$_{01}$ **HE$_{11}$**

LP$_{11}$

TE$_{01}$

TM$_{01}$

HE$_{21}$

LP$_{21}$

EH$_{11}$

HE$_{31}$

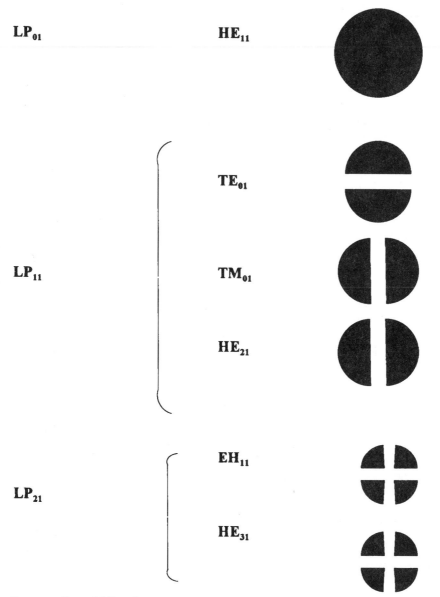

Figure 3.3 Several LP mode patterns.

2. The field pattern for each LP$_{\ell m}$ mode constituent is nearly identical (if we ignore rotation). In practice, when a particular LP$_{\ell m}$ mode is stimulated in a fiber, it is virtually impossible to observe only one of the constituents (provided there is more than one). They all will propagate as a result of coupling between the modes created by imperfections in the core–cladding interface.

3. The LP modes are only approximate solutions to the field equations in cylindrical coordinates, but they're a very good approximation for practical optical fibers.

4. It is useful to note that the HE modes with $\ell' = \ell + 1$ have very nearly the same β as do EH modes with $\ell' = \ell - 1$. Furthermore the superposition of these modes has only four non-zero-field components, which makes this combination very much like a TE or TM mode.

3.1.7 Mode propagation factors

In this section we shall examine a relatively simple graphical technique for determining which modes propagate and their corresponding propagation constant β. The main theme of this text is ultra-high-speed data communications, so we will consider only LP-type modes here.

The X and Y parameters that appear in Eq. (3.16) are connected to each other by the relation $V \doteq \sqrt{X^2 + Y^2}$; thus, for a given normalized frequency, V, Eq. (3.16) is really an equation in a single variable (say, X). Soon we shall employ this equation to find the propagation constant for the fundamental mode LP_{01}. We note here, however, that when $V > 2.405$ there *may* be multiple solutions to this equation. For each value of the mode parameter ℓ, each of these multiple solutions corresponds to different values of m beginning with $m = 1$.

The numerical solutions plotted in Fig. 3.4 display the normalized propagation constant b versus the normalized frequency V for several $LP_{\ell m}$ modes. The first step in using this graph is to compute the normalized frequency V using the parameters of the fiber and the free-space wavelength λ_0 of the optical field:

$$V \approx \frac{2\pi}{\lambda_0} \, an_1 \, \sqrt{2\Delta}$$

The coordinates of the intersections of vertical lines drawn at the specified V with the various $LP_{\ell m}$ mode curves yield the normalized propagation constant for each $LP_{\ell m}$ mode that will propagate, and this, in turn, determines the propagation constant β for each mode. In Gloge's paper [7] the ordinate b is given by Eq. (3.17a), and it presents a relation between b and β that differs from Eq. (3.10). Namely

$$b \approx \frac{(\beta/k_0) - n_2}{n_1 - n_2} \tag{3.17a}$$

This difference can be reconciled by noting that

$$b = \frac{(\beta/k_0)^2 - n_2^2}{n_1^2 - n_2^2} = \left\{ \frac{(\beta/k_0) - n_2}{n_1 - n_2} \right\} \left\{ \frac{(\beta/k_0) + n_2}{n_1 + n_2} \right\} \tag{3.17b}$$

Furthermore, both κ^2 and γ^2 are defined for cylindrical waveguides exactly

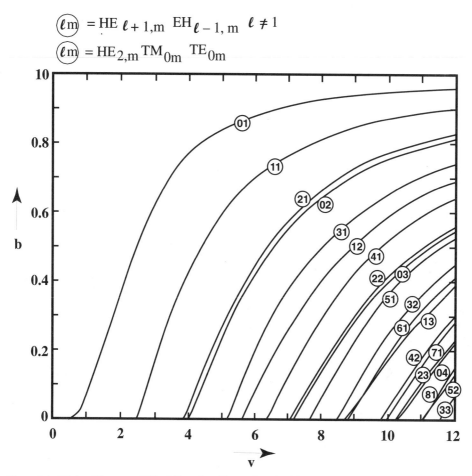

$$\textcircled{\ell m} = \text{HE}\,_{\ell+1,m}\ \text{EH}\,_{\ell-1,m}\ \ell \neq 1$$

$$\textcircled{\ell m} = \text{HE}\,_{2,m}\ \text{TM}\,_{0m}\ \text{TE}\,_{0m}$$

Figure 3.4 Plots of b versus V for LP modes. (*Reproduced with permission from Ref. 2.*)

the same way as they were in the slab waveguide analysis. Therefore we have once again the relation

$$n_2 < \frac{\beta}{k_0} < n_1$$

Now the $n_1 \approx n_2$ aspect of the weakly guided modes condition makes the second bracket in Eq. (3.17b) approximately equal to unity, thereby resulting in the simplified form of Eq. (3.17a).

The preceding analysis also shows that cutoff conditions occur when $\beta \rightarrow n_2 k_0$. This results in the following trends:

1. When this limit is reached, the field in the cladding region will no longer be an evanescent field and energy will flow outward in the radial direction.

Thus the field in the cavity will rapidly lose energy as it propagates through the fiber.

2. There are a few "leaky" modes which propagate despite being outside the cutoff bounds because of an angular momentum barrier at the boundary that can reflect light;

3. For modes with $\beta \approx n_2 k_0$ a significant fraction of the power travels in the cladding in the z direction (not the r direction).

4. On the other hand, as β increases, the modes tend to congregate toward the center of the core and very little power travels in the cladding.

Example 3.1   Let $a = 4.2$ μm, $n_1 = 1.463$, $\Delta = 0.01$, and $\lambda_0 = 1.55$ μm. The first step is to use this specification to find V. Thus we have

$$V = \frac{2\pi}{1.55 \times 10^{-6}} (1.463)\sqrt{0.02}\,(4.2 \times 10^{-6}) \approx 3.52$$

The next step in the process is to draw a vertical line on the graph at $V \approx 3.52$. This is displayed in Fig. E3.4. Clearly this line will intersect the curves only for the LP_{01} and LP_{11} modes. This means that this particular system will support only those two

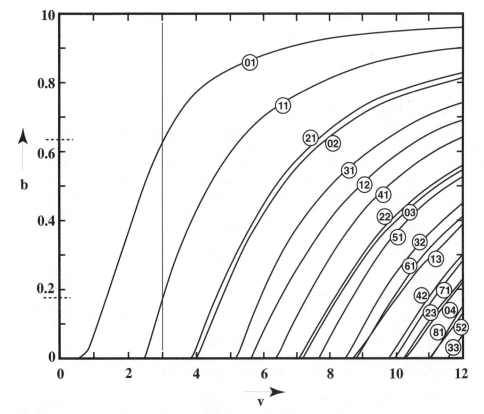

Figure E3.4   Plots of b versus V for LP modes with $V = 3.52$ as in example 3.1.

particular modes; i.e. these two modes are the only modes that will propagate under the low-loss condition. Other modes may be stimulated by the optical source driving the fiber, but the other modes will be lossy and die out exponentially as they travel down the fiber. Furthermore, if the fiber is illuminated by an optical source operating at 1.55 μm, both of these modes will be excited regardless of the structure of the field pattern entering the fiber. Even if it were possible to excite only one of the modes at the fiber input, imperfections in the core–cladding interface would soon distribute energy (not necessarily equally) among all the modes that could propagate.

We next determine numerical values for the propagation parameter β and the phase velocity of each mode that propagates.

1. The Y projections of the intersections in the graph are approximately $b_{01} \approx 0.62$ and $b_{11} \approx 0.16$.

2. Next we find β from Eq. (3.17a) with the result

$$\beta = k_0[b(n_1 - n_2) + n_2] \qquad (3.17c)$$

Also recall that $n_2 = n_1(1 - \Delta) \approx 1.448$, which leads to $n_1 - n_2 \approx 0.0146$. For the given wavelength

$$k_0 = \frac{2\pi}{1.55 \times 10^{-6}} = 4.054 \times 10^6$$

We finally obtain $\beta_{01} \approx 5.93 \times 10^6$ and $\beta_{11} \approx 5.81 \times 10^6$ and also

$$v = \frac{\omega}{\beta} = \frac{2\pi c}{\lambda_0 \beta} \rightarrow v_{01} = 2.05 \times 10^8 \qquad \text{and} \qquad v_{11} = 2.09 \times 10^8$$

3.1.8 Single-mode propagation conditions

We can see from Table 3.1 that the lowest order mode turns out to be the HE_{11}, which happens to be a "corkscrew" mode. It is also apparent from Fig. 3.4 that there is no cutoff condition for this mode, which means that this mode can propagate at any optical wavelength. On the other hand, it can be seen that all the other modes will not propagate for fiber systems where $V < 2.405$ and only the LP_{01} will propagate under that condition. Thus the single-mode propagation condition is expressed as

$$2.405 > V \approx \frac{2\pi}{\lambda_0} a n_1 \sqrt{2\Delta}$$

which leads to the following constraint core radius

$$a < \frac{2.405\lambda_0}{2\pi n_1 \sqrt{2\Delta}} \qquad (3.18a)$$

and the following constraint free-space wavelength for the optical field:

$$\lambda_0 > \frac{2\pi a n_1 \sqrt{2\Delta}}{2.405} \qquad (3.18b)$$

We note that, using Eq. (3.18a) on the fiber system described in Example 3.1, we find that $a_{max} = 2.87$ μm. Thus the optical fiber used in Example 3.1 would not work for single-mode propagation at $\lambda_0 = 1.55$ μm. The minimum wavelength for single-mode operation of that fiber is

$$\lambda_{min} = \frac{2\pi a n_1 \sqrt{2\Delta}}{2.405} = 2.27 \text{ μm}$$

The data depicted in Fig. 3.4 demonstrate that, in general, as V increases, the number of modes M that can propagate also increases. It can be shown [3] that for values of $V > 20$ a ballpark estimate for the number of modes that will propagate for a step index optical fiber is approximately

$$M_s \approx \frac{V^2}{2}$$

where the subscript on M_s stands for step index fiber.

Example 3.2 As a pair of simple examples of the preceding formula

1. Let $n_1 = 1.45$, $\Delta = 0.01$, $\lambda_0 = 0.82$ μm, and $a = 50$ μm. Then $V = 78.4$ and $M_s = 3.07 \times 10^3$.
2. Let $n_1 = 1.45$, $\Delta = 0.01$, $\lambda_0 = 1.55$ μm, $a = 50$ μm, $V = 41.6$, and $M_s = 864$.

Thus we see that as λ_0 increases (frequency decreases), the number of modes that can propagate decreases and vice-versa.

In this text we shall be primarily interested in high-speed digital transmission, which, in turn, means that we shall focus on single-mode propagation through fiber-optic channels. We use Eq. (3.16) to simplify the calculation of the propagation constant for the LP_{01} mode. The results of these calculations are presented in Fig. 3.5 as a plot of b_{01} versus V. Once b_{01} has been found, it is a straightforward to find the propagation constant for the fundamental mode β_{01}.

3.2 Some Special Topics

3.2.1 Numerical aperture

A readily measurable parameter that is often used in describing an optical fiber is the *numerical aperture* (NA), which measures the light-collecting ability of the fiber and is one of the primary parameters that one specifies when ordering optical fibers from a manufacturer. The previously encountered parameter V as well as the single-mode conditions are compactly expressed in terms of numerical aperture. The basic definition arises from the structure depicted in Fig. 3.6. We define the NA parameter by

$$\text{NA} = n_0 \sin \theta_a = n_1 \sin \bar{\phi}_c$$

where $\bar{\phi}_c$ is the complement of the critical angle ϕ_c. The last equality follows from Snell's law. Clearly, if $\theta \geq \theta_a$, the ray inside the fiber will not be totally

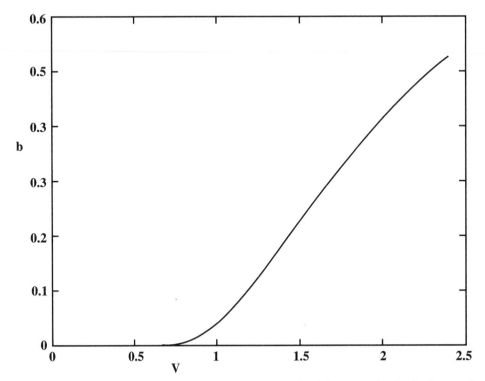

Figure 3.5 Normalized propagation constant versus normalized frequency for the fundamental mode.

reflected. Thus θ_a is the largest angle θ for which the ray will be totally internally reflected inside the fiber (hence propagate with low loss).

We note that $\sin \bar{\phi}_c = \cos \phi_c$ and that

$$\cos \phi_c = \cos \theta_a = \sqrt{1 - \left\{\frac{n_2}{n_1}\right\}^2} = \frac{\sqrt{n_1^2 - n_2^2}}{n_1}$$

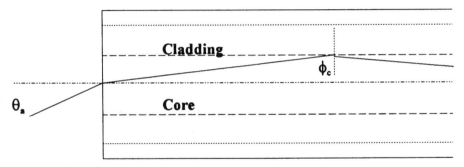

Figure 3.6 Numerical aperture parameters.

Then we get

$$NA = \sqrt{n_1^2 - n_2^2} \tag{3.19a}$$

Also $n_2 \approx n_1 (1 - \Delta)$ (for $\Delta \ll 1$) and

$$NA \approx n_1 \sqrt{2\Delta} \tag{3.19b}$$

We can now express the normalized frequency as

$$V \approx \frac{2\pi}{\lambda_0 \, a n_1 \sqrt{2\Delta}} = \frac{2\pi}{\lambda_0 \, a \, NA} \tag{3.19c}$$

and the single-mode propagation conditions in terms of the numerical aperture:

$$a < \frac{2.405 \, \lambda_0}{2 \, \pi n_1 \sqrt{2\Delta}} = \frac{2.405 \, \lambda_0}{2 \, \pi \, NA} \tag{3.20a}$$

$$\lambda_0 > \frac{2\pi a n_1 \sqrt{2\Delta}}{2.405} = \frac{2\pi \, NA}{2.405} \tag{3.20d}$$

3.2.2 P_{clad} versus P_{core}

In Chap. 2, when we derived Snell's law and the phenomenon of total internal reflection, we noted that $|\mathbf{E}|$ decays exponentially as it progresses into the transmission side of the interface (*the cladding region in optical fibers*). We found that the component of the Poynting vector perpendicular to the intersection and inside the cladding region was pure imaginary. We also found that there existed a real component of the Poynting vector, and hence propagation in the direction parallel to the interface within the cladding region. In the cylindrical fiber we have a similar situation. The E_r component is an evanescent field in the cladding region and S_r is pure imaginary there, but S_z is real and flows in the z direction. The total power that flows through each region can be determined from

$$P_{core} = \int_0^a \int_0^{2\pi} S_z \, r \, dr \, d\phi$$

$$P_{clad} = \int_a^\infty \int_0^{2\pi} S_z \, r \, dr \, d\phi$$

For weakly guided modes (LP and $\Delta \ll 1$)

$$\frac{P_{core}}{P_{clad}} \approx \left\{1 - \frac{X^2}{V^2}\right\} \left\{1 - \frac{J_l^2(Xa)}{J_{l+1}(Xa)J_{l-1}(Xa)}\right\} \tag{3.21}$$

A plot of this ratio for individual LP modes is presented in Fig. 3.7.

The curves displayed in Fig. 3.7 were generated by computer using the values of β obtained from the preceding Gloge plot of b versus V. Recall that β is needed to evaluate the parameter X in Eq. (3.21). Some general trends in these figures are

1. For each mode we note that as V increases, P_{clad}/P_{tot} decreases. This may be interpreted as showing that as V increases, the field pattern for each mode is concentrated toward the principal axis of the fiber.

2. In particular, consider the single-mode case. Two specific examples are

$$V = 2 \Rightarrow \frac{P_{clad}}{P_{tot}} = 0.28 \quad \text{and} \quad V = 1 \Rightarrow \frac{P_{clad}}{P_{tot}} = 0.70$$

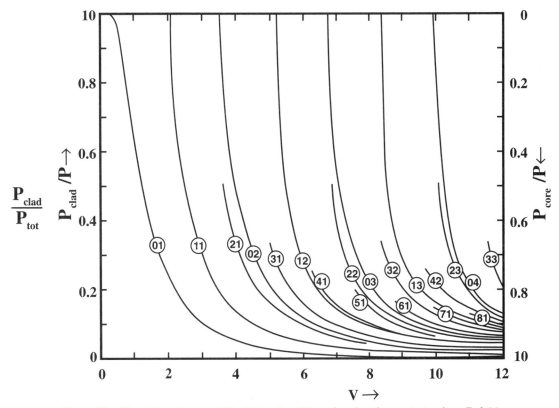

Figure 3.7 Plots of P_{clad}/P and P_{core}/P versus V for LP modes. (*Reproduced with permission from Ref. 2.*)

These results were obtained from the LP_{01} mode curve of the Gloge graph. If we imagine a fiber illuminated by a variable-wavelength source, we see that the minimum cladding propagation occurs for the wavelength that yields $V = 2.405$. Longer wavelengths will produce smaller values of V and hence will have greater proportions of the energy propagating in the cladding. *The main point here is that high-grade glass must also be used for the cladding.*

3.2.3 Graded index fibers

As an introduction to this subject, let's consider two examples of single-mode step index fibers:

1. Let $n_1 = 1.45$, $\Delta = 0.01$, and $\lambda_0 = 1.3$ μm. The single-mode condition on the core radius for this example is $a < 2.43$ μm.

2. Similarly, $n_1 = 1.45$, $\Delta = 0.01$, and $\lambda_0 = 1.55$ μm $\Rightarrow a < 2.9$ μm.

In both examples we see that the core radius is less than 3 μm for the two most frequently used wavelengths. Today this does not present a serious problem for launching optical signals down this narrow fiber, but it does complicate the process of splicing (necessary on long cables). At an earlier stage in the development of fiber-optic communication systems, launching optical signals down very narrow cores was a problem and the graded index fiber was considered as one way to solve the launching problem.

Snell's law demonstrates that light rays always bend toward the medium with the higher index of refraction. One means of concentrating light near the axis of the core is to employ graded index fibers which result in a gradual bending or light rays toward the axis of the fiber. A typical graded index fiber is depicted in Fig. 3.8. The mathematical form for the index of refraction is given by

$$n(r) = \begin{cases} n_1\sqrt{1 - 2\Delta\left(\dfrac{r}{a}\right)^p} & \text{(for } r < a \text{ core)} \\ n_2 \approx n_1\sqrt{1 - 2\Delta} & \text{(for } r > a \text{ clad)} \end{cases} \qquad (3.22)$$

The analysis for propagation in these fibers is extremely complicated, and we shall not pursue it at all. Some consequences of the analysis are

1. Let M_g = number of modes in the graded fiber. Then we get for large V

$$M_g \approx \frac{p}{p} + 2\,\frac{V^2}{2}$$

Now $p = 2$ provides a distinct advantages for multimode dispersion (see discussion of dispersion), and for that case we have

$$M_g = \frac{M_g}{2}$$

that is, half as many modes.

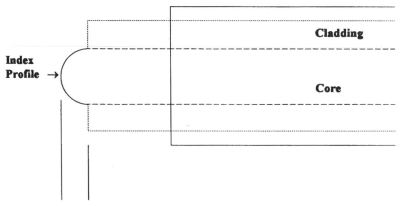

Figure 3.8 Graded fiber with index profile.

2. It also turns out that graded index fibers have a single-mode cutoff at a larger V than the step index fiber and hence a larger core radius.

$$V_{\text{cutoff}} = 2.405 \sqrt{1 + \frac{2}{p}}$$

For example, assume that $p = 2$ increases V_{cutoff} by a factor of $\sqrt{2}$ while $p = 1$ leads to an increase in V_{cutoff} by a factor of $\sqrt{3}$. It would seem that the concomitant increase in core radius for single-mode propagation would provide some relief to the splicing alignment problem. However, the need to align the graded profiles accurately creates even more stringent requirements for splicing fibers. Additionally there is the problem of manufacturing long graded index fibers with uniform profile parameters.

3.3 Transmission Characteristics of Optical Fibers

In this section we shall discuss two principal physical phenomena that cause the signal to deteriorate as it travels down the optical fiber. First we note that the optical signal is weakened by a variety of propagation losses and second, the structural integrity of the optical signal is distorted by the effects of dispersion (i.e., pulse spreading). We shall begin by briefly describing the loss mechanisms (without equations) and concentrate on dispersion, which, in practice, creates the serious constraints on data rates in high-speed fiber-optic communication systems.

3.3.1 Transmission losses

The primary loss mechanisms in optical fibers fall into the following categories:

1. Absorption losses
2. Scattering losses (linear)
3. Scattering losses (nonlinear)
4. Bending losses

The numerical aspects of these loss mechanisms are usually described by the loss coefficient parameter α expressed in decibels per kilometer (dB/km). The total decibel loss in a fiber of length L (in kilometers) is then αL, which results in (see Chap. 1).

$$\frac{P(L)}{P(0)} = 10^{-\alpha L/10}$$

Although all the loss mechanisms are important, the most important is the absorption of light by the materials that constitute the optical fiber. If this is too large, then the fiber is unsatisfactory for communications and the other loss mechanisms are irrelevant. We noted in Chap. 1 that the high-quality glass used in communication-grade optical fibers has very low losses at specific wavelengths (see Fig. 1.2 [4,5] as well as Figs. 3.9 and 3.10). In order to

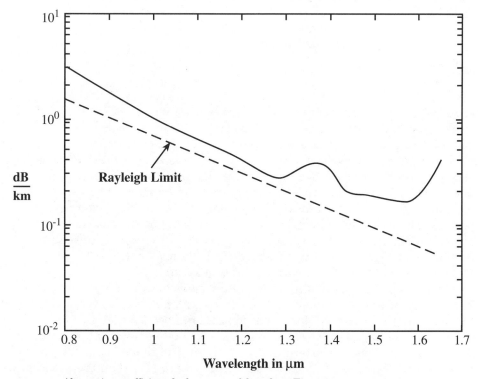

Figure 3.9   Absorption coefficient for best case of data from Fig. 1.2.

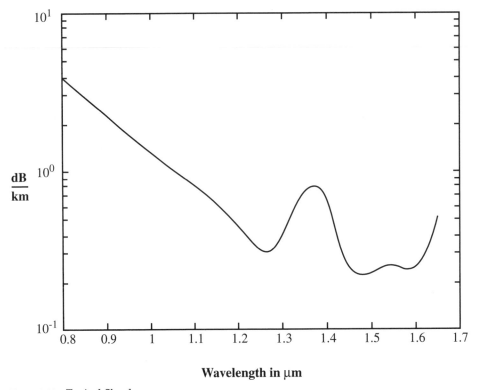

Figure 3.10 Typical fiber loss curve.

examine the loss coefficients in greater detail, we now examine Fig. 3.9. The solid-line curve approximates the loss coefficient versus wavelength for the best optical fiber from the production-run data depicted in Fig. 1.2. Strong absorption at wavelengths above ~1.6 μm are caused by the tail edges of very strong absorption lines at even further into the infrared band. These lines are produced by oxides in the glass. As such, they can be reduced by using nonoxide compounds such as fluorides and chlorides in the fiber manufacturing process. There is also some ultraviolet absorption, but it is masked by other effects (in particular, Rayleigh scattering losses). In any case, the ultraviolet portion of the spectrum is not utilized in modern fiber communication systems, and we shall omit this portion of the spectrum.

Most metallic impurities are adequately removed these days by modern materials processing techniques. The OH radical, however, is responsible for the absorption peak near 1.4 μm that appears in Figs. 3.9 and 3.10. It is more difficult to eliminate this absorption line since it arises from water vapor, and thus humidity must be carefully controlled in the manufacturing process. When the OH has been reduced to less than 1 part 10^{-7} then the absorption (loss) curve that is shown in the Fig. 1.2 accrues. We note that absorption minima occur near $\lambda = 1.3$ and 1.55 μm. In particular, at 1.55 μm the loss is almost equal to the lower limit set by Rayleigh scattering displayed as a dashed line in Fig. 3.9. We also include a loss curve for a more typical fiber in Fig. 3.10. This particular loss curve approximates the center of the production-run data depicted in Fig. 1.2, and it has been included for problem-solving purposes.

Scattering losses occur when some of the optical energy propagating through the fiber is converted into modes and/or wavelengths that do not propagate well through the fiber. These can be subdivided into linear and nonlinear scattering phenomena. The linear type consists of Rayleigh and Mie scattering. Both are caused by random inhomogeneities in the index of refraction in the fiber. The effects of Rayleigh scattering are displayed as the dashed curve in Fig. 3.9. For Rayleigh scattering these inhomogeneities are much smaller than the wavelength of the light, while Mie scattering involves scattering from inhomogeneities that are comparable (or larger) than the wavelength. The small inhomogeneities in the index of refraction that cause Rayleigh scattering were solidified into the glass during the cooling phase of the manufacturing process. Although some improvements have been made in reducing the density of these inhomogeneities, there is a fundamental lower bound to this density which is caused by the vibrations in the constituents of the glass as it solidifies at high temperatures. This process must take place gradually during this annealing time, and the accompanying high temperatures cause relatively large molecular vibrations during solidification. Rayleigh scattering is responsible for the upper edge of the intrinsic scattering region of Fig. 1.2, and it places a lower bound on the fiber loss coefficients. The magnitude of this scattering is proportional to λ^{-4}, and therefore it causes less distortion in infrared systems than in visible light systems. The effects for Mie scattering are more severe, but fortunately they can be *eliminated by appropriate processing techniques.*

Losses are also incurred as a result of nonlinear effects generated by strong optical fields in fibers. One of the nonlinear scattering processes is *Brillouin scattering,* which is generated at high power levels by modulation of the propagating optical signal by molecular vibrations within the medium (glass) with vibrational energy units called *phonons* (acoustical photons). This modulation generates new frequencies and thereby scatters energy into these frequencies and in some cases into unwanted modes. The energy scattered into these undesired modes will dissipate, and this represents a loss mechanism to the propagating optical field. The other form of nonlinear scattering that is important in fiber optics is *Raman scattering,* which is generated at even higher levels than Brillouin and is produced by interaction with optical frequency phonons as opposed to acoustic photons for Brillouin scattering. Both of these effects are characterized by thresholds dependent on energy density [6]. Below threshold the effects are insignificant, while above threshold these effects become significant and grow extremely rapidly with increasing input level.

We note that nonlinear scattering effects are of particular concern in single-mode fiber-optic systems since the smaller core radius results in higher energy densities for a given input power level. Therefore nonlinear scattering may occur at much lower input power levels in single-mode fibers than in the larger-core multimode fibers. Fortunately, single-mode fiber-optic communication systems require less initial transmitter power to achieve desired error rates than do multimode systems. One consequence of nonlinear scattering is that it is not possible to overcome system defects by continually increasing source power. These nonlinear effects place an upper bound on useful source signal power. Before we leave this topic, we note that both these nonlinear effects can be used to advantage in several types of optical amplifiers

One loss mechanism that we have not as yet considered is losses due to bends in the fiber. Fiber bend losses are characterized by

$$\text{Loss} \propto \exp\left\{\frac{-R}{R_c}\right\}$$

where R_c is the critical radius. If $R > 10\,R_c$ the losses become insignificant. Typically $R_c \approx 1$ mm for a 1.55-μm system. If the radii of the microbends are kept greater than, say, 10 mm, the losses will be negligible. Thus those rolled-up spools, on the order of about 0.5 m in diameter, which are often used in laboratory tests of fiber-optic communication systems, will not cause significant losses due to bends. On the other hand, we shall see in the next chapter that some practical optical amplifier structures employ small-diameter spools with fiber lengths on the order of 25 m, and bending losses are a definite concern in the design of these devices.

3.3.2 Dispersion effects (introduction)

An optical signal propagating down a fiber can be resolved into a linear combination of component parts in several ways. For example, the signal can be expressed as a sum of modes, and/or it can be expanded as a linear combination of its Fourier components. The modal portion of the expansion will be limited to those modes that propagate and may be limited to only one term for single-mode fiber systems. The frequency decomposition of the optical signal is given by the Fourier transform of the modulated optical carrier being transmitted over the fiber-optic system. When the individual frequency components of the signal travel at different velocities down the fiber, then the reconstituted signal at the receiving terminus will be distorted. This process, known as *dispersion* [7,8], is one of the primary factors that limit data rates in fiber-optic channels. In digital communication system dispersion causes each pulse to spread in time (disperse). When these pulses spread sufficiently, intersymbol interference (ISI) will be generated and thereby increase error rates. The three primary sources of dispersion in fiber-optic cables are:

1. *Material dispersion,* which occurs whenever the index of refraction is a function of frequency. In most optical materials the permitivity $\epsilon(f)$ is the

source of this frequency dependence. It is this property of glass that causes the splitting of white light into a rainbow of colors by glass prisms.

2. *Waveguide dispersion,* which results from the fact that the mode propagation factor for a given mode is a function of frequency. Thus, frequency variation is accompanied by wavelength variation, which, in turn, affects the self-consistency property of the ray description of modes (see Chap. 2). Specifically, it causes the ray angle to vary with frequency, which, in turn, causes further pulse spreading.

3. *Intermodal dispersion,* which is caused by the unique propagation factors that accompany each separate LP mode. Clearly this particular type of dispersion is absent from single-mode systems.

We shall discuss each of these types of dispersion further and present some quantitative formulas for estimating the amount of pulse spreading soon, but first we shall take a brief look at group velocity versus phase velocity.

3.3.3   Group velocity

As an introduction to the concept of group velocity, let us consider the following simple monochromatic plane wave traveling in the $+z$ direction:

$$\psi = A \cos(\omega t - kz)$$

Now, as we previously demonstrated

$$\omega t - kz = \text{const} \Rightarrow v_p = \frac{dz}{dt} = \frac{\omega}{k}$$

Next we consider the superposition of two such waves, each propagating at slightly different frequencies:

$$\psi_1 = A \cos[(\omega + \Delta\omega)t - (k + \Delta k)z]$$

$$\psi_2 = A \cos[(\omega - \Delta\omega)t - (k - \Delta k)z]$$

$$\psi = \psi_1 + \psi_2$$

We shall use $2 \cos X \cos Y = \cos(X + Y) + \cos(X - Y)$ and $Y = \omega t - kz$ and $X = \Delta\omega t - \Delta kz$ to arrive at $\psi = 2A \cos(\omega t - kz) \cos(\Delta\omega t - \Delta kz)$:

$$\psi = \left\{ \underbrace{2A \cos (\Delta\omega t - \Delta kz)}_{\text{envelope}} \right\} \left\{ \underbrace{2A \cos (\omega t - kz)}_{\text{carrier}} \right\} \qquad (3.23)$$

Students with a background in communication theory will recognize this signal as a DSB/SC (double sideband/suppressed carrier)-modulated carrier where the underlying carrier propagates according to the second { } factor while the envelope propagates according to the first { }. At this point we do not intend to envelope-detect this signal here, but we shall refer to Eq. (3.23) when we discuss optical heterodyne detection in Chap. 8. Now the envelope velocity is obtained from $\Delta\omega t - \Delta kz = $ const, and the group velocity v_g is given by

$$v_g = \frac{dz}{dt} = \frac{\Delta\omega}{\Delta k} \rightarrow \frac{d\omega}{dk}$$

It was pointed out in Chap. 2 that, because of the linearity of Maxwell's equations, frequency is the same in all materials, and that the wavelength changes as the material composition changes. Specifically, we have $\lambda = \lambda_0/n$, and for many materials, including glass, the index of refraction is a function of the frequency of the light. Thus, since

$$k = \frac{2\pi}{\lambda} = \frac{2\pi n(\omega)}{\lambda_0} = k(\omega)$$

the frequency ω is the independent variable while $k(\omega)$ is the dependent variable. It turns out that an important parameter that will be used in deriving a formula for material dispersion in the next section is the group delay τ_g. It is related to group velocity by the simple formula distance = velocity $\times$ time. This results in

$$\frac{\tau_g}{L} = \frac{1}{v_g} = \frac{\Delta k}{\Delta\omega} \rightarrow \frac{dk(\omega)}{d\omega}$$

which is the proper form for differentiation. We next obtain $dk(\omega)/d\omega$ in terms of the index of refraction. Now in a material of index of refraction n, we have that

$$k = nk_0 = \frac{2\pi}{\lambda_0} n = \frac{\omega}{c} n \qquad (\text{where } c = f\lambda_0)$$

Thus

$$\frac{dk(\omega)}{d\omega} = \frac{1}{c}\left[n + \omega\frac{dn(\omega)}{d\omega}\right]$$

Now it turns out that the measured data for index of refraction is typically graphed with the index of refraction plotted versus an abscissa in terms of λ_0. The preceding equation can be put in a format that is only a function of λ_0 by noting

$$\frac{dn}{d\omega} = \frac{dn}{d\lambda_0}\frac{d\lambda_0}{d\omega}$$

and that

$$\lambda_0 = \frac{2\pi c}{\omega} \Rightarrow \frac{d\lambda_0}{d\omega} = -\frac{2\pi c}{\omega^2}$$

Thus

$$\tau_g = \frac{L}{c}\left[n - \frac{2\pi c}{\omega}\frac{dn}{d\lambda_0}\right] = \frac{L}{c}\left[n - \lambda_0\frac{dn}{d\lambda_0}\right] = \frac{L}{c}\left[n - \lambda_0\frac{dn}{d\lambda_0}\right] \qquad (3.24a)$$

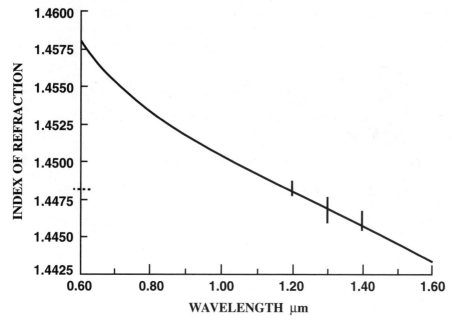

Figure 3.11 Index of refraction n versus wavelength for fused silica. (*Adapted from Ref. 3.*)

Figure 3.11 displays $n(\lambda_0)$ for fused silica. We next employ this graph to work a numerical example of group velocity.

Example 3.3 Consider the following parameters:

$$L = 1 \quad \text{and} \quad \lambda_0 = 1.3\mu \Rightarrow n(\lambda_0) = 1.4465$$

Note the tick marks on Fig. 3.11 for numerical values used below:

$$\frac{\Delta n}{\Delta \lambda_0} = \frac{1.4457 - 1.4482}{0.2\mu} = -\frac{0.0012}{\mu}$$

$$v_g = \frac{1}{\tau_g} = \frac{c}{1.4465 + (1.3)(0.0012)} = +\frac{c}{1.4494} = 2.0709 \times 10^8 \text{ m/s}$$

$$v_p = \frac{c}{n} = \frac{c}{1.4465} = 2.0740 \times 10^8 \text{ m/s} > v_g$$

If the index of refraction is not a function of wavelength, then the second term in the denominator of the v_g equation above vanishes and the group velocity equals the phase velocity.

3.3.4 Material dispersion

Signals employed in high-speed communication systems consist of a series of short-duration pulses. The Fourier transform of such pulses imply that a broad range of frequencies are transmitted through the fiber rather than a single frequency. Let $y(t)$ be the response of a linear system whose input is $x(t)$. Furthermore, $y(t)$ will be considered as an undistorted version of the input signal $x(t)$, provided

$$y(t) = Ax(t - t_d) \qquad (3.24b)$$

In other words, we do not consider simple magnification and/or a reasonable time delay to be distortion. Thus we have

$$Y(f) = X(f)H(f) \qquad (3.24c)$$

where $X(f)$ and $Y(f)$ are Fourier transforms of $x(t)$ and $y(t)$, respectively, while $H(f)$ is the transfer function of the linear system in the frequency domain. Now from Eqs. (3.24b) and (3.24c), along with the shifting theorem for Fourier transforms, we have

$$Y(f) = X(f) \{A \exp [-j2\pi t_d f]\} \qquad (3.24d)$$

Thus the transfer function for this "ideal" filter is given in polar form by

$$|H(f)| = A \qquad \text{and} \qquad \angle 2\pi t_d f = \text{constant} \times f \qquad (3.24e)$$

We also recall that the phase propagation constant β is related to frequency by

$$\beta = \frac{\omega}{v_p} = \frac{2\pi}{v_p} f$$

Now as long as the index of refraction, hence the phase velocity v_p, is a constant with respect to frequency (or λ_0), propagation through this material will be free of distortion. If, on the other hand, $v_p(f)$, then there will be dispersion. Furthermore, since $v_p(f)$ hence $n(f)$ are properties of the glass that make up optical fibers, this type of distortion is called *material dispersion*.

In this subsection we are concerned with the effect of $n(f)$ on dispersion. We can and will concentrate on uniform plane waves in homogeneous media. Thus k and β are identical since there are no boundary value waveguide effects in homogeneous media. In this case we define the material delay as a function of wavelength $\tau_m(\lambda)$ as

$$\tau_m(\lambda_0) = \tau_g(\lambda_0) = \frac{L}{c} \left[n - \lambda_0 \frac{dn}{d\lambda_0} \right] \qquad (3.24f)$$

Now let there be two separate plane waves propagating in the same direction; one at wavelength λ_1 and the other at wavelength λ_2. Furthermore, we take the separation between these two wavelengths to be $\Delta\lambda$ in the manner prescribed below:

$$\lambda_1 = \lambda_0 - \frac{\Delta\lambda}{2} \quad \text{and} \quad \lambda_2 = \lambda_0 + \frac{\Delta\lambda}{2}$$

Then we define difference in the propagation time through the fiber as $\Delta\tau_m$ and obtain the following relations:

$$\Delta\tau_m = \tau_m(\lambda_2) - \tau_m(\lambda_1) = \tau_m\left(\lambda_0 + \frac{\Delta\lambda}{2}\right) - \tau_m\left(\lambda_0 - \frac{\Delta\lambda}{2}\right)$$

Using this formula as a guide, we further define material dispersion as the time spread, or pulse broadening, due to material properties of the optical fiber, and we denote this as σ_m. Now the preceding equation suggests the following format for material dispersion:

$$\sigma_m \doteq \Delta\tau_m \approx \Delta\lambda \left|\frac{d\tau_m}{d\lambda}\right|_{\lambda=\lambda_0} \quad \text{(for } \Delta\lambda << \lambda_0)$$

Since σ_m represents the time spread, we shall represent the spectral spread of the optical signal as $\sigma_\lambda = \Delta\lambda$. Thus σ_λ is the bandwidth of the propagating radiation expressed in terms of the wavelength:

$$\sigma_m = \sigma_\lambda \left|\frac{d\tau_m}{d\lambda}\right|_{\lambda=\lambda_0} \tag{3.24g}$$

Now by using Eq. (3.24f), we obtain

$$\frac{d\tau_m}{d\lambda} = \frac{L}{c}\left[\frac{dn}{d\lambda} - \frac{dn}{d\lambda} - \frac{d^2\lambda}{d\lambda^2}\right] = -\frac{L}{c}\lambda_0 \frac{d^2n}{d\lambda_0^2} \tag{3.24h}$$

Then from Eqs. (3.24g) and (3.24h) we have

$$\sigma_m = \sigma_\lambda L \left|\frac{\lambda_0}{c}\frac{d^2n}{d\lambda_0^2}\right| \tag{3.24i}$$

This formula may be expressed in a more compact form by using the dispersion coefficient D, defined as

$$D = \frac{\lambda_0}{c}\frac{d^2n}{d\lambda_0^2} \rightarrow \text{ps nm}^{-1}\text{ km}^{-1}$$

$$\sigma_m = \sigma_\lambda L |D| \tag{3.24j}$$

A graph of D for a typical fiber is displayed in Fig. 3.12.

We use this form so that the parameter D clearly separates the material properties from the source characteristics. As an example, consider the following.

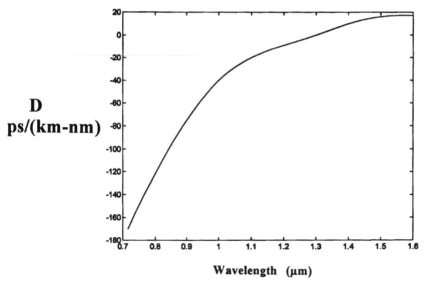

$$\mathbf{D}$$
$$\mathbf{ps/(km\text{-}nm)}$$

Wavelength (μm)

Figure 3.12   Dispersion coefficient versus wavelength.

Example 3.4 $L = 1$ km and $\lambda_0 = 1.55$ μm $= 1550$ nm. Then $D \approx 17$ ps km^{-1} nm^{-1}

Case 1.   For an LED source, a reasonable value for σ_λ/λ_0 is 0.01. Thus

$$\sigma_\lambda \approx \lambda_0\,(0.01) = (1550 \text{ nm})\,(0.01) = 15.5 \text{ nm}$$

and

$$\sigma_m = 15.5(1)\,(17) = 263.5 \text{ ps/km}$$

$$L = 100 \text{ km} \rightarrow \sigma_m = \Delta\tau = 26.3 \text{ ns}$$

Case 2.   Consider a modulated laser source to be used for ultra-high-speed data transmission. Let $\Delta f = 10^{10}$. Before we can use this parameter, we need to relate σ_λ to Δf. We begin by noting that although Δf is very large, it is still a narrowband system because the optical carrier frequency is $c/\lambda_0 = 1.936 \times 10^{14}$. Then

$$\frac{\Delta f}{f} = \frac{f_1 - f_2}{f} = \frac{(c/\lambda_1) - (c/\lambda_2)}{c/\lambda} = \frac{\lambda}{\lambda_1\lambda_2}\,(\lambda_2 - \lambda_1)$$

For narrow bandwidth $\lambda \approx \lambda_1 \approx \lambda_2$, which yields

$$\frac{\Delta f}{f} = \frac{\sigma_\lambda}{\lambda} \tag{3.25}$$

Then for this case

$$\frac{\Delta f}{f} = \frac{10^{10}}{1.936 \times 10^{14}} = 5.167 \times 10^{-5} = \frac{\sigma_\lambda}{\lambda_0}$$

Finally we obtain

$$\sigma_m = (5.167 \times 10^{-5})\,(1550)\,(17) = 1.362 \text{ ps/km}$$

and for

$$L = 100 \text{ km} \quad \text{and} \quad \Delta\tau_{\text{m}} = 136.2 \text{ ps} << 1 \text{ ns}.$$

On the basis of this dispersion coefficient, this system could probably be employed for a 1-Gbit/s data rate, but it would run into serious difficulty at 10 Gbits/s (as implied by this Δf). It does not take much intersymbol interference to prevent one from obtaining error rates $\sim 10^{-9}$, which is usually the benchmark for high-speed systems.

It is interesting to note that in the vicinity of $\lambda_0 \sim 1.3$ μm the material dispersion virtually disappears. This dispersion minimum can be shifted to longer wavelengths by doping the fiber materials with GeO and/or selection of core diameter. A degree of dispersion cancellation can occur between material and wavelength dispersion and the latter is affected by fiber geometry.

3.3.5 Waveguide dispersion

The derivation of waveguide dispersion is complicated, and we won't go into any details here. It arises because the mode angles, thus the path lengths, are wavelength-dependent. This type of pulse spreading is very small, and it is important only in single-mode systems for $\lambda_0 \sim 1.3$ μm, where the material dispersion virtually vanishes. It has been shown by Gloge that

$$\sigma_{\text{wg}} = \frac{nL\sigma_\lambda\Delta}{c\lambda}\left[V\frac{d^2(Vb)}{dV^2}\right]$$

Plots of the bracketed term can be found in the literature (e.g., see Fig. 3.16, p. 106 of Keiser [13]). We note also that the completely separate treatment of these two spreading mechanisms is in itself an approximation [9]. They do interact with one another, but the type of treatment described above yields reasonable estimates of dispersion effects. In most cases one type of dispersion dominates the pulse spreading. As we shall see in the next subsection, multimode dispersion dominates pulse spreading except for the case of single-mode propagation. Material dispersion dominates pulse spreading in single-mode fibers, except for wavelengths in the vicinity of the material dispersion minimum (see Fig. 3.12), where waveguide dispersion dominates.

3.3.6 Intermodal dispersion

An easily understood estimate of the pulse spreading due to intermodal effects utilizes the travel-time difference between the slowest mode and the fastest mode. Thus

$$T_{\text{min}} = \frac{\text{distance}}{\text{velocity}} = \frac{L}{c/n_1} = \frac{Ln_1}{c} \quad \text{(straight-down axis)}$$

$$T_{max} = \frac{Ln_1}{c \cos \theta} = \frac{Ln_1^2}{cn_2} \quad \text{(at critical angle)}$$

where we have used $\cos \theta = \sin \phi_c = n_2/n_1$. Now let

$$\partial T_s \doteq T_{max} - T_{min} = \frac{Ln_1^2}{cn_2} - \frac{Ln_1}{c} = \frac{Ln_1}{c}\left[\frac{n_1}{n_2} - 1\right]$$

$$\partial T_s = \frac{Ln_1}{c}\frac{n_1 - n_2}{n_2} \approx \frac{Ln_1\Delta}{c}$$

where $n_2 = n_1(1 - \Delta)$, and where $\Delta \ll 1$ were used to obtain the preceding formula. This analysis, however, does not take into account the distribution of energy among the propagating modes. One approach to this is to assume a uniform distribution of energy into the modes that do propagate and computing the variance of a random variable uniformly distributed between $\pm \partial T_s/2$. By using this approach, one arrives at the following estimate for intermodal dispersion:

$$\sigma_s = \frac{Ln_1\Delta}{2c\sqrt{3}}$$

where the subscript s on σ_s stands for step index fiber. This formula will be different for graded index fibers. Note that here σ_s does depend on Δ.

As an example, we shall find the multimode dispersion for a fiber with the following characteristics. Let $L = 1$ km, $\lambda_0 = 1.55$ μm, and $n_1 \approx 1.4478$. We also shall assume that $\Delta = 0.01$ (a typical value for Δ), and note that the parameter Δ is not required for the analysis of material dispersion. On the other hand, we will not require the spectral width parameter σ_λ here since, for multimode transmission (with a large number of modes), the propagation velocity extremes are determined by fiber structure parameters. Then for the assumed numerical values for this example we get

$$\sigma_s = \frac{(1)(1.4478)(0.01)}{2\sqrt{3}\,3 \times 10^5} \approx 13.9 \text{ ns/km}$$

where $c = 3 \times 10^5$ km/s. This is much larger than our material dispersion result. For the λ_0 of interest these days, 1.3 and 1.55 μm, the material dispersion is orders of magnitude smaller than intermodal dispersion.

One further point to be made here is that σ_s as represented by the *preceding formula is actually too large*. There are several effects that tend to reduce intermodal pulse spreading:

1. Mode coupling due to imperfections tends to redistribute energy among slow and fast modes as they progress down the fiber. This creates a kind of averaging effect on the speed of mode travel.

2. There are many more fast modes than slow modes, thus tending to compress toward the front of the pulse period. This effectively reduces pulse spreading.

Nevertheless, there is a great deal more pulse spreading in multimode step index fibers than in single-mode fibers, and ultra-high-speed data transmission over moderate to long distances is feasible only with single-mode fibers.

3.3.7   Further discussion of dispersion

The dispersion analysis of graded index fibers is very complicated, and we shall present only a few of the more important results:

1. The optimum index profile (with regard to dispersion characteristics) exponent is

$$p_{0,p} = 2 - \frac{12\Delta}{5} \quad \text{and} \quad \sigma_g \approx \frac{L n_1 \Delta^2}{20\, c\sqrt{3}}$$

   For $\Delta = 0.01$, this leads to an improvement of a factor of 1000 over the step index multimode fiber.

2. This improvement often cannot be realized because it is difficult to manufacture kilometer-length fibers with the proper profile. It turns out that the dip in the dispersion factor versus *p is extremely sharp,* indeed, and small variations in the manufacture of the fiber will seriously affect system performance. In practice, however, an improvement over step index fibers of 100 can be readily achieved.

3. A qualitative explanation for this improvement can be obtained by noting that the modes with the longer ray paths travel further out from the core center where the index of refraction is lower than at the center. Thus they travel faster than the shorter-path modes that stay near the core center. This tends to make all modes travel at nearly the same speed down the fiber.

4. A formula for combining the pulse spreading due to intramodal and intermodal dispersion is

$$\sigma_T \approx \sqrt{\sigma_{intra}^2 + \sigma_{intermod}^2} \tag{3.26}$$

In practice, a good estimate for the amount of pulse spreading in multimode fibers can be obtained from the intermodal dispersion term alone. For single-mode fibers, material dispersion provides a sufficiently accurate estimate, except for the case where this term is very small and waveguide dispersion dominates dispersion. Before we leave this topic, we review some further characteristics of single mode fibers:

1. Dispersion is due to three effects. Two of them, σ_m and σ_{wg}, have already been discussed; the third effect, which is proportional to $d\Delta/d\lambda$, is called *profile spreading*. These three effects are not truly separable, and when carefully considered as a whole, some interesting consequences are observed.

2. In some situations the various dispersion contributors can cancel each other. This is how the dispersion coefficient $D \to 0$ near $\lambda_0 \sim 1.3$ μm (see Fig. 3.11).

3. As noted earlier, these dispersion minima can be shifted to longer wavelengths by doping the fiber with GeO. We note, however, that this increases the Rayleigh scattering losses.

4. In practice, some high-order nonlinear effects set lower bounds on the dispersion factor. For fused silica, this is $\sim 2.5 \times 10^{-2}$ ps nm^{-1} km^{-1} at $\lambda_0 = 1.273$ μm (in the absence of GeO doping).

3.3.8 Solutions

Glass manifests some nonlinear properties that if properly taken advantage of can counter the effects of dispersion [10,11]. In particular, intense optical fields can affect the index of refraction within the fiber, which, in turn, affects the propagation velocity and phase of the propagating light. The interaction between the optical field and the fiber material is part of the polarization vector **P** as indicated in the following two equations

$$\mathbf{D} = \epsilon_0 \mathbf{E} + \mathbf{P}$$

$$|\mathbf{P}| = \epsilon_0(\chi_1 \xi + \chi_2 \xi^2 + \chi_3 \xi^3 + \cdots)$$

where ξ in the magnitude of the electric field. In linear media only the coefficient $\chi_1 \neq 0$, and in most materials χ_2 and χ_3 are very small. In glass one does not experience any noticeable nonlinear effects until ξ is very large such as may be encountered at the input of a single-mode fiber. Recall that the single-mode propagation condition leads to very small-diameter cores and even a few milliwatts can yield very large energy densities.

The wave equation for electromagnetic propagation in glass ($\mu = \mu_0$, $\chi_2 \approx 0$, and $\chi_3 \ll 1$) takes the form

$$\nabla^2 \mathbf{E} - \frac{1}{\mu_0 \epsilon_0} \frac{\partial^2 \mathbf{E}}{\partial t^2} = \mu_0 \frac{\partial^2}{\partial t^2} (\mathbf{P}_{\mathrm{L}} + \mathbf{P}_{\mathrm{NL}}) \tag{3.27}$$

where $P_{\mathrm{NL}} = 4 \chi_3 \xi^3$ is the nonlinear portion of the electromagnetic polarization density while P_{L} is the linear portion. In general, the polarization density is related directly to the electric dipole density within the medium, and this, in turn, determines the index of refraction. (See Chap. 2 of Siegman [14] for an excellent description of this process.)

Equation (3.27) is a nonlinear differential equation, and we shall not attempt to solve it here. The student interested in delving further into this topic will find a very readable presentation in Section 19.8 of the Saleh–Teich book listed at the end of Chap. 1. This equation has a number of eigensolutions called *solitary waves* or *solitons*. For the *envelope* of the traveling wave $A(z,t)$, the lowest-order (fundamental) solution is

$$A(z,t) = A_0 \, \text{sech} \left\{ \frac{t - (z/v)}{\tau_0} \right\} \exp \left\{ \frac{jz}{4z_0} \right\}$$

where τ_0 is a measure of pulse width and z_0 is a constant proportional to τ_0^2. If the fiber is illuminated with a pulse whose envelope has the shape

$$A(z,t) = A_0 \, \text{sech} \left\{ \frac{t}{\tau_0} \right\}$$

it will retain this shape without dispersing as long as the intensity of the field remains at an appropriate level. Since the field decays as it propagates, it is necessary to boost the signal level with optical amplifiers at frequent intervals. For example, one of the major telephone companies has proposed a 9000 km undersea fiber-optic cable using optical amplifiers spaced at 30-km intervals. This cable has been designed for soliton transmission at a 5-Gbits/s transmission rate, which explains the rather small (30-km) spacing between amplifiers.

Problems

1　Let a particular optical fiber have the characteristics $n_1 = 1.45$ and $\Delta = 0.05$. Find
  (a)　Numerical aperture
  (b)　Acceptance angle θ_a
  (c)　Critical angle
2　Let a particular fiber have the same characteristics as the cladded fiber of problem 1. Now suppose that a GaAs laser ($n = 3.6$) is put perfectly flush (no gap at all) to the end of the fiber. Do the following:
  (a)　Find θ_a as viewed from inside the laser for this set up.
  (b)　Compare the preceding acceptance angle with that obtained for the air–fiber interface for this fiber.
3　Let a ray of $\lambda_0 = 1.3$-μm light strike the core–cladding boundary of the fiber specified in problem 2, at an angle of 87° with regard to the normal of the boundary.
  (a)　Determine the exponential loss coefficient (in nepers per micrometer) for penetration into cladding.
  (b)　What is the loss in decibels for a point 4λ into the cladding?
4　Describe some similarities and differences between the modal solutions for slab and cylindrical waveguides. Include in your discussion
  (a)　The method by which the modal solutions are found
  (b)　Exponential decay in cladding

(c) Oscillatory nature of spatial distribution inside core

(d) How single-mode cutoff is found

(e) TE, TM, and other modes

5 Let $n_1 = 1.446$, $\Delta = 0.005$, and $a = 3$ μm.

    (a) Find λ_c for single-mode operation.

    (b) Let $\Delta = 0.01$, and repeat part **a.**

    (c) Let $\Delta = 0.002$, and repeat part **a.**

    (d) How does this trend agree with intuition (if, in fact, it does)?

6 Let $a = 25$ μm in problem **2** above, and determine the approximate number of modes that propagate for

    (a) $\lambda_0 = 0.83$ μm

    (b) $\lambda_0 = 1.3$ μm

    (c) $\lambda_0 = 1.55$ μm

7 It is desired to design an optical fiber that will operate as a single-mode fiber for *any and all* wavelengths within the range 1.2 μm $\leq \lambda_0 \leq 1.6$ μm.

    (a) Find the largest a_{max} that will accomplish this task when $n_1 = 1.45$ and $\Delta = 0.005$.

    (b) Find the propagation constant β for the wavelengths at both ends of the wavelength region specified in part **a** above.

    (c) Suppose that this fiber is now illuminated by radiation at $\lambda_0 = 0.8$ μm. Which LP modes will be excited?

Fiber losses

8 Let a fiber-optic communication system using a fiber whose loss characteristics is given by Fig. 3.11 have the specifications power in = 10 mW and power received 40 nW fiber loss. Give the maximum length of the cable that will satisfy these conditions at

    (a) $\lambda_0 = 0.8$ μm

    (b) $\lambda_0 = 1.3$ μm

    (c) $\lambda_0 = 1.55$ μm

9 Explain how the Rayleigh scattering loss curve shown in Fig. 3.10 is related to the discussion in the paragraph following Fig. 3.11.

10 Consider the setup shown below in Fig. P3.10. The laser output incident on the lens is 1 mW.

Now the laser output power is measured separately and found to be 10 mW. The power exiting the fiber is observed to be 12.5 μW.

    (a) What is the loss coefficient for this particular fiber?

    (b) Note that the wavelength of the light has not as yet been specified. Use Fig. 3.10 to estimate the wavelength of the optical radiation based on the measurement.

    (c) Use Fig. 3.10 to explain why this particular measurement scheme cannot be used to measure wavelength in general.

Dispersion

11 Use Fig. 3.11 to find the dispersion for a single-mode fiber-optic channel with the following properties: $\Delta = 0.004$, $\lambda_0 = 0.83$ μm, $L = 100$ km, and source bandwidth = 6×10^8 Hz.

12 Now suppose that the core radius of the preceding problem is expanded so that it becomes a multimode fiber. Find the total dispersion for this fiber (ignore waveguide dispersion).

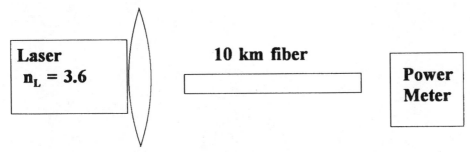

Figure P3.10 The index of refraction for both lens and optical fiber is $n_0 = 1.45$.

13 Suppose that we now have an optical fiber consisting of a core of radius 2.5 μm with $n = 1.463$ surrounded by an air cladding. Let this fiber be illuminated by radiation at $\lambda_0 = 1.3$ μm. Determine the following quantities for this fiber:

(a) V

(b) NA (numerical aperture)

(c) An estimate of how many modes can propagate

(d) An estimate of the dispersion per kilometer

14 Consider the peculiar fiber described in the previous problem. Indicate which of the equations and figures *are no longer valid* for this fiber.

References

1. M. Abramowitz and I. A. Stegun, *Handbook of Mathematical Functions with Formulas, Graphs, and Mathematical Tables,* National Bureau of Standards, 1964, p. 374.
2. D. Gloge, "Weakly guiding fibers," *Appl. Opt.* **10,** 2552–2558 (1971).
3. A. Cherin, *An Introduction to Optical Fibers,* McGraw-Hill, New York, 1983.
4. D. B. Keck, "Fundamentals of optical waveguide fibers," *IEEE Commun. Mag.* **23,** 18 (May 1985).
5. T. Miya, Y. Terunuma, H. Hosaka, and T. Miyashita, "Ultimate Low-loss Single-Mode Fibre at 1.55 μm," *Electron Lett.* **15,** 1979.
6. R. H. Stolen, *Nonlinearity in Fiber Transmission, Proc. IEEE* **68,** 1232–1236 (1980).
7. D. Gloge, "Dispersion in weakly guiding fibers," *Appl. Opt.* **10,** 2442–2445 (1971).
8. J. W. Fleming, "Material dispersion in lightguide glasses," *Electron. Lett.* **14,** 326–328 (1978).
9. D. Marcuse, "Interdependence of waveguide and material dispersion," *Appl. Opt.* **18,** 2930–2932 (1979).
10. A. Hasagawa and T. Tappert, "Transmission of stationary nonlinear optical pulses in dispersive dielectric fibers. I Anomalous dispersion," *Appl. Phys. Lett.* **23,** 142–144 (1973).
11. B. J. Hong, L. Wang, and C. C. Yang, "Using nonsoliton pulses for soliton-based fiber communications," *J. Lightwave Technol.* **8,** 568–575 (1990).
12. D. Gloge, "Weakly guiding fibers," *Appl. Opt.* **10,** 2553 (1971); see also B. E. A. Saleh and M. C. Teich, *Fundamentals of Photonics,* Wiley, New York, 1991, p. 280 ff.
13. G. Keiser, *Optical Fiber Communications,* 2d ed., McGraw-Hill, New York, 1991.
14. A. E. Siegman, *An Introduction to Lasers and Masers,* McGraw-Hill, New York, 1971.

4

Lasers and Optical Amplifiers

4.1 Introduction

4.1.1 Historical background

Experimental measurement techniques developed around the late 1890s and early 1900s demonstrated that there were phenomena involving the interaction between electromagnetic fields and matter that were not adequately described by classic electromagnetic theory. Several prime examples of these anomalies that concerned physicists at that time are:

1. Blackbody radiation

2. Photoelectric effect

3. Gas discharge phenomena

Blackbody radiation in particular was an enigma because there existed seemingly accurate theoretical models for the spectral rolloff on both the infrared and ultraviolet sides of the central portion of the blackbody emission spectrum. Neither of these two separate models matched observed spectral behavior for the central portion of the blackbody spectrum, nor were they compatible with the central region on an individual basis. Planck derived a single equation that fit the entire blackbody spectrum by decomposing the field into discrete modes and furthermore postulating a model that divided the electromagnetic energy of each mode into the discrete units called *photons*. The density of the number of modes as a function of frequency was a very important aspect of this model. The success of Planck's model as well as the assumptions used to derive the blackbody spectrum were milestones in the development of modern quantum mechanics. In this chapter we shall discuss the following subjects:

1. Definition and characteristics of photons

2. Interaction of light with atoms

3. Optical amplifiers

4. Lasers in general

5. Diode lasers

6. Amplifiers of optical frequency signals

The most successful and complete theoretical model for the interaction between light and atoms is a very complicated theoretical model is known as *quantum electrodynamics* (QED). It accurately describes the phenomena observed when electromagnetic fields interact with matter. Among its more prominent successes of QED is the remarkable prediction by Dirac of the existence and properties of the positron before it was observed experimentally. To a degree the laser is also a triumph for the theory, but QED is a more complicated theoretical model than we need to describe the devices that are utilized in fiber-optic communication systems (FOCS). We shall use the much simpler *photon-optics model* (POM) [1a] that was postulated by Planck and others, including Einstein. It doesn't do all that QED can do, but it accurately describes all the optical phenomena of interest in this course. The POM forms a readily comprehensible bridge between classic electromagnetic and the quantum-based particle-like properties of light (photons).

4.1.2 Photon optics

Electromagnetic radiation in a cavity can be expressed as a linear superposition of modes in the format

$$\widetilde{\mathbf{E}}(\mathbf{r},t) = \sum_q \sum_k \widetilde{A}_q U_q(\mathbf{r}) \exp\{j2\pi f_q t\}\hat{\mathbf{P}}_k \qquad (4.1)$$

where q is the index for the spatial aspect of the mode (usually three parameters) while k is the polarization index. The spatial portion of the mode is represented by $U_q(\mathbf{r})$. Each of the $U_q(\mathbf{r})$ is individually normalized to have unit energy, fits the shape of the cavity, satisfies the wave equation, and fits the boundary conditions. The phasor $\widetilde{A}_q$ is the "complex amplitude" for the qth mode, and it specifies the magnitude and phase of that mode, while f_q denotes the resonant frequency of the qth mode. The coordinate system is usually selected to match the shape of the cavity, and $\mathbf{r}$ represents the three spatial coordinates in that coordinate system.

We note that one of the primary differences between the classic and quantum-mechanical representations of electromagnetic fields is the nature of the oscillators used to represent the time development of the modal solutions. In the classic model a simple (classic) mechanical harmonic oscillator is used for each mode while in the quantum-mechanical description, the solution of Schroedinger's equation for the quantum mechanical oscillator is the basis for each mode. One major difference between the two oscillator models is that while classic oscillators have a continuous range of possible oscillator energies, the allowed energy levels for quantum-mechanical oscillators are discrete with allowed energy levels states given by

$$E_q = (m + \tfrac{1}{2})\, hf_q \qquad (\text{where } m = 0,1,2,3,4,...)$$

The energy difference between any two states of the quantum-mechanical oscillator are integer multiples of the photon energy, which is

$$hf_q = \hbar\omega_q \qquad \text{where} \qquad \hbar = \frac{h}{2\pi}$$

and where $h = 6.626 \times 10^{-34}$ J $\cdot$ s (Planck's constant). (*Note:* Dimensionality $hf = $ J $\cdot$ s/s $=$ joules, as it should.)

Although it has no bearing on the design of FOCS, it is interesting to note that photons have momentum, and in particular that the photons in circularly polarized light exhibit angular momentum. The Star Wars concept of laser cannons is based on impulse momentum doing the damage.

The extra half-photon energy in each mode is known as the *zero-point field*. The existence of this zero-point oscillation is manifested by the inability of hydrogen to solidify at absolute zero temperature because the zero-point energy molecular vibration prevents solidification at atmospheric pressure. The effects of the zero-point field have been observed in other interactions between electromagnetic fields and atoms, but these effects are not related to FOCS, so we shall not pursue them here, but some aspects of zero-point fields will arise in Chap. 10. In Chap. 4 we are concerned primarily with transitions of atoms from one quantum state to another as well as the corresponding transitions of the energy states of the electromagnetic fields which occur in discrete steps. Since energy is conserved in the interactions of fields and matter, the energy difference between two states of the atom must equal the corresponding photon energy for both absorption or emission processes. Thus, for example, the energy of the incident electromagnetic field is reduced by integer multiples of the corresponding photon energy. Furthermore, the number of absorbed photons will equal the number of atoms that have made the corresponding upward transition. We shall soon employ this type of reasoning to investigate the basic processes involved in lasers.

There is a tendency to view photons as solid particles much like shotgun pellets. Photons, however, have wavelike properties which are clearly displayed in various optical interference experiments. Let $\widetilde{A}_{1q}$ and $\widetilde{A}_{2q}$ be the complex mode amplitudes in Eq. (4.1) for two different fields within the same cavity. If both fields are superposed in the cavity, then the net field may be expressed by

$$\widetilde{A}_q = \widetilde{A}_{1q} \oplus \widetilde{A}_{2q}$$

where [cpl] implies phasor addition. Now the energy of the qth mode of the superposed field is proportional to

$$\propto |\tilde{A}_q|^2 = |\tilde{A}_{1q} \oplus \tilde{A}_{2q}|^2$$

The net energy of the qth mode after superposition will still be quantized to the discrete levels of the quantum-mechanical oscillator levels, but the probability distribution of finding a mode at a particular energy will depend on the phase relation between the complex mode amplitudes. This is a form of optical interference, and it clearly is a manifestation of the wavelike properties of photons. We shall study this phenomenon in some detail when we study optical heterodyne detection systems in Chap. 8.

In general, the particle properties of photons become more noticeable at high frequencies than at low frequencies, where the wavelike properties are more prevalent. A dilemma that confronted physicists during the early years of the quantum age was the question of why so many commonly observed phenomena, particularly those involving macroscopic bodies, did not manifest quantization phenomena. This question led to the formulation of the *Bohr correspondence principle* (BCP), which stated that *when the quantum number (number of photons) of a system such as the simple harmonic oscillator was large, it behaved much as a classic oscillator.* The details of the BCP and its place in quantum mechanics is beyond the scope of this text, but we note here that BCP will arise again in Chap. 10. For now, though, we consider the following two important aspects of the quantum numbers for a mode that will be used later on:

1. First we note that if we compare the number of photons in a mode with a specified energy for $\lambda_0 = 0.2$ μm with, say, $\lambda_0 = 2.0$ μm. Since the energy of the photons in the latter case is one-tenth that of the $\lambda_0 = 0.2$ μm photons, there will be 10 times as many photons for the latter case for a specific input power.

2. An even more dramatic difference is observed when one considers the number of modes density factor at the two wavelengths. An integral part of Planck's derivation of the blackbody radiation spectrum was the assumption of the following form for the mode density $M(f)$:

$$M(f) = \frac{8\pi f^2}{c^3} \tag{4.2}$$

where $M(f)$ is the mode number density per unit volume and per unit bandwidth at center frequency f. Now consider the situation described in part 1 above. The mode density will be ~ 100 times larger for the shorter wavelength, which means that the available photons will be distributed among ~ 100 times as many modes. Thus, including the effect described in part 1 above, the net dilution of photon numbers for this case is a factor of ~ 1000. If we now apply this analysis to a comparison of photon densities for a 1-GHz microwave signal and a typical optical frequency of 10^{15} Hz, we would find an enormous photon dilution factor of $\sim 10^{18}$ per mode at the microwave frequency. Therefore it is not surprising that quantum effects are seldom observed at microwave frequencies. Nevertheless, some quantum-mechanical effects have been observed at microwave frequencies, as is evidenced by the maser, but as of now no one has observed the photoelectric effect at microwave frequencies.

4.1.3 Photon streams

The process by which photons are generated by excited atoms is inherently random. Thus both the number and arrival times of photons are random variables. We shall see in Chap. 6 that those random phenomena which are based on quantum effects will be manifested as shot noise in the detection process. Now consider a monochromatic beam of light at time t and position $\mathbf{r}$, with optical intensity $I(\mathbf{r},t)$ (in watts per unit area). Then the mean photon-flux density denoted by $\phi(\mathbf{r},t)$ is given by

$$\phi(\mathbf{r},t) = \frac{I(\mathbf{r},t)}{hf} = \text{average number of photons per second per area}$$

The total flux $\Phi(t)$ through a surface at time t may then be obtained from

$$\Phi(t) = \int_A \phi(\mathbf{r},t) \cdot dA = \frac{1}{hf} \int_A I(\mathbf{r},t) \cdot dA \tag{4.3a}$$

For a spatially uniform steady-state source, the average number photons [ba]n that flow through a cross section A in T seconds is

$$\bar{n} = \Phi AT = \frac{\text{energy}}{hf} \tag{4.3b}$$

Now, to obtain an intuitive feel for the randomness of photon statistics, we shall examine a few simple examples. This material will be supplemented considerably in Chaps. 6 and 7 when we study the detection statistics for FOCS receivers. The randomness of the photon streams is initially generated by the source of the optical radiation. These initial photon statistics, however, may be modified by the communication channel. For example, slight pressure changes on the optical fiber can cause random changes in the polarization in single-mode systems as well as the random reconfiguration in the energy distribution among the modes in multimode fibers. We now consider a few important examples of photon statistics in optical sources.

For a steady-state light such as an ideal steady-state laser, the photon statistics are a simple homogeneous Poisson process. The number of photons that pass through a particular surface during a time interval T is random and the probability of exactly k photons in T seconds $p(k)$ is given by the simple Poisson distribution

$$p(k) = \frac{(\bar{n})^k e^{-\bar{n}}}{k!} \tag{4.4}$$

where $\bar{n}$ is as given by Eq. (4.3b). An important property of the simple Poisson distribution is the fact that the mean and variance are equal:

$$\mathbf{E}[k] = [\bar{n} \qquad \text{and} \qquad \text{var}[k] = \sigma^2 = \bar{n} = \text{mean}$$

Thus, even for a constant optical source, the number of photons in a specified time is a random variable. Next we note that for a steady-state source the average number of photons observed is directly proportional to the intensity, and it can be shown that a time-dependent optical source will generate a mean number of photons given by

$$\bar{n} \propto \int_0^T I(t)\, dt$$

In this text we shall often define the signal-to-noise ratio (SNR) as

$$\text{SNR} = \frac{(\text{mean})^2}{\text{Variance}} = \text{which for simple Poisson} = \frac{\bar{n}^2}{\bar{n}} = \bar{n}$$

As a second example, we next consider the photon number distribution for a single mode of thermal (natural) light. It is well known that the distribution of energy for a system of particles in thermal equilibrium at temperature T is

$$P(E_n) \propto \exp\left\{ -\frac{E_n}{k_B T} \right\} \tag{4.5}$$

where $k_B = 1.38 \times 10^{-23}$ J/K (Boltzmann's constant). This equation also applies to the probability distribution of the number of photons and hence the probability distribution of energy in a single mode of natural light. We note that photons are bosons; and, unlike electrons, several photons can occupy the same state simultaneously. In the following analysis the principal goal is to obtain the photon distribution, and therefore we shall ignore the zero-point field. Let $\zeta = hf/k_B T$ and $E_n = nhf$. Therefore $p(n) \propto e^{-n\zeta} = [e^{-\zeta}]^n$, and note that

$$\sum_{n=0}^{\infty} q^n = \frac{1}{1-q} \qquad \text{for} \qquad |q| < 1 \Rightarrow \sum_{n=0}^{\infty} \{e^{-\zeta}\}^n = \frac{1}{1 - e^{-\zeta}}$$

Then, properly normalized, we have

$$p(n) = \frac{(e^{-\zeta})^n}{[1 - e^{-\zeta}]^{-1}} = [1 - e^{-\zeta}](e^{-\zeta})^n$$

Next let

$$a = \frac{1}{(e^{\zeta} - 1)} \Rightarrow \frac{1}{a} = e^{\zeta} - 1$$

$$e^{\zeta} = \frac{1}{a} + 1 = a + \frac{1}{a} \qquad \text{or} \qquad e^{-\zeta} = \frac{a}{a+1}$$

Finally we arrive at what is called the *Bose–Einstein distribution*

$$P(n) = \frac{1}{a+1}\left\{ \frac{a}{a+1} \right\}^n \tag{4.6}$$

When we cover these and related photon statistics more carefully in Chap. 6, we shall demonstrate that $\bar{n} = a$ and $\sigma_n^2 = a + a^2$.

The variance for a single mode of thermal light is clearly larger than the variance for a mode of ideal laser light, which, as it turns out, may be modeled by a simple Poisson distribution for the photon number. The excess variance in photon number observed for a single mode of natural light is manifested in a phenomenon known as *photon bunching,* and, as we shall see in Chap. 6, photon bunching results in excess shot noise at the receiver. This tendency to cluster is in stark contrast to the simple Poisson process, wherein photon arrival times are statistically independent of each other.

4.2   Interactions between Light and Atoms

4.2.1   Introduction

It is well known from spectroscopic experiments of absorption and/or emission of light that most atoms and molecules exhibit sharp line spectra. These phenomena can be explained quantum-mechanically by assigning discrete allowable energy levels (E_i) to the natural states (eigenstates) of an atom or molecule. Let E_m and E_n represent two such energy levels for a typical atom or molecule. The energy diagram for a photon emission event is depicted in Fig. 4.1, while the energy diagram for an absorption event depicted below in Fig. 4.2. The conservation of energy requires that the photon energy match the energy difference between the upper and lower levels of that particular transition.

The following relations between energy levels and photon frequency apply to both emission and absorption processes:

$$E_m - E_n = hf_{mn} \Rightarrow f_{mn} = \frac{E_m - E_n}{h}$$

and

$$\lambda_{mn} = \frac{c}{f_{mn}} = \frac{hc}{E_n - E_m}$$

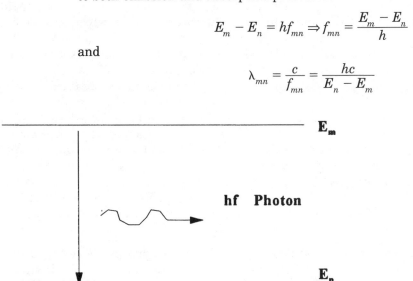

E_m

hf   Photon

E_n

Figure 4.1   Photon emission process.

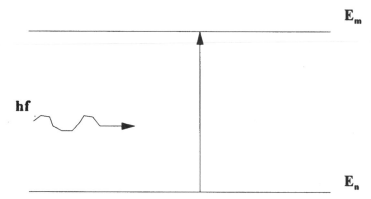

Figure 4.2 Photon absorption process.

Many of the energy levels in an atom or molecule are actually composed of several overlapping levels at virtually the same energy. The various energy states associated with these "degenerate" energy states will split into distinct and spectrally discernible energy levels when the atoms are exposed to strong electric or magnetic fields. One example of this phenomenon that may be familiar to the reader is how the two spin states—which have nearly the same energy level—of an electron within an atom split into separate energy levels separated by microwave frequency transitions when atoms are immersed in a strong magnetic field. The number of levels associated with these degenerate energy levels plays a role in some of the equations that we are about to develop, but to keep this development as simple as possible, we shall omit the level degeneracy parameter from our coverage of the topic. This will not affect the main thrust of our coverage here, which is aimed at understanding the operation of the semiconductor laser. When we get to this topic toward the end of this chapter, we shall see that the Pauli exclusion principle permits only solo occupation of any quantum state by an electron.

In theory, the specific energy levels for an atom could be obtained by solving Schroedinger's equation for that particular species of atom. Unfortunately, these equations are very complex, and hydrogen is the only species of atom (or molecule) for which complete solutions of the Schroedinger equation have been found. Thus, while theoretical analyses have yielded some subsets of the possible energy levels for other types of atoms, the wavelength, bandwidth, and line strength of most spectral lines have been determined by experiment, and tables of these may be obtained from the National Bureau of Standards. Some general aspects of these transitions and spectral lines are

1. All atoms in excited states, that is, those with energy above the ground state (the lowest energy state), exhibit a natural tendency to drop down to a lower state. In this process the energy difference is carried off by either a *photon,* if the emission is in the form of an electromagnetic radiation, or an

acoustic photon known as a *phonon,* if the emission is in the form of vibrational or rotational energy.

2. The decay from an upper state to a lower state is random in nature; thus the time spent in the upper state by a particular atom is a random variable. The decay process is characterized by an exponential decay constant called the lifetime (*not half-life*), and the reciprocal of the decay constant represents the average time an atom in the upper state remains there. Thus, the larger the decay constant, the less likely you are to find an atom selected at random in the upper state. These finite decay times broaden the energy-level lines via Fourier transform concepts; thus, the spectral lines emanating from the atom are not monochromatic. There is a nonzero spectral bandwidth and lineshape that applies to both emitted and absorbed radiation. The most frequently encountered spectral shape is given by

$$S(f) = \frac{A}{(\Delta f)_{mn}^2 + (f - f_{mn})^2} \tag{4.7}$$

where f_{mn} is the center frequency of the particular transition and $(\Delta f)_{mn}$ is the 3-dB bandwidth for that transition. The latter parameter is often known in the literature as the *full width at half maximum* (FWHM). The size of the FWHM is inversely proportional to the lifetimes associated with the energy-level states involved in the transition. The spectrum represented by Eq. (4.7) is known as the *lorentzian spectrum,* a spectral shape that should be familiar to electrical engineers since it represents the power spectral response of a high-Q single-tuned RLC (resistance, inductance, capacitance) bandpass filter.

3. Each transition within an atom and separately for each species of atom, has a different affinity for making the transition to the other state. A parameter that represents this affinity is the oscillator strength S_{mn}. It is a unique parameter for each atom and for each transition within that atom. Most oscillator strengths are also obtained by experiment.

4. There is a direct connection between oscillator strength and linewidth. The lifetime of an upper-level state is decreased by all the possible transitions to lower states. Transitions with large oscillator strengths will lead to short lifetimes for the upper state, and all transitions from such a level will have large bandwidths. Energy states that are characterized by very weak transitions to lower state, and hence have long lifetimes, are called *metastable* states, and they play a very important role in the design of lasers. Transitions starting from metastable states will be very narrow.

4.2.2 Two-level model

In the following material we shall employ two constants called the *Einstein A and B coefficients* [1b]. They both depend on oscillator strength, while B alone is also a function of the number of modes density factor expressed by Eq. (4.2). To simplify the derivation of the key equations, we shall assume that all atoms (or molecules) are of the same species, and we focus our attention on the two energy levels associated with one particular transition between the energy levels denoted by E_1 and E_2. Now let

N_2 = volume density of the atoms in the upper state (E_2)

N_1 = volume density of the atoms in the lower state (E_1)

This assignment of specific numbers to the population density of each state is really an oversimplification of the quantum-mechanical picture of the interaction of light with atoms. In quantum mechanics the state of the atom is determined by the wavefunction Ψ. For the two-level atom model, Ψ is a linear superposition of the only two energy states considered in this simplified model:

$$\Psi(\mathbf{r},t) = a_1 \, \Psi_1(\mathbf{r},t) + a_2 \, \Psi_2(\mathbf{r},t)$$

Where the Ψ_k are the normalized wavefunctions for an atom in the kth state. If we now measure the energy of a single atom in the mixed state represented by Ψ, we will observe either E_1 or E_2 with probabilities given by

$$\mathrm{Prob}(E_1) = |a_1^2| = \frac{N_1}{N_1 + N_2} = \frac{N_1}{N}$$

and

$$\mathrm{Prob}(E_2) = |a_2^2| = \frac{N_2}{N_1 + N_2} = \frac{N_2}{N}$$

where $N = N_1 + N_2$, $N_1 = |a_1|^2 2\,N$, and $N_2 = |a_2|^2\,N$. This forms a connection between the formal quantum treatment of the status of the atoms and the simplified picture using N_1 and N_2, which is the picture that we shall employ here. On the other hand, if we measure the energy level of many such atoms, each of which has the same wavefunction Ψ, the number of atoms observed at each energy level is random. The parameters N_1 and N_2 are proportional to the ensemble average of the number of times each energy level is observed in an experiment consisting of $N_1 + N_2$ trials.

We next examine each basic interaction process involved in the interaction of light with atoms, namely, absorption, spontaneous emission, and stimulated emission. These processes play a crucial role in the operation of lasers and optical amplifiers.

4.2.3 Absorption

The absorption process for a two-state single atom is displayed in Fig. 4.2. It occurs when a photon, whose energy coincides with the difference between energy levels for the transition in question, collides with an atom in the lower state. The term *collision* here is used in the quantum-mechanical sense, meaning that the photon and the atom interact in such a way that both are noticeably affected by the interaction. In this case the photon is annihilated while the atom is excited to a higher energy state. This type of collision should not be viewed like two billiard balls colliding. In fact, the wave nature of the photon is such that the photon and the atom never touch in the ordinary sense. During the absorption interaction one quantum of energy (hf) is removed from the optical field. This photon can come from any mode whose frequency is within the transition bandwidth, although it will more likely come from photons with frequencies near the line center than out in the wings of the response spectrum [see Eq. (4.7)]. Simultaneously with the photon annihilation, one lower-state atom will be elevated to the upper energy state. Thus this interaction conserves energy, and in a more subtle way it also conserves momentum.

When an atom and a photon come into close proximity, the probability of a quantum-mechanical collision taking place (in this case absorption) depends on a parameter called the *collision cross section*. This parameter is directly proportional to the oscillator strength for that particular transition. These parameters, as well as the mode density factor expressed by Eq. (4,2), are embedded into Einstein B coefficient, which is one of the basic factors that appear in the absorption rate equation:

$$r_{abs} = BN_1\Phi \qquad (4.8a)$$

where Φ = flux density of the electromagnetic radiation incident upon an atom expressed in photons per second per unit area.

Clearly the quantum collision rate is proportional to the product of the number of atoms in the lower state [as represented by N_1 in Eq. (4.8a)] and the number of photons per second incident on the atoms (represented by Φ). Simply stated, the number of hits (close encounters) is proportional to the number of shots (photons) times the number of targets (atoms). The Einstein B coefficient is the constant that connects the close encounter rate with the absorption rate as expressed by Eq. (4.8a).

4.2.4   Spontaneous emission

Spontaneous emission is a symptom of the tendency of all systems to migrate toward a lower energy state. In this process an atom in the upper state *spontaneously* drops down to a lower state and simultaneously emits a photon whose frequency matches the energy level difference. The spontaneous emission process is depicted in Fig. 4.1, and the rate for this process is given by

$$r_{spont} = AN_2 \qquad (4.8b)$$

where A is the Einstein A coefficient. It is related to the B coefficient by

$$\frac{A}{B} = \frac{8\pi h}{\lambda^3} \approx 1.665 \times 10^{-8} \qquad \text{(for } \lambda_0 = 1 \text{ } \mu\text{m)}$$

Thus for the range of wavelengths employed in fiber optics, $A << B$. Spontaneous emission is dependent only on the number of atoms in the upper state. It is not affected directly by the incident radiation, but the spontaneous emission rate may be affected indirectly because strong incident light beams can alter the population levels. Atomic transitions between states with different angular momenta will generate circularly polarized photons. We note that this class of transitions includes most first-order transitions in the optical region.

The spontaneous photon is emitted with equal probability [modified by the lineshape function of Eq. (4.7)] into all the optical modes within the atomic transition linewidth. The emitted photon has random phase and direction with respect to an incident optical field in the vicinity, and, as such, spontaneous emission is a source of noise in optical fiber communication systems. We shall see in later chapters that one very important aspect of this noise is the random-phase walk generated by the spontaneous emission process in the output of lasers. It is called *phase noise,* and in some systems phase noise sets lower limits on the data rates. We note here that requiring something is not the same as enabling it to happen, and we shall see in Chap. 8 that this lower bound on data rate definitely is not an advantage.

4.2.5 Stimulated emission

Stimulated emission plays a key role in the physical process that makes optical amplification possible, and optical amplification, in turn, makes laser action possible. Einstein first proposed the existence of stimulated emission in 1917 based on the results obtained from his application of the principle of detailed balance to a system in thermodynamic equilibrium. His main contribution here was recognition that stimulated emission was necessary to bring the equations into balance. It wasn't until the advent of the maser in 1955 that stimulated emission was utilized in a practical device.

Stimulated emission results in atoms being stimulated by an incident optical field to descend from the upper state to the lower state and emitting a photon of the proper energy in the process. The process is depicted in Fig. 4.3.

It is generated by direct interaction (collisions) between photons and upper-state atoms. Thus the rate for stimulated emission r_{stim} is also proportional to the product of the number of photons and the number of atoms in the upper state. This leads to the following equation for the stimulated emission rate:

$$r_{stim} = BN_2\Phi \tag{4.8c}$$

where Φ is the incident photon rate and B is exactly the same Einstein B coefficient that we encountered when we discussed absorption. The primary difference between the stimulated emission rate formula and the absorption rate is the presence of the upper-state population density N_2 in Eq. (4.8c) in place of the lower-state population density N_1 that appears in Eq. (4.8a). The

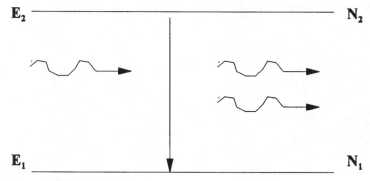

Figure 4.3 Stimulated emission

relative strength of these two processes will depend on which of the two states has the larger population. If $N_1 > N_2$, then absorption dominates, while if $N_2 > N_1$, stimulated emission will be the dominant rate. Clearly the latter condition is a minimum requirement for lasing, but as we shall soon see, it is not a sufficient requirement to achieve lasing.

Figure 4.3 displays a simplified version of the process where only one incident photon generates one other photon. Thus the figure does not show the proportionality of r_{stim} to the rate of incident photons. Probably the most important aspect of the photons generated by stimulated emission is that these photons will have the same polarization, direction, and phase (except for a frequency-dependent nonrandom phase shift) as the incident light. Furthermore, each stimulated photon will be stimulated into the same mode as the incident photon. This is the process that leads to optical amplification and lasing action.

4.3 Optical Amplification

4.3.1 Population inversion (PI)

We shall now investigate how these processes affect a light beam passing through a uniform medium. Specifically, consider a uniform monochromatic light beam passing through an infinitesimally thin plane within a uniform dielectric.

Let $\Phi(z)$ be the incident intensity and $\Phi(z + \Delta z)$ be the exit intensity of a single-mode light crossing a plane perpendicular to the z axis.

Now, as is shown in Fig. 4.4, $\Phi(z + \Delta z) = \Phi(z) + d\Phi$, where $d\Phi \doteq$ change in photon flux encountered by the light beam in the Δz wide region. Now

$$d\Phi = \underbrace{\frac{-BN_1\, \Phi(z)\, \Delta z}{\text{absorption}}} + \underbrace{\frac{BN_2\, \Phi(z)\, \Delta z}{\text{stimulated emission}}}$$

where we assume that spontaneous emission is negligibly small. Then we have

$$\phi(z + \Delta z) - \phi(z) = -B(N_1 - N_2)\phi(z)\Delta z = -\alpha\phi(z)\Delta z$$

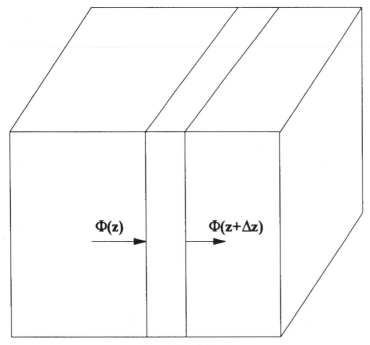

Figure 4.4 Intensity change through a slab.

where

$$\alpha = B(N_1 - N_2) \tag{4.9a}$$

This clearly leads to

$$\frac{d\phi}{dz} = -\alpha\phi = \Rightarrow \frac{dI}{dz} = -\alpha I$$

Thus we have

$$I(z) = I_0 e^{-\alpha z} \tag{4.9b}$$

Now from Eq. (4.9a) we see that when

$$N_1 > N_2 \qquad \alpha > 0 \Rightarrow \text{attenuation}$$

and when

$$N_2 > N_1 \qquad \alpha < 0 \Rightarrow \text{amplification}$$

Equation (4.9b) shows that both attenuation and amplification effects vary exponentially with distance. As light of an appropriate wavelength passes through a population-inverted medium, it will be amplified. We shall discuss optical amplifiers in Sec. 4.6, but we note here that one practical device employing the traveling-wave amplification process is the Erbium-doped fiber-optic amplifier. While traveling-wave amplifiers generally have superior gain–bandwidth products over cavity amplifiers, it is traveling-wave amplification of light bouncing back and forth between two end reflectors that forms the basis for lasers (optical oscillators).

The situation in which $N_2 > N_1$ is called *population inversion*. Population inversion does not occur naturally in thermal equilibrium because the population of the various energy levels satisfy

$$\frac{N_2}{N_1} = \exp\left\{-\frac{E_2 - E_1}{K_B T}\right\} = \exp\left\{-\frac{hf}{k_B T}\right\} < 1 \qquad (4.9c)$$

Thus, in order to achieve population inversion (PI), a non-thermal-equilibrium situation must be established. Note that this does not rule out the possibility of a steady-state situation. A number of schemes now exist for obtaining PI, and we shall now describe the three- and four-level laser pumping schemes.

4.3.2 Laser pumping schemes

The first oscillator device based on stimulated emission was the ammonia beam maser. This was actually a two-level system which took advantage of quadrupole moment charge symmetry of these molecules to physically remove the lower-state atoms from the particle beam. Very few optical frequency masers (lasers) operate with only two levels.

The population of the three states in the three-level laser for both before and after pumping are shown below in Figs. 4.5a and 4.5b, respectively [2a]. Pumping is usually accomplished by strongly irradiating the atoms with light whose frequency is determined by the energy of levels for the transition from E_1 to E_3. Thus we have

$$f_{13} = \frac{E_3 - E_1}{h}$$

A three-level laser has been designed by selecting atoms with states such that there is a strong connection between levels 2 and 3 which causes atoms at E_3 to quickly decay to level E_2. The atoms for this laser are chosen to also have the property that level E_2 is a metastable level (relatively long lifetime). Thus a traffic jam (in terms of atoms) forms at level 2, which will create a population inversion between levels 2 and 1.

Now for levels separated by an optical photon, virtually all the atoms are in the lower state prior to pumping [see Eq. (4.9c)]. Thus it is necessary to raise at least half of all the atoms into the upper state in order to achieve population inversion. Because of this requirement, relatively few substances have energy levels and suitable lifetimes that allow three-level lasing. Nevertheless, the very first laser was a three-level laser using ruby (an Al_2O_3 crystal doped with chromium). Since our main interest in this text will be diode lasers, I will outline only a few salient points about the ruby laser:

1. Chromium atoms are the sources of laser radiation;

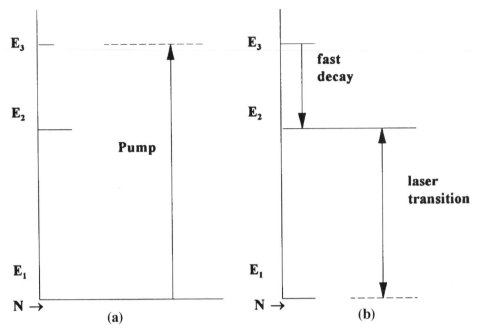

Figure 4.5 Three-level population diagrams: (*a*) before pumping; (*b*) after pumping.

2. The upper pumping level (E_3) is really a broad collection of levels separated from the ground level (denoted here by E_1) by transitions in the blue and green portions of the spectrum. The pumping is achieved with flash lamps filled with argon, xenon, and krypton mixtures selected to maximize the optical output in the green and blue regions.

3. The fast decay from level 3 to level 2 is greatly enhanced by the Al_2O_3 crystal host, and it takes place in many small steps using phonons for each transition. The upper lasing level E_2 (really two very close levels) is determined by an inner-shell electron in the chromium atom, and therefore it is shielded to a great extent from the outside world. This makes E_2 a metastable level, but it does not prevent lasing.

By incorporating a fourth level into the pumping scheme, one can eliminate the need to raise at least half of the atoms out of the ground state. The four-level pumping scheme is shown in Fig. 4.6. Note that the labeling of the energy levels in Fig. 4.6 is such that once again the lasing takes place between the E_2 and E_1 levels. Level E_0 represents the ground state, which is the lowest level in the atom. Exciting ground-state atoms to level E_3 can be achieved by many means, one of which is strong illumination at f_{30},

$$\text{where } f_{30} = \frac{(E_3 - E_0)}{h}.$$

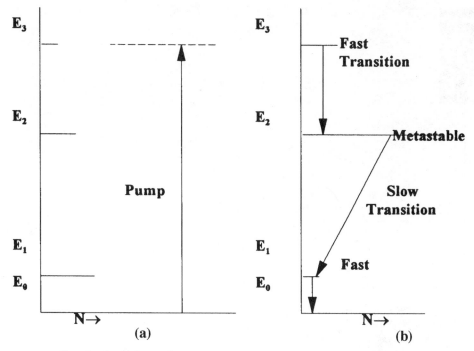

Figure 4.6 Four-level population diagrams: (*a*) before pumping; (*b*) after pumping.

A very prominent example of a four-level laser is Nd-YAG (neodymium–yttrium aluminum garnet), whose energy-level diagram is shown in Fig. 4.7. The top energy level E_3 employed in the pumping scheme for this laser is really a broad set of states, all of which are connected to the ground state by radiation in the green and blue portions of the spectrum. Flash lamps are often used for pulsed operation, while tungsten filament blackbody sources are often used to pump the Nd-YAG laser to achieve steady-state laser operation. The population inversion occurs here for the Nd atoms, while the YAG crystal acts as a host for the Nd. The YAG crystal lattice is responsible for the rapid decay of atoms excited to level E_3 down to the upper lasing level E_2. Note that many other hosts are possible for Nd, including glass, other crystals, and some organic liquids.

While the four-level Nd-YAG laser will operate as a steady-state source of light, it is very difficult to operate the three-level ruby laser in a steady-state mode of operation. Ruby lasers that have been built for steady-state operation have output powers on the order of a few milliwatts. This stands in stark contrast to pulse operation ruby lasers whose output pulse energies were at one time measured by the number of razor blades that the laser pulse could blast through. On the other hand, powerful four-level Nd-YAG lasers are available in both pulsed and steady-state versions, sometimes in the same device. We note here that the combination of Nd-glass lasers and optical amplifiers was

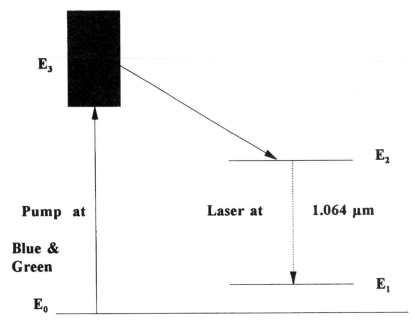

E_3

E_2

Pump at

Laser at   1.064 μm

Blue &
Green

E_0

E_1

Figure 4.7   Level diagram for Nd-YAG.

one of the prime candidates for laser-driven fusion reactors when that area
was a hot topic for research a few years ago.

4.3.3 Optical amplification

The preceding section demonstrated several general methods for achieving population inversion. We now consider some important characteristics of lasers:

1. The laser output is not completely monochromatic, and even a so-called
   single-frequency laser has an output spectrum of the form given by Eq.
   (4.7). This lineshape occurs frequently and is known as a *lorentzian spectrum*. We shall encounter it again in Chap. 8 when we discuss a laser phenomenon known as *phase noise.*

2. As is common in tuned electronic amplifiers, there is also a frequency-
   dependent phase shift encountered by the electromagnetic field as it passes
   through the optical amplifier. This phase shift causes an effect on the laser
   known as *frequency pulling,* which affects the laser output frequency. In
   most lasers it is a relatively small effect that is of use mainly in studying
   laser operation; however, it plays an important role in tuning diode lasers.

   In the remainder of this section we shall concentrate on optical amplification in connection with lasers. However, we note here the existence of a
nonoscillatory optical amplifiers that have assumed great importance in

fiber-optic system design. We shall describe such devices in Sec. 4.6, and note here that one such device is an erbium-doped fiber [1c,3] with a useful gain profile that may be as broad as 400 GHz has become a very important component in practical high-speed FOCS.

4.3.4   Laser cavity

Thus far in this chapter we have described the optical amplification process by itself. The laser, however, is an optical frequency oscillator that employs some form of positive feedback to stabilize operation of the device. This is accomplished for most lasers with an optical frequency cavity of the form shown in Fig. 4.8. The active medium refers to the population-inverted atoms (or molecules) that we have been discussing in this section. This cavity is enclosed at both ends by optical reflectors; usually made with layers of dielectric coatings to achieve desired reflectivities. Typically one reflector is only partially reflecting and serves as laser output port while the other is made as close to 100 percent reflection as possible. The other sides of the laser cavity are not part of the feedback path and are usually nonreflecting. This open-sided optical cavity arrangement is known as a Fabry–Perot cavity. For masers, the cavity is fully enclosed with a few relatively small ports to connect it with the outside world. A Fabry–Perot cavity for a maser would result in large diffraction losses. Diode lasers employed in FOCS, however, have cavity dimensions that are on the order of several wavelengths of light. We shall see in Sec. 4.4.7 that many FOCS grade diode lasers have confinement in the top and bottom directions.

The laser oscillation process begins with a spontaneous emission photon propagating down the principal axis of the laser cavity. This photon, in turn, stimulates more photons to join the original photon and thereby initiates the buildup of the field strength in the mode. Now, in order to have gain, the *population inversion must be sufficient to overcome losses within the cavity*. This includes the desired output, which also represents energy lost to the cavity. A dynamic analysis of the process yields an exponential buildup of energy in

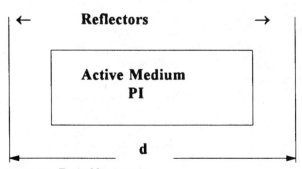

Figure 4.8   Typical laser cavity.

the optical field. Clearly this process cannot continue indefinitely, and eventually a level of stimulated emission is reached which just balances the pumping rate at which atoms are elevated to the upper lasing state. This process usually—but not always—yields a steady-state laser output.

The resonant frequency of the optical cavity is determined by the physical condition that the *round-trip length (2d) be such that round-trip signal adds constructively to the primary beam.* This requires that the wavelength of oscillation be such that

$$2d = q \lambda_q \quad \text{or} \quad \lambda_q = \frac{2d}{q}$$

where q is an integer. This type of mode is known as an *axial mode.* In order for oscillation to occur at wavelength λ_q, its associated frequency f_q must lie within the gain profile [see Eq. (4.7)] of the lasing transition. Since for most lasers $2d >> \lambda_q$, it is possible for several axial modes to have a resonant frequencies within the gain response of the laser, and therefore several modes may be able to oscillate at the same time. Each axial mode has a separate wavelength, hence a corresponding separate lasing frequency. These are denoted by q as above, and the spacing between axial modes in the frequency domain is given by

$$\Delta f = f_{q+1} - f_q = \frac{c}{n\lambda_{q+1}} - \frac{c}{n\lambda_q} = \frac{c}{2nd}\left[q + 1 - q\right] = \frac{c}{2nd} \qquad (4.10)$$

The mode spacing is somewhat affected by the frequency pulling effect in such a way that the resonant wavelength for each axial mode is pulled slightly toward the center frequency of the transition. Therefore the axial mode frequencies are not exactly uniformly spaced.

In addition to axial modes, there is another class of modes known as *transverse modes.* There are higher-order resonant solutions to the wave equation inside the laser cavity with field patterns that display multiple bright spots. A few such modes are displayed in Fig. 4.9. If the laser were operating in a single transverse mode, the shape of mode pattern would be the same (except for a size factor) for any cross section in the cavity. It is, however, extremely difficult to operate a laser in a single higher-order mode for any length of time. Furthermore, each transverse mode can support a number of axial modes. One vital aspect of these modes evident in Fig. 4.9 is that higher-order transverse modes have broader mode patterns. For most practical applications, laser operation is desired with the single solid spot of the fundamental mode. One can achieve this pure fundamental mode operation by placing an optical constriction such as a camera iris at a strategic location inside the cavity, and then reducing the opening until only the solid dot of the fundamental mode is seen in the output. The iris constriction causes the higher modes to experience too much attenuation to achieve lasing. We note here, however, that this process alone won't eliminate axial modes, and therefore it does not achieve single-frequency operation.

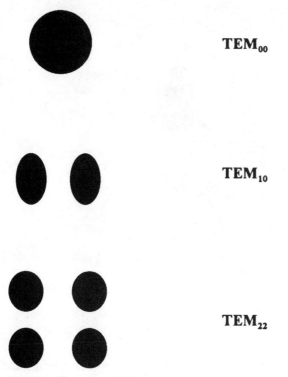

TEM_{00}

TEM_{10}

TEM_{22}

Figure 4.9 Transverse Modes.

To reduce the output to a single axial mode, two routes are possible. In the first method the mirror spacing is reduced until only one axial mode frequency fits under the useful gain spectrum of the pumped medium. Recall that the axial mode spacing is inversely proportional to mirror separation. Unfortunately, this shortening of the laser cavity also reduces the laser output power.

A second general method consists of placing a frequency-dependent passive element inside the cavity. One such arrangement that has been used for a number of commercial lasers is to use a high quality optical prism and take advantage of the $n(f)$ properties of glass shown in Fig. 4.10. Typically only one pertinent wavelength will be deflected at just the right angle to line up with the cavity axis and, provided this line is within the gain curve, the device will oscillate at a single frequency. This can also be achieved by employing an appropriately designed diffraction grating as reflector. Both of these realizations of the second method enable the user to select any one of a number of specific wavelengths that can be made to oscillate in a particular device, or on a finer scale provide some fine-tuning capability within the usable bands for one of the aforementioned laser lines.

In Fig. 4.10 the dashed line represents the path of the light ray in the opti-

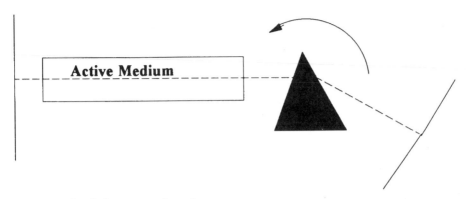

Figure 4.10   Single frequency using prism.

cal cavity. Obviously one cannot put a prism inside a diode laser cavity. Nevertheless, as we shall see later on in this chapter, a wavelength-dependent substructure can be manufactured within the diode cavity using well-known semiconductor manufacturing techniques.

4.3.5   Gain threshold condition

To obtain a steady-state laser output, the *net round-trip gain must equal unity*. Otherwise the output will either continue to grow if it's greater than unity or shrink if it is less. We introduce now the following notation. Let

$$\bar{g} = B(N_2 - N_1) = \text{exponential gain coefficient of the pumped medium}$$

$$\bar{\alpha} = \text{exponential loss coefficient (oversimplified) due to scattering and other imperfections}$$

$$R_1, R_2 = \text{the fraction of power (photons) reflected at each mirror}$$

Then

$$R_1 R_2 \exp\{2d(\bar{g}_{\text{th}} - \bar{\alpha})\} = 1$$

where $\bar{g}_{\text{th}}$ is the value of the gain coefficient *[ba]g* that is required to just barely reach the lasing threshold. We now take the natural logarithm of this equation to obtain

$$\bar{g}_{\text{th}} = \bar{\alpha} + \frac{1}{2d} \ln \left\{ \frac{1}{R_1 R_2} \right\} \tag{4.11}$$

The physical implication of Eq. (4.11) is that the pumping source which maintains the population inversion must be sufficient to support all energy-consuming mechanisms within the cavity. This includes the internal loss mechanisms as well as both the desired and undesired outputs.

The population inversion that generates unity net round-trip gain defines the minimum population-inversion density $(N_2 - N_1)_{th}$ required to just barely begin lasing. This is known as the *threshold condition,* which explains the subscript notation in Eq. (4.11) and above. Note, however, that Eq. (4.11) determines $\bar{g}_{th}$ for steady-state lasing regardless of the output power level. We shall see in the next subsection that this trend will have important consequences to the spontaneous emission rate as well as the nature of laser light below threshold. For a semiconductor laser diode (SLD), the threshold is manifested by a lower bound on the current required for lasing. We now investigate briefly the nature of the laser output below and above lasing threshold.

4.3.6 Steady-state equations for single-mode laser

In this subsection we shall find the steady-state characteristics of a single-mode laser operating below threshold as well as the same laser operating above threshold. Eventually we shall examine the remarkable transition laser in operation that occurs as the laser passes through threshold. In the following analysis we shall employ a simplified model for the laser which focuses on the two energy states E_1 and E_2 associated with the lasing transition. Here we shall assume a single mode whose frequency is central to the transition spectrum. To further simplify the analysis, we may define some additional terms to make the following additional assumptions:

1. $N_1 \approx 0$.
2. n_p = number of photons in the lasing mode of the cavity.
3. r_p = rate (atoms per second per unit volume) that are excited to E_2 state by the pumping mechanisms.
4. Total *spontaneous decay rate* for upper-level atoms, including both photon and phonon decay mechanisms = (N_2/τ_{21}), where τ_{12} is the characteristic lifetime for the $E_2 - E_1$ transition.
5. *Stimulated emission rate* = $Kn_p N_2$, where K is the Einstein B coefficient expressed in terms of the number of photons in the cavity mode of interest n_p, instead of the photon flux Φ.
6. Spontaneous radiative decay rate into a single mode = KN_2. The appearance of the same K for spontaneous decay as well as stimulated decay is a result of Einstein's work. Note, however, that the spontaneous decay does not depend on the incident field.

Equation (4.12a) expresses the rate of change for the upper-level population density, while Eq. (4.12b) is the number of photons in the lasing mode of the cavity in terms of the rate of stimulated emission, absorption, and the pumping and cavity loss rates.

$$\frac{dN_2}{dt} = -\underbrace{Kn_pN_2}_{\text{stimulated emission}} - \underbrace{\frac{N_2}{\tau_{21}}}_{\text{spontaneous decay}} + \underbrace{\frac{r_p}{\text{pumping}}}_{} \quad (4.12a)$$

$$\frac{dn_p}{dt} = +\underbrace{Kn_pN_2}_{\text{stimulated emission}} + \underbrace{KN_2}_{\text{spontaneous decay}} - \underbrace{\frac{n_p}{\tau_p}}_{\text{losses}} \quad (4.12b)$$

where the term n_p/τ_p [Eq. (4.12b)] is the rate at which photons are lost from the cavity field by all the loss mechanisms that appear in Eq. (4.11). Thus the photon decay lifetime parameter τ_p includes the effects of the exponential loss coefficient $\bar{\alpha}$ from Eq. (4.11) as well as the output losses through the mirrors at the end of the cavity. Now, in order to apply steady-state conditions, *we set all $d/dt = 0$* to arrive at the following two equations:

$$0 = -KN_2n_p - \frac{N_2}{\tau_{21}} + r_p \quad (4.13a)$$

$$0 = KN_2(n_p + 1) - \frac{n_p}{\tau_p} \quad (4.13b)$$

These are two coupled nonlinear algebraic equations where *the two unknowns are N_2 and n_p*. We now examine each of these equations separately to see what they reveal about the behavior of the laser above and below lasing threshold. From Eq. (4.13b) we obtain both of the following relations between N_2 and n_p:

$$n_p = \frac{N_2}{N_{2\text{th}} - N_2} \quad (4.14a)$$

where $N_{2\text{th}} = 1/K\tau_p$ and

$$N_2 = \frac{1}{K\tau_p}\left\{\frac{n_p}{n_p + 1}\right\} = N_{2\text{th}}\left\{\frac{n_p}{n_p + 1}\right\} \quad (4.14b)$$

The following trends are discernible from Eqs. (4.14a) and (4.14b):

1. As $N_2 \rightarrow N_{2\text{th}}$, $n_p \rightarrow \infty$.
2. As $n_p \rightarrow \infty$, $N_2 \rightarrow N_{2\text{th}}$, *but never exceeds it*. Thus the population inversion of a steady-state laser never exceeds its value required to just barely reach threshold.

If we had not assumed that $N_1 = 0$ and used a third equation for (dN_1/dt), we would have obtained the following variation on part 2 above:

$$\text{As} \quad n_p \rightarrow \infty, \quad (N_2 - N_1) \rightarrow (N_2 - N_1)_{\text{th}}$$

$$\text{For} \quad N_2 \ll N_{2th}, \qquad n_p \approx \frac{N_2}{N_{2th}} \ll 1$$

Equations (4.14a) and (4.14b) justify the choice of subscript on N_{2th}. Even when the laser is operating well above threshold and n_p is very large, the upper-state population does not exceed the value required to just barely begin lasing action. We now use Eq. (4.13a) to relate N_2 and n_p to the upper-state pumping rate r_p.

$$N_2 = \frac{r_p \tau_{21}}{1 + K\tau_{21}n_p} \tag{4.15}$$

$$n_p = \frac{r_p \tau_{21} - N_2}{K\tau_{21}N_2} \tag{4.16}$$

Now *well below threshold* $(n_p \ll 1)$ *we find from Eq. (4.15) that*

$$N_2 = r_p \tau_{21} = \text{constant} \times r_p \tag{4.17}$$

Next we define the pumping threshold parameter $r_{p,th} = N_{2th}/\tau_{21}$. Then, using this definition of $r_{p,th}$ [Eq. (4.14a)] and Eq. (4.16) for pumping rates below threshold, specifically, $r_p < r_{p,th}$, we find that

$$n_p = \frac{N_2}{N_{2th} - N_2} = \frac{r_p \tau_{21}}{r_{p,th}\tau_{21} - r_p \tau_{21}} = \frac{r_p}{r_{p,th} - r_p} \tag{4.18a}$$

We now make the reasonable assumption that $n_p \gg 1$ for lasers operating above threshold (i.e., $r_p > r_{th}$). We also know that above threshold $N_2 \approx N_{2th}$, so we substitute this into Eq. (4.16) and utilize the definition of $r_{p,th}$ to obtain

$$n_p \approx \frac{1}{K\tau_{21}} \left\{ \frac{r_p}{r_{p,th}} - 1 \right\} \tag{4.18b}$$

Plots of both N_2 and n_p versus the pumping rate are displayed in Fig. 4.11. The linear growth of N_2 below threshold comes from Eq. (4.17), while the saturation of this parameter above threshold comes from Eq. (4.14b). The number of photons in the single mode that will eventually provide the laser output grows slowly well below threshold as indicated by Eq. (4.18a) but grows very quickly once r_p passes threshold. Well above threshold $(r_p \gg r_{p,th})$, this growth of the output power is an approximately linear function of the pumping rate.

To clarify the basic nature of these processes (particularly near threshold), we introduce the number of modes parameter p:

$$p = \frac{8\pi V f_{21}^2 (\Delta f)_{21}}{(c/n)^3} \tag{4.18c}$$

This formula is derived using the mode density factor $M(f)$ [Eq. (4.2)] and represents the number of modes within the transition linewidth $(\Delta f)_{21}$ and cavity of volume V. Typically p is an extremely large number for most lasers, and some typical values are $p \sim 10^{12}$ for a ruby laser and $p \sim 10^9$ for a helium-

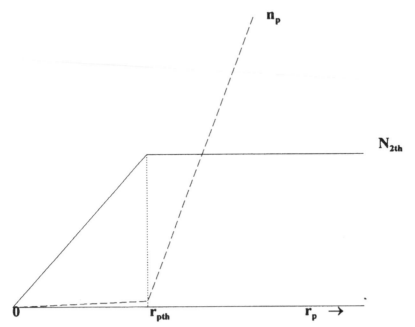

Figure 4.11 Plots of N_2 and n_p versus the pumping rate.

neon laser. Diode lasers have very small cavities; nevertheless we still have $p \sim 10^6$ for these devices.

To examine the dynamic changes that take place inside the laser cavity near threshold, we shall gradually increase the upper-level pumping rate r_p from below achieving population inversion (here in this simplified model, it means N_2) right up through lasing threshold. Below threshold the radiation is primarily spontaneous emission and the photons are distributed more or less equally over all the p modes. This is natural light, called *fluorescent radiation,* and it is a form of thermal radiation. The electric field for steady-state fluorescent radiation is a zero-mean stationary gaussian random process. The output power is determined by the variance, which turns out to be directly proportional to N_2. The saturation of the upper-level population density N_2 as $r_p \rightarrow r_{p,th}$ can be verified experimentally by observing the radiation emitted from the side of the laser at a wavelength with an optical transition that has the same upper-level E_2 as the laser transition. The brightness of this spectral line will be almost constant as the laser output is varied over a large range, thereby confirming that N_2 remains constant when r_p exceeds threshold. It should be noted, however, that optical transitions that don't feed off the E_2 level will most likely brighten as pumping is increased.

Now as we increase r_p beyond threshold, the excess pumping must be going directly into the single lasing mode. This is evidenced by the sudden and sharp increase in laser output that takes place at threshold. The threshold level for a particular laser cavity depends on cavity losses and mirror reflectivities as represented by Eq. (4.11). It takes a pumping level of $r_{p,th}$ to raise N_2 just to the threshold level. Suppose, for example, that we let r_p be twice $r_{p,th}$. We then have as much power in the single-laser mode as was spread among all the other $p - 1$ modes put together. Thus the power density in the laser mode is $\approx p \times$ the power in each of the other nonlasing modes. Another way to see this is to rewrite Eq. (4.18b) as

$$K\tau_{21}n_p = \frac{\text{stim.}}{\text{spont.}} \approx \frac{r_p}{r_{p,th}} - 1$$

In the preceding example we selected $r_p = 2 \times r_{p,th}$, and when we use this value for r_p in the revised form of Eq. (4.18b) above, we find that the stimulated emission (stim.) equals the spontaneous emission (spont.). We shall soon see how this single-laser mode differs dramatically from thermal light, but first let us examine what happens in the vicinity of threshold.

Siegman [2b] has employed Eqs. (4.13a) and (4.13b) to demonstrate that near threshold we have

$$n_p \approx \frac{p}{2}\left\{(\gamma - 1) + \sqrt{(\gamma - 1)^2 + \frac{4\gamma}{p}}\right\} \tag{4.18d}$$

where $\gamma = (r_p/r_{p,th})$. Now, because of the extremely large value of p, the transition from just below threshold to just above threshold is very sharp, indeed, as indicated by Fig. 4.12.

We now define a parameter ε where $\varepsilon \ll 1$. Then we can show that

$$\gamma = \begin{cases} 1 - \varepsilon \to n_p \approx \dfrac{1}{\varepsilon} & \text{just below threshold} \\ 1 + \varepsilon \to n_p \approx \varepsilon\, p & \text{just above threshold} \end{cases}$$

As a numerical example, let $\varepsilon = 0.01$ and $p = 10^{10}$

$$n_p \approx \begin{cases} 100 & \text{just below threshold} \\ 10^8 & \text{just above threshold} \end{cases}$$

Thus, in going from 1 percent below to 1 percent above threshold, the number of photons changes by a factor of 10^6. The source of this sudden change is the nonlinear nature of the basic equations and the very large size of parameter p. In fact, it is very difficult to observe the transition region, and most experiments demonstrating the transition through threshold are performed with diode lasers because of the smaller value for p.

4.3.7   What's special about laser light?

Thermal or chaotic radiation is characterized by equiphase wavefronts which have random shapes. Such radiation cannot be focused to produce a higher

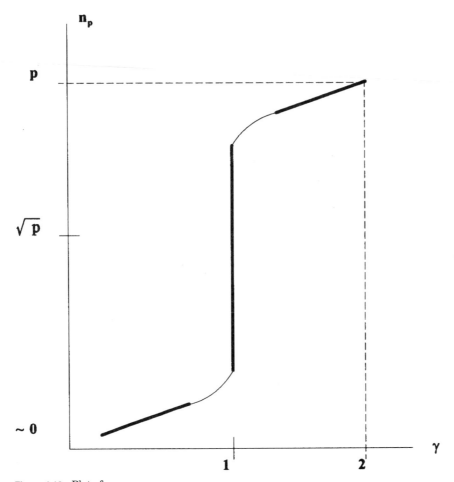

Figure 4.12 Plot of n_p versus γ.

temperature than the temperature of the blackbody that generated the light to begin with. On the other hand, the specific phase relation between the stimulated emission photons and the cavity field generates coherent light which has orderly equiphase wavefronts. These can be focused to yield considerably higher temperatures than the ambient temperature of the laser itself. Now, to further emphasize the difference between coherent and chaotic radiation, we shall calculate the temperature required by a blackbody source to yield the same number of photons per second and within the same degree of optical confinement as a single mode of a very modest-size laser propagating in a single-mode optical fiber (see Fig. 4.13 for an explanation of the hypothetical experiment).

Let n_T be the average number of thermal photons emitted from a blackbody source whose frequencies lie within a bandwidth Δf. Planck showed that for blackbody radiation we have

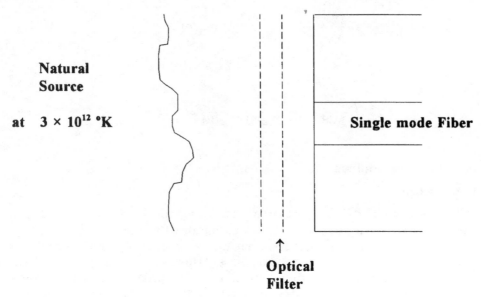

**Natural
Source**

at 3×10^{12} °K

Single mode Fiber

↑
**Optical
Filter**

Figure 4.13   Hypothetical diagram showing propagation of laser in a single-mode optical fiber.

$$n_{\mathrm{T}} = \frac{\Delta f}{\exp\{hf/kT\} - 1} \approx \frac{\Delta f k T_{\mathrm{eq}}}{hf} \qquad (4.19)$$

where we arrived at the approximation in Eq. (4.19) by using the soon-to-be-verified fact that

$$T_{\mathrm{eq}} >>> 1 \qquad \text{so} \qquad \frac{hf}{kT_{\mathrm{eq}}} << 1$$

(*Note:* The transparent heat shield is not feasible at the temperatures shown in Fig. 4.13, but it helps us to visualize the process.)

Now, to get the same number of thermal photons in the same bandwidth and the same directional constraints as a laser, we use Eq. (4.19) to obtain

$$P_{\mathrm{laser}} = n_{\mathrm{T}} hf = \Delta f k T_{\mathrm{eq}}$$

and arrive at

$$T_{\mathrm{eq}} = \frac{n_{\mathrm{T}} hf}{k\,\Delta f} = \frac{P_{\mathrm{laser}}}{k\,\Delta f}$$

As a numerical example, we consider a single-frequency, 1.0-mW diode laser operating at $\lambda_0 = 1.3$ μm. We also assume a bandwidth-to-center-frequency ratio of 10^{-7}. Thus, from our work on material dispersion in Chap. 3, we have

$$\frac{\Delta f}{f} = \frac{\Delta \lambda}{\lambda} = 10^{-7} \qquad \text{(single mode)}$$

$$\Delta f = f \times 10^{-7} = \frac{10^{-7}c}{1.3\ \mu} = 2.31 \times 10^{7}\ \text{Hz}$$

$$T_{eq} = \frac{10^{-3}}{1.38 \times 10^{-23}(2.31 \times 10^{7})} \approx 3.14 \times 10^{12}\ \text{K}$$

4.4 Semiconductor Diode Sources

4.4.1 General introduction

References 1d and 4–7 are particularly useful for the following material. The same basic physical processes that we encountered in the preceding description of lasers—namely, absorption, spontaneous emission, and stimulated emission—are all present in the diode laser. However, instead of the concentrating on the states of individual atoms as in the previous material, we are concerned here with the states of electrons in a very dense collection of atoms. In this regard we note that one of the major differences between the types of lasers we have studied so far and semiconductor lasers is the nature of the energy levels. Individual atoms of the chromium in ruby and the Nd in Nd-YAG were almost completely separate systems (with only small interactions between Cr atoms via the optical field). The Pauli exclusion principle applied to isolated atoms states that no two electrons in an individual atom can occupy the same state at the same time. On the other hand, the atoms in semiconductors are so closely packed that *a great many atoms share the same outer-shell electrons*; thus these electrons cannot be associated with any particular atom. The *energy levels are determined not by individual atoms but by an extended system consisting of an enormous number of densely packed atoms.* Some salient features of this structure are

1. The Pauli exclusion principle applies to these electrons also, which means that no two electrons can occupy the same state in this new, greatly expanded version of a closed system.

2. Energy levels are so closely packed that they essentially form continuous groups of energy levels known as *bands.*

3. Because of the repetitive structure of the semiconductor crystal, there is a spatially periodic structure to the potential well in the Schroedinger equation. In general, partial differential equations with periodic boundary conditions lead to discrete eigenvalues. In semiconductors these densely packed states result in the energy levels breaking up into densely packed, distinct bands of energy levels.

4. The *conduction band* is the highest of the energy bands. It consists of electrons which are free to respond to externally applied electric fields. The

next-highest band, known as the *valence band,* consists of bound-state electrons. These electrons are shared among multiple atoms and can move around throughout the structure in this sharing mode, but not in the same way as electrons in the conduction band. Actually it is vacancies in the valence band (holes) that act like positive charges, and they move in a manner analogous to electrons.

5. The energy difference between the bottom of the conduction band and the top of the valence band is called the *bandgap energy,* which we denote as E_g. It is the minimum energy required to excite an electron from the valance band to the conduction band. The region between these two levels is called the "forbidden region," for there are no levels in this region (except for those due to impurities)

The energy of an electron is composed of potential energy represented by its energy state plus kinetic energy. In Fig. 4.14 we present some simplified plots E versus k for Si and GaAs. In these diagrams E_c represents the bottom of the conduction band, E_v represents the top of the valence band, and $E_g = E_c - E_v$ is the bandgap.

For silicon, the bottom of the conduction band and the top of the valance band are offset from each other. This arrangement is known as an *indirect bandgap,* and the emission process for such a material requires three particles simultaneously; an electron, a photon, and a phonon. This process is a two-stage or second-order process, and it is a much rarer event than a first-order transition event as is encountered in direct-bandgap transitions. For GaAs, on the other hand, the bottom of the conduction band and the top of

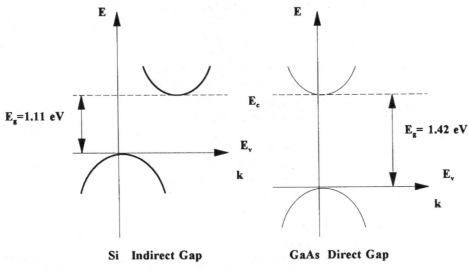

Figure 4.14 Approximate *E–k* diagrams for Si and GaAs.

the valance band lie vertically above each other. This permits direct photon emission without momentum change, and therefore it is called a *direct bandgap.*

The preceding explains why Si is not used for lasers while GaAs and InP (and mixtures of such materials) are used for semiconductor lasers. Silicon, however, can be used in detectors. Silicon photodiodes are much cheaper than GaAs, and silicon works well for absorption because it requires only that the photon have enough energy to cross the elevated bandgap. The phonons needed for the upward transition are provided in abundance by the crystal lattice vibrations (crystal momentum) to the valence band electrons. On the other hand, free electrons in the conduction band are not bound to atoms, and therefore they don't vibrate within the crystal structure. Thus the phonon matching is much harder to come by for the photon emission process than the absorption process. Finally, we note that Si cannot be used to detect the $\lambda_0 = 1.3$ and 1.55 μm because the bandgap is too large to be bridged by such low-energy photons.

4.4.2   Fermi levels

To obtain a better understanding of what population inversion means in diode lasers, we need to introduce the concept of Fermi levels. The probability that a particular energy level E is occupied is given by the Fermi–Dirac distribution

$$f(E) = \frac{1}{1 + \exp\{(E - E_f)/k_B T\}} \tag{4.20}$$

where E_f is called the *Fermi level.* It can also be seen from this equation that $f(E_f) = \frac{1}{2}$ at all temperatures. Although Eq. (4.20) is known as the *Fermi–Dirac distribution* and $0 \le ff(E)$, $f(E)$ is not normalized in the variable E and *therefore it is not a probability density in E.* Also, it can be seen from Eq. (4.20) that at $T = 0$ K all the energy levels below E_f are filled while all the energy levels above E_f are empty. This phenomenon is depicted in Fig. 4.15. As the temperature increases, kinetic energy excites several electrons across the energy gap to the conduction band. For each electron so emitted, there is a vacancy left behind in the valence band which we call a *hole.* The Fermi level plays an important role in understanding the operation of diode lasers. In particular, we shall use Fermi levels to explain why modern diode lasers are far more efficient in the conversion of electrical energy into optical energy than any other type of laser, but first we need to discuss doped semiconductors and p–n junctions.

4.4.3   Doped semiconductors

The basic semiconductor without doping is said to be *intrinsic.* For an intrinsic semiconductor, the Fermi level E_f lies in the center of the bandgap.

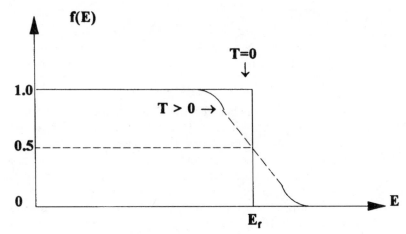

Figure 4.15 Fermi–Dirac distribution.

The electrical properties of semiconductors can be altered enormously by imbedding specific impurities in the semiconductor crystal called *acceptors* or *donors*. A donor has one more outer-shell electron than the host semiconductor (HS), while an acceptor has one less outer-shell electron than the HS. The key to the sharing of electrons in a crystal is a strong tendency for two atoms to share a total of eight electrons. Thus Si pairs contribute four outer-shell electrons each to form a Si–Si structure while Ga As forms a 3–5 structure, with Ga providing three outer-shell electrons and arsenic providing five.

To see how doping operates, let's use GaAs as an example. If some of the As atoms in the HS are replaced by elements from the VI (six outer-shell electrons) of the Periodic Table of Elements (PTE), then these dopant atoms will provide an extra electron to the outer-shell-sharing pair. The force retaining this electron in the vicinity of the dopant atom is weak, and it takes very little kinetic energy to lift this electron to the conduction band. A semiconductor doped with these *donor* atoms is said to be an *n*-type semiconductor. Similarly, replacing some Ga atoms in the semiconductor with elements from the II column of the PTE leads to outer-shell pairings with only seven electrons. There is a strong tendency for this pair to capture an electron from an adjoining pair, leaving that pair short one outer-shell electron. This vacancy is called a *hole,* and it is passed through the crystal by the sharing exchanges described above. It behaves as though it were a free charge with a positive charge equal in magnitude to that of an electron. A hole responds to an applied electric field much like an electron, except that the force is in the opposite direction. Note, however, that in reality no positive charge actually moves through the crystal and the motion of the hole is really a flow of bound-state electrons in the opposite direction. A few donors and acceptors for IV–IV and III–V semiconductors are

	Donors	Acceptors
For IV–IV semiconductors (C, Si)	As P from V	Al, Ga, In, from III
For III–V GaAs, InP	S, Se, Te from IV	Zn, Cd, Hg from II

Donor atoms carry a residual + e charge because the charge balance of the atom has been disturbed by the loss of an electron to the conduction band, while acceptor atoms carry an equivalent residual negative charge due to the extra electron that has been taken from a nearby atom. These residual charges are not mobile; nevertheless, they play a vital role in the behavior of a *p–n* junction. The electric properties of the doped semiconductors are quite different from those of the intrinsic semiconductor. In particular, the Fermi levels are altered as is shown in Fig. 4.16. In Fig. 4.16 E_C is the energy level at the bottom of the conduction band and E_v is the energy level at the top of the valance band.

Now the Fermi level of the *n*-type semiconductor *t* is also known as the *donor level,* and similarly the Fermi level of the *p*-type semiconductor is called the *acceptor level.* We now define $E_D = E_C - E_f$ and $E_A = E_f - E_v$,

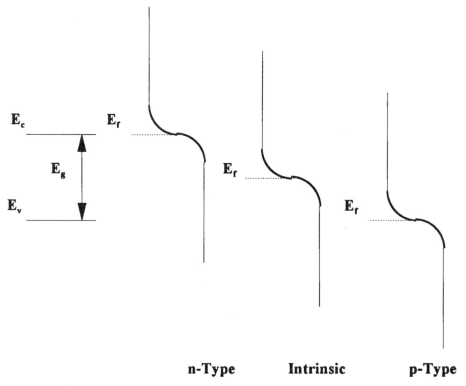

Figure 4.16 Fermi distribution for doped semiconductors.

where E_D represents the minimum energy required to raise an electron from the donor level up to the conduction level. For n-type material, this is much less than the bandgap energy, and as a consequence, the transition is easily achieved from kinetic energy available at room temperature. Similarly, it takes only $E_A << E_g$ to raise an electron whose energy level is near the top of the valence band to an acceptor level, thereby creating a vacancy in the valence band called a *hole*. Thus we see that doping a semiconductor with donor or acceptor atoms alters the conductivity significantly.

There are other processes that create holes and conduction band electrons simultaneously in pairs. In two such processes the energy to accomplish this can come from either kinetic energy (typically from phonons provided by the vibrating atoms) or from the absorption of a photon whose energy is sufficient to bridge the bandgap. In both cases an electron in the valence band is elevated up to the conduction band, leaving a hole behind in the valence band. Recombination is another related process that is very important to recognize at this point in the development of laser diode theory. It is the inverse of the hole–electron pair production process. In recombination, a conduction band electron drops down to a vacancy in the valence band, thereby annihilating a hole and a conduction band electron simultaneously. Energy is conserved in this process by the emission of a photon or a phonon. These transitions are subject to stimulated emission just as much as are transitions in isolated atoms, and, as expected, stimulated emission is the engine that drives optical amplification and lasing in semiconductors. Population inversion is achieved in laser diodes by the application of a strong forward bias to a suitably designed p–n junction diode which includes very heavy doping of the p–n junction. This will bring together a large number of holes and electrons in a small region of the device known as the *active region*. We next review some salient features of p–n junctions.

4.4.4 p–n junction

Hole–electron pairs are continually being generated by thermal agitation. Now let n be the free-electron charge carrier density, p be the hole charge carrier density, and $n_i = p_i$ be charge carrier densities in an intrinsic semiconductor generated by thermal excitation. Now it can be shown (see, e.g., Sec. 2.5 of Ref. 17) that

$$n_i^2(T) = np = C \exp\left(-\frac{E_g}{kT}\right) = n_i^2$$

Note that the product np is independent of the Fermi level, and that as T increases, the negative exponent decreases and the np product increases exponentially. At sufficiently high temperatures, even a doped semiconductor behaves like an intrinsic semiconductor. The charge carrier density, however, is affected by the doping concentrations, and at lower temperatures, including normal room temperature, we have

$$n \approx N_d \qquad \text{and} \qquad n \gg p \qquad \text{in n-type semiconductors}$$

$$p \approx N_a \qquad \text{and} \qquad p \gg n \qquad \text{in p-type semiconductors}$$

and

$$n = N_D \qquad \text{and} \qquad p = \frac{n_i^2}{N_D} \qquad \text{in n-type material}$$

$$p = N_A \qquad \text{and} \qquad n = \frac{n_i^2}{N_A} \qquad \text{in p-type material}$$

Although it is not the way that p–n junction diodes are manufactured, imagine that we place a block of n-doped semiconductor flush up against a p-doped block of the same type of semiconductor. Initially (nanoseconds) there is a flow of majority carriers across the junction where they become minority carriers. The immobile dopant atoms in the immediate vicinity of the junction retain the opposite charge of the majority carrier and thereby set up an electric field which opposes the flow of majority carriers across the junction (see Fig. 4.17). In effect, a drift current is formed which exactly cancels the diffusion current across the junction, and a static equilibrium is maintained. This

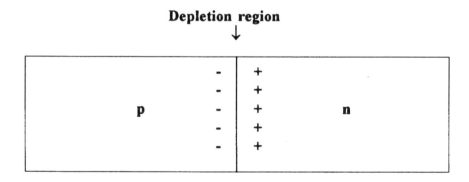

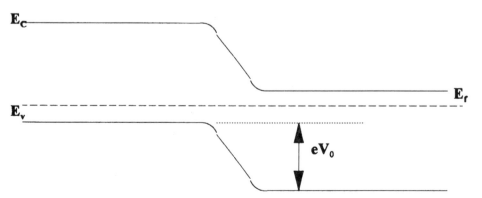

Figure 4.17 Unbiased p–n junction.

equilibrium is such that the Fermi level is the same on both sides of the junction. The E-field that has been set up in the depletion region produces a potential barrier V_0 which opposes majority carrier flow across the junction. This potential barrier, known as the *contact potential,* exists in one form or another when dissimilar materials are in contact with one another. At a $p-n$ junction this contact potential is given by

$$V_0 = \frac{kT}{e} \ln\left(\frac{N_a N_d}{n_i^2}\right)$$

Consider the unbiased $p-n$ junction shown in Fig. 4.17.

The field at the junction removes free electrons and holes from the neighborhood of the junction. This explains the term *depletion region* for this part of the device and why the depletion region is characterized by a high impedance. If an external voltage is applied to the diode, *most of the voltage drop across the diode occurs across the very small depletion region.* We'll come back to this point soon.

Now consider the forward-biased $p-n$ junction depicted in Fig. 4.18. The voltage drop across the depletion region has been reduced by the applied volt-

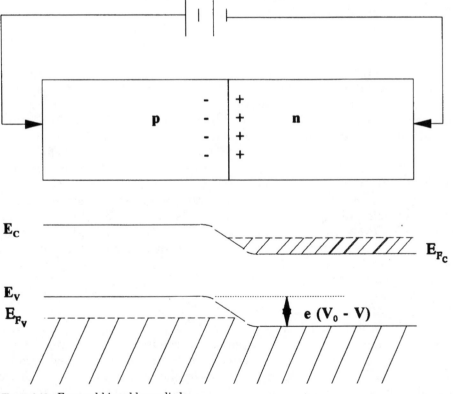

Figure 4.18 Forward-biased laser diode.

age, and there now exists a region containing large concentrations of both conduction band electrons and holes in close proximity. Some of these recombine to generate photons, and some recombine to generate phonons (and thereby heat in the process). Thus the recombination rate as well as the photon production rate are increased by the forward bias applied to the device. An external power supply provides the energy to drive this process by replenishing the charge carriers annihilated by the recombination process. Recall that the recombination rate is considerably larger for direct-gap transitions than for indirect-gap transitions.

Stimulated emission also drives the recombination process; electrons in the conduction band are stimulated to drop into a vacancy (hole) in the valence band by the optical field in the cavity. If the forward bias on the laser diode is large enough, then a level of population inversion can be attained that will support optical amplification as well as optical oscillation (laser action). To enhance population inversion, both the p and n regions of the laser diode are heavily doped. It so happens that diode lasers can be constructed that are more efficient in converting electrical energy into optical energy than any other type of laser. Some diode lasers are so efficient that a new class of lasers known as *diode-pumped lasers* based on this phenomenon has been developed. We next examine the basic reason why an appropriately designed laser diode is so efficient.

4.4.5 Semiconductor lasers

We now tackle the question of why semiconductor lasers are so efficient compared to other lasers. Our quest begins by noting that Fermi levels apply to thermal equilibrium conditions and even a steady-state diode laser is not in equilibrium. Figure 4.19 shows the bandgap diagram at $T = 0$ K for a heavily doped diode laser operating in a steady-state mode. The presence of a vacant region at the top of the valence band can occur because of the pumping and the particular relaxation times. Within each band any disturbance from equilibrium will right itself very quickly with a time constant of $\sim 10^{-12}$ s. This is much faster than the band-to-band relaxation times and gives the levels labeled $E_{F,c}$ and $E_{F,v}$ the appearance of equilibrium Fermi levels. These levels are called *quasi-Fermi* levels. At $T = 0$ K, all energy levels are filled within each band up to the corresponding quasi-Fermi level, and none are filled beyond that level. To see why diode lasers are so efficient, consider Fig. 4.19.

In Fig. 4.19 $E_q = E_{F,c} - E_{F,v}$. Now when a photon enters the active region whose energy lies in the range defined by $E_g < hf < E_q$, then absorption is inhibited because the upper levels of transitions in this frequency range (i.e., from top of valence band to lower portion of conduction band) are already occupied. On the other hand, stimulated emission is not so impeded because of all the vacancies available in the empty region at the top of the valence band. At normal laser operating temperatures these effects are somewhat

Relaxation time ~ 10^{-12}
↓

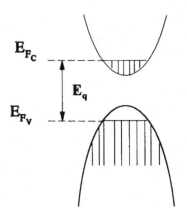

Band-to Band Relaxation time ~ 10^{-9}

Figure 4.19 Quasi-Fermi levels.

muted, but they are strong enough to produce lasers with conversion efficiencies that exceed 50 percent.

To obtain oscillation, we require optical feedback. This may be achieved in semiconductor lasers by cleaving the ends of the laser cavity (output face and direct opposite face). Cleaving provides flat and parallel reflecting surfaces. The reflection at the cleaved edges is due to the different indices of refraction at the interface between air or glass and the semiconductor material. The fraction of power reflected at the interface is given by [see also Eq. (2.31)]

$$R = \left\{ \frac{n_2 - n_1}{n_2 + n_1} \right\}^2$$

As an example, consider that an air–GaAs interface yields $R = 0.32 = 32$ percent where we used $n_1 = 1$ for air and $n_2 = 3.6$ for GaAs. In practice, the side opposite the output is usually coated with dielectric layers to make it ~ 100 percent reflecting. This complex coating, however, is one of the principal failure mechanisms for semiconductor lasers.

The simple diode structure shown in Fig. 4.20 is called a homojunction as opposed to the more complicated structures called *heterojunctions* that we shall encounter soon.

One of the most important parameters for laser diodes is the current densi-

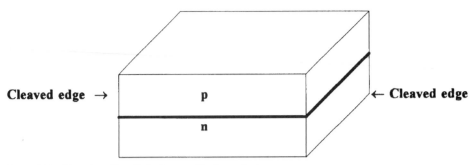

Figure 4.20 Homojunction diode laser with cleaved edges.

ty at threshold, denoted by J_t. Since J_t is the primary source of the charge carriers injected into the active region, it is the primary pumping mechanism for achieving population inversion in a semiconductor laser diode (SLD). It is reasonable, therefore, to assume that the population inversion threshold hence $\bar{g}_{th}$ is directly related to J_t by

$$\bar{g}_{th} = \bar{\beta} J_t \qquad \text{or} \qquad J_t = \frac{\bar{g}_{th}}{\bar{\beta}}$$

where $\bar{\beta}$ is a constant appropriate to a specific device. It is well known that within limits the threshold current density is inversely proportional to the thickness of the active region and furthermore that J_t can be decreased by confining the internal laser beam within this region. We shall return to these aspects when we describe heterojunction SLDs and quantum-well lasers. Now, substituting this into Eq. (4.11), we obtain

$$J_t = \frac{1}{\bar{\beta}} \left\{ \bar{\alpha} + \frac{1}{2d} \ln \frac{1}{R_1 R_2} \right\} \tag{4.21}$$

This equation is a somewhat simplified version of the formula for the threshold current density; however, it will suffice for the purposes of this text. More details can be found in the references listed at the end of this chapter. Actually the current density is internal to the device and not measurable directly. On the other hand, the external threshold current I_t is a more easily observed measure of how much power is required to drive the laser. These two quantities are simply related by $I_t = A J_t$, where A is the cross-sectional area of the active region of the SLD. The diode current at threshold is a measure of how much heat is generated in the SLD. To stabilize SLD operation[1] it is necessary to maintain a constant ambient temperature, a task that is sim-

---

[1]Figures 4.25 and 4.27 display some of the effects of temperature variations. These figures will be discussed in later subsections.

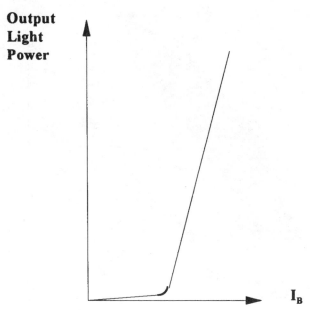

Figure 4.21   Laser output.

plified by reducing I_t as much as possible. In the next section we shall see several schemes for reducing I_t substantially; some by reducing J_t directly and others by adjusting the cross-sectional area of the active region and thereby maintaining the threshold current density J_t while at the same time reducing the total I_t.

The output power versus drive current for a good SLD is shown in Fig. 4.21. This figure is in good agreement with the discussion associated with Fig. 4.11. For currents below laser threshold, the optical output is extremely small. Once threshold is reached, however, the output power in terms of photons generated per second rises rapidly and almost linearly with increasing laser drive current. This type of device can be readily modulated in both analog and digital (ON/OFF) modes directly without resorting to devices external to the SLD by modulating the injection current I_t.

4.4.6   Single-mode and single-frequency lasers

For many applications it is desirable to obtain single-mode and single-frequency operation. A laser diode such as that shown in Fig.4-20 would almost surely oscillate in numerous modes and frequencies simultaneously. We also noted in Sec. 4.3.4 that the spacing between axial modes is given by $\Delta f = c/2nd$. Now d is very small in a SLD; thus the frequency spacing between axial modes is quite large (on the order of 45 GHz). This will lead to large dispersion and therefore will not suffice for ultra-high-speed data transmission. To achieve single-mode operation, we shall use some of the techniques dis-

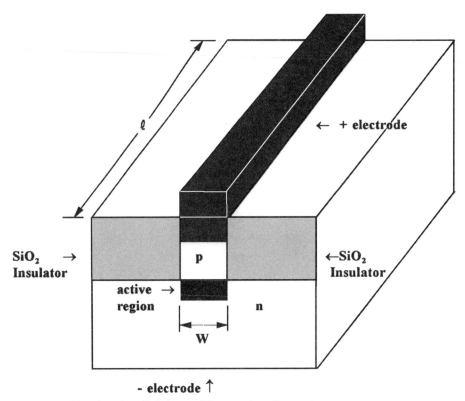

Figure 4.22 Homojunction diode laser with current confinement

cussed earlier in this chapter. One method is to reduce the cross section of the active region and thereby eliminate higher-order transverse modes. This will produce an effect akin to the effect of an iris (although this constriction is not adjustable). Unwanted axial modes can be eliminated by building into the SLD cavity a periodic structure in the doping levels which acts as a tuned element. Variations of this. concept have yielded single-frequency tunable SLDs.

Consider the stripe geometry lasers such as that shown in Fig. 4.22. The current is confined to only a small part of the width of the diode structure, and only those modes traveling close to the central axis of the laser cavity can achieve lasing.

In homojunction junction lasers the indices of refraction both inside and outside the pumped region are virtually identical. As a result, there is no noticeable optical confinement of light to the "active" region. We shall encounter one remedy to this situation in Sec. 4.4.7 when we discuss heterojunction lasers. The primary variable that determines lasing threshold is the current density J_t. The various parameters that affect J_t are shown in Eq. (4.21). Note that J does not explicitly depend on cross section. Thus, by constraining the current flow to a narrow stripe, we are able to reduce the external threshold current I_t while maintaining the value of J_t. This current confinement yields a considerable reduction in heat generated in the SLD; furthermore, current confinement eliminates many transverse modes by pinching the cavity in a manner analogous to an iris in open-structure macroscopic lasers. Another advantage of this design is that by eliminating mode hopping among transverse modes, it eliminates "kinks" from the linear portion of the laser output power–current curve. Actually the diode laser output curve depicted in Fig. 4.21 represents the response of a laser with current confinement. The SLD output curve of a simple laser such as that depicted in Fig. 4.20 would display irregularities known as kinks in the SLD power–current response curve. Current confinement alone, however, does not eliminate axial modes.

We now describe three SLD structures that have been employed to achieve single-frequency operation by reducing laser oscillation to a single axial mode. These are depicted in Figs. 4.23 to 4.25.

The *cleaved coupled cavity* or C³ laser (see Fig. 4.23), as it is also known, requires that the lasing mode satisfy the boundary conditions of both cavities simultaneously. This condition severely limits oscillation to the axial mode that first reaches threshold. The output of this type of laser is fairly stable, but tends to mode-hop (make discontinuous jumps in frequency) when being tuned. Although this type of laser is used mostly in laboratories as opposed to operating fiber-optic communication systems, there is one variation of the C³ laser that is very important for applications where there is a requirement for broad tuning without mode hopping. We shall discuss this variation at the end of this section.

The most successful design for single-frequency diode lasers embeds one or more frequency selective elements in the SLD cavity. In one realization of this concept, we build into the ends of the SLD cavity tuned feedback mecha-

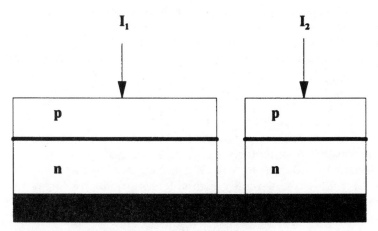

Figure 4.23   Cleaved coupled cavity laser (side view).

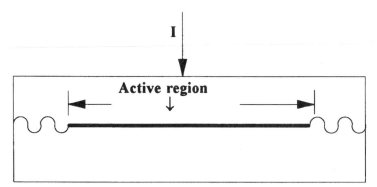

Figure 4.24   DBR laser (side view).

nisms in the form of periodically corrugated reflectors as depicted in Fig. 4.24. This particular configuration, termed the *distributed Bragg reflection* (DBR) laser, uses the Bragg effect to select a single frequency for oscillation. For a mode to be resonant in this compound cavity structure, the various reflections from each section of the corrugated ends of the cavity must be in phase to add constructively. This will maximize the feedback active region. The mode which best matches this condition will reach laser threshold first and thereby prevent other modes from reaching threshold (recall how $N_{2\text{th}}$ stops growing once threshold is reached). This type of mode competition effect happens in all lasers, but the DBR portions sharpen the frequency selectivity enough to enable single-frequency operation. The DBR portion acts very much like a diffraction grating.

One drawback of the DBR as shown in Fig. 4.24 is the fact that the corrugated portions are embedded into the optical cavity itself. Since these regions are not in the pumped region of the cavity, they represent a loss mechanism for the SLD, thereby lowering the efficiency of the device. A useful variation on the theme of inserting a tuned element to the cavity is the DFB arrangement shown in Fig. 4.25.

In the SLD displayed in Fig. 4.25 the corrugated section runs the length of the cavity, but it is physically located outside the active region. In effect, there is a periodic boundary condition structure set up along the interface of the active region and its surroundings. The boundary conditions imposed by this periodicity are very selective in choosing the resonant frequency of the cavity. The DFB diode laser is better suited to our next topic (heterojunction lasers). We note here that, as we saw when we studied optical fibers, at the boundaries between materials of different indices of refraction, the boundary conditions are affected by what's going on outside the boundary. Finally we note that most commercially available semiconductor lasers are of the DFB variety.

DFB lasers have been mass-produced for CD (compact-disc) players and optical computer memory readers. This type of laser is usually employed in

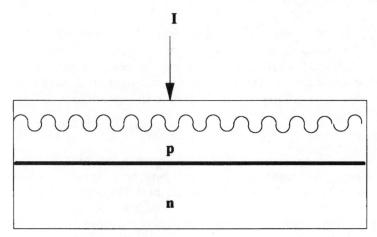

Figure 4.25   DFB laser (side view).

conjunction with current confinement, and "kinks" are seen only in devices that are breaking down. The threshold current for an SLD is a function of temperature, as can be seen from Fig. 4.26. It is clear from this figure that the threshold can be very sensitive to the ambient temperature. This can cause serious difficulties when diode lasers are used in ON/OFF-keyed (OOK)-type systems. For example, if the diode is biased to operate at, say, 30°C and heats up to 50°C, then the SLD output may remain below threshold for both binary states. One means of stabilizing laser outputs is to monitor the light output from the nonoutput end, where the end is often coated to be ≈ 99 per-

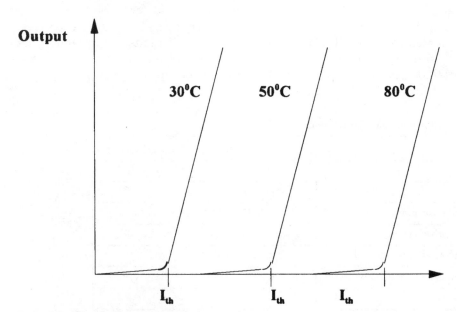

Figure 4.26   Threshold current versus temperature.

cent reflecting, and use this output in a feedback control system to adjust the bias current. This technique is often employed in repeaters for long-range PCM (pulse-code-modulated) OOK systems. Often the repeater contains a control system that accounts for the number of ON pulses in a long data sequence to avoid creating a data-driven imbalance in the system.

4.4.7   Heterojunction lasers (HJLs)

We have not as yet considered how an SLD can be constructed to oscillate at a particular frequency. The transmission properties of modern optical fibers require the use of $\lambda_0 = 1.3$ or 1.55 μm. However, a pure GaAs homojunction would operate at $\lambda_0 \sim 0.87$μm. A somewhat more complex laser diode can be constructed using *multiple layers of closely matching yet different materials*. To see how this works, consider the structure displayed in tabular form in Table 4.1.

The host semiconductor for AlGAs lasers is GaAs, while the host for InGaAsP lasers is InP. In Table 4.1 the x and y subscripts represent mole fractions of the atoms used in the SLD. Thus, for example, $Al_{0.4}Ga_{0.6}$ means that in that region 40 percent of the Ga atoms have been replaced by Al atoms in that layer. The bandgap, and hence the lasing frequency for each type of laser may be specified by selecting the appropriate combination of the mole fractions represented by x and y. AlGas lasers operate in the 0.78- to 0.85-μm range. They are mass-produced for compact-disc and optical computer memory devices, and they have been used in early versions of fiber-optic communication systems. Heterojunction lasers of the InP type can be built to operate in the 1.0- to 1.6-μm range. These multiple-layer devices are grown on top of the host crystal. In order for this to occur, the lattices for each layer must match the adjoining layers. This requirement limits the useable range of mole fractions x and y (see Table 4.1). Fortunately it is possible to construct lasers that operate in the 1.3- and 1.55-μm wavelength ranges.

TABLE 4.1   Some Heterojunction Laser Structures*

Function of layer	AlGaAs laser	InGaAsP laser
Metal contact		
Improve metal-semiconductor contact	p^+-type GaAs	p^+-type InP
Charge carrier and mode confinement	p-type $Al_xGa_{1-x}As$	p-type InP
Recombination or active region ~0.1 μm	n-type $Al_yGa_{1-y}As$	$In_{1-x}Ga_xAs_yP_{1-y}$
Charge carrier and mode confinement	n-type $Al_xGa_{1-x}As$	n-type InP
Substrate and metal-semiconductor contact	n^+-type GaAs	n^+-type InP
Metal contact		

*The notations p^+ and n^+ implies heavy doping in respective charge carrier.

We begin our discussion of the confinement regions which surround the active region by noting that only recombination that occurs in the active region can contribute to the laser output. A potential barrier is formed at each boundary between the dissimilar materials that form the active and the confinement layers (note additional space between these regions in Table 4.1). These potential barriers help reduce the bandgap in the active region to the desired value, but they also tend to repel the free electrons (*at upper end*) and holes (*at lower end*) as they try to exit the active region. The abrupt material change at these interfaces results in a change of index refraction on the order of 5 to 10 percent, providing optical confinement inside the active region. Furthermore, mode confinement is made possible because the index of refraction in the active region is larger than the indices of the confinement layers, and this process retains the lasing mode inside the tiny active region by total internal reflection. The principal consequence of all these effects is to lower the threshold current density J_t, especially when current confinement techniques are also applied. One possible arrangement for a device with these properties is shown in Fig. 4.27.

The buried heterojunction laser depicted in Fig. 4.27 has an active region that is completely surrounded by materials with lower indices of refraction, a property that helps confine the optical beam to the active region. The high-

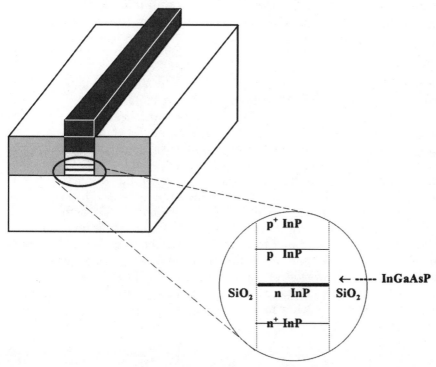

Figure 4.27 Buried heterojunction laser.

lighted region of Fig. 4.27 is a head-on view of a heterojunction laser (HJL). To obtain single-frequency operation, a distributed, feedback element is often introduced into one of the HJL confinement regions where it is aligned along the length of the active region. It can also be seen from Fig. 4.27 that this laser has current confinement by the SiO_2 insulator that forces the current to flow in a narrow strip. Thus the HJL is capable of single-frequency operation at a specified wavelength while reducing threshold current.

We noted earlier in this section when discussing Eq. (4.21) that the threshold current density could be reduced by decreasing the thickness of the active region of the SLD. Typically the active region in heterojunction lasers is between 0.1 and 0.3 μm in width. When one attempts to reduce the active-layer thickness beyond ~0.1 μm, one encounters a significant reduction in optical field confinement. The net outcome of these opposing processes is that instead of reducing J_t, the current density required to achieve lasing is actually increased. However, when the thickness of the recombination region is made even smaller, say, less than ~10 nm (more specifically, the thickness approximates the DeBroglie wavelength[2] of a conduction band electron in thermal equilibrium), then the junction region no longer behaves physically as bulk semiconductor; instead it takes on the physical properties of a *quantum well,* meaning that the properties of this very tiny region take on properties of a quantum-mechanical potential well. One consequence of this behavior is that the conduction band breaks up into several subbands. The technology for growing such thin layers has been solved, and the specialized equipment for growing such ultrathin layers are now available. The literature is full of reports of a large variety of devices that employ quantum wells.

The primary advantage of quantum-well lasers (QWLs) over bulk semiconductor lasers is that they greatly reduce the threshold current. Threshold currents I_t under 1 mA are commonly obtained with these devices. In fact, some of the lowest reported threshold currents have been attained using quantum-well lasers [5]. One of the principal advantages of quantum-well lasers is that the sharp potential barriers at the edges of the well tend to trap holes and conduction band electrons inside the wells and the active region more strongly than in bulk semiconductor heterojunctions. This results in higher recombination rates in the active region, which yields greater amplification and lower J_t. In many practical applications of single quantum-well lasers the single well is too small to yield the desired output level. To remedy this problem, many QWLs employ multiple quantum-well structures to achieve larger output power levels (up to ~100 mW). However, the author of the text has heard some engineers refer to quantum-well devices somewhat scornfully with the phrase "they make good fuses."

---

[2]DeBroglie wavelength $= h/p,$ where $h =$ momentum.

4.5 Tunable Lasers

Single-frequency lasers play a major role in optical heterodyne receivers. As is the case with most rf and microwave heterodyne receivers, it is necessary to be able to tune the local oscillator over a broad range of frequencies. Such is also the case with optical heterodyne detection (OHD) receivers, but the diode lasers that we have described so far do not meet the stringent requirements for transmitters and local oscillators in OHD systems. Diode lasers may be tuned by exercising temperature control and/or current control. It turns out that both the bandgap and diode dimensions are altered as temperature changes while the SLD drive current affects the index of refraction in the active region; thus, the free charge carrier densities affect n. Some typical plots of SLD output wavelength versus temperature and versus driving current) are shown in Fig. 4.28.

As can be seen from these plots, the output wavelength can change abruptly as either current or temperature is altered. This phenomenon, called "mode hopping," restricts the useful tuning range of these diodes. If one attempts to tune these lasers over too large a range, there will be a sudden change in the output wavelength. In an OHD receiver, this change can cause the difference frequency between local laser and incoming signal to change suddenly from a manageable tens of gigahertz to an unmanageable f_{if} on the order of hundreds of gigahertz. Fortunately there are variations on the types of diode lasers described previously in this chapter that permit continuous tuning over broad wavelength ranges without mode hopping. One variation combines the DBR version of an SLD with a C^3 version. This type of multiple-section laser is

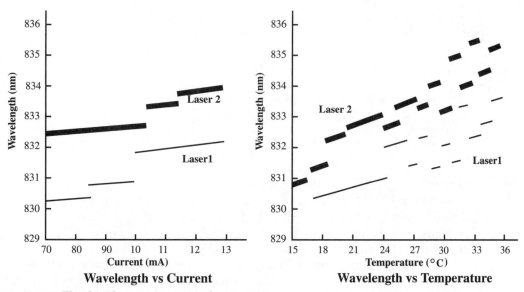

Figure 4.28 Wavelength versus current and temperature.

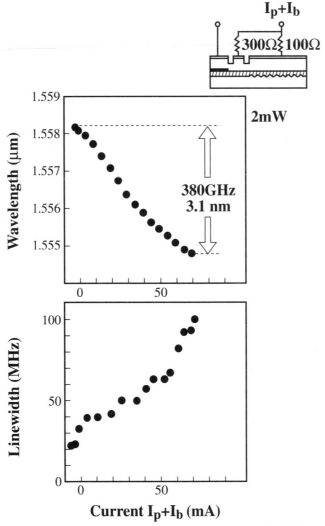

Figure 4.29 Multiple-section tunable single-frequency diode laser. (*From Ref. 8.*)

shown in Fig. 4.29. The three-terminal structure separates the various tasks to be accomplished in the cavity. The laser derives its output power from the active region which is pumped by I_a. Tuning and stability come from I_p and I_b sections. It is difficult to see from Fig. 4.29, but the laser depicted there has a Bragg reflector built in to the cavity below the I_b terminal.

The type of tuning shown in Fig. 4.29 is more effective here because the dependence of the index of refraction on free charge carrier density is far more effective in the Bragg section than it would be in the active region.

Some versions of this type of laser are capable of continuous tuning over nearly one terahertz. Furthermore, these lasers can be switched to oscillate at a new wavelength in 0.5 μs. This flexibility is important in some optical heterodyne detection networks that require rapid reconfiguration (see Sec. 9.2, on wide area networks, for an example of such a network).

4.6   Optical Amplifiers

4.6.1   Introduction

The relatively recent advent of practical optical amplifiers (OAs) has greatly altered the structure of fiber-optic systems being deployed today. In particular, the development of erbium-doped fiber-optic amplifiers (EDFAs) with relatively low noise figures as well as low-loss input and output coupling capabilities has had a profound effect on system design. The three primary applications for OA in fiber-optic communication systems are

1. *Preamplifier.*   The primary function of the preamplifier is to boost the optical signal that arrives at the receiver prior to optical detection by a photodiode. The advantages of "predetection gain" will be discussed in subsequent chapters, but we note here that the preamplifier does not amplify the postdetection electrical noise generated in the receiver electronics.

2. *Repeater.*   Repeaters enable transmission over long distances. The OA version of a repeater amplifies the signal in optical form in a FOCS and thereby avoids the need to convert optical to electrical followed by regenerating the optical signal.

3. *Booster.*   A booster amplifier may be used to increase transmitter power.

Our primary interest in this text will be on the preamplifier version of optical amplifiers.

The primary mechanism for optical amplification is stimulated emission, which, along with population inversion and optical feedback, forms the basis of laser oscillation. When an optical signal is inserted into an oscillating laser, the output will fluctuate wildly, but the response as observed as the laser output will not be a linearly (or nonlinearly) amplified replica of the inserted optical signal. Suppose, however, that we pass an optical signal through a laser-type device with population inversion that is powered below threshold. The subthreshold condition could be achieved by a reduction of the pumping rate and/or reducing the reflectivity of the end mirrors. One possible version of this type of optical amplifier is an AlGaAs heterojunction laser without any coatings on the ends of the cavity (hence $\sim$ 32 percent reflecting) and driven by a current strong enough to create population inversion but not strong enough to reach threshold [9]. In this section we shall briefly discuss some of the salient features of this type of OA. We shall also discuss the properties of semiconductor-based single-pass traveling-wave optical amplifier

(TWOA) devices without any end reflectors at all. Finally, we examine the erbium-doped single-pass TWOA, which has assumed a dominant role in present-day FOCS. In this chapter we shall describe only the operational aspects of OA and postpone discussion of the noise aspects of these devices until Chaps. 6 and 7.

4.6.2   Cavity-type semiconductor OA

A simple semiconductor optical amplifier is presented in Fig. 4.30. The current is confined to the central thin black strip which represents the active region, and it flows perpendicular to the plane of the page. To confine the current to this narrow strip, some of the techniques depicted in Fig. 4.27 can be employed. One of the greatest assets of this type of OA is that it may be designed to operate at any one of a broad range of center frequencies by employing an appropriate combination of heterojunction layers. To increase the gain in this OA version, a Fabry–Perot cavity is formed by the reflectors at each end of the active region with power reflectivities denoted by R_1 and R_2.

The overall power gain $G_A(v)$ for such an OA with power reflectivities (as indicated in Fig. 4.30) may be expressed as [9,10]

$$G_A(v) = \frac{(1 - R_1)(1 - R_2)G_s(v)}{\left[1 - G_s(v)\sqrt{R_1 R_2}\right]^2 + 4G_s(v)\sqrt{R_1 R_2}\,\sin^2\{\pi(v - v_0)(2nL/c)\}} \quad (4.22)$$

where v = frequency of input radiation
v_0 = resonant frequency of the OA cavity, including the phase shifts of the active medium

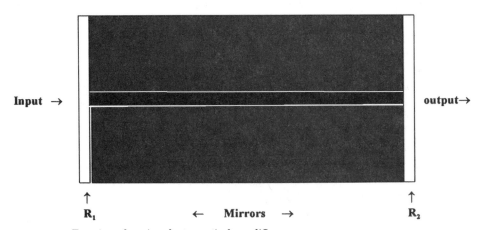

Figure 4.30   Top view of semiconductor optical amplifier.

$G_s(v)$ = frequency-dependent single-pass gain

L = spacing between reflectors (not necessarily the same as the length of the active region in some variations)

n = index of refraction in the active region

For 3–5 semiconductors used for fiber-optic communications, the index of refraction is actually fairly constant with respect to frequency, but as we saw when we studied tunable lasers, n may be function of the pumping level.

The axial mode spacing in a Fabry–Perot cavity is given by $c/2nL$ (see Sec. 4.3.4). This quantity appears (inverted) in the denominator of Eq. (4.22). In this application it is known as the *free spectral range* (FSR) of the Fabry–Perot cavity formed by the reflectors.

One important characteristic common to all cavity-type amplifiers is that they display a fixed gain–bandwidth product. This trend is displayed in Fig. 4.32, in which the gain curves are plotted in terms of a normalized frequency ξ, and where ξ is defined as the frequency deviation from v_0 divided by the axial mode spacing. The 3-dB bandwidth associated with each curve is represented in the figure by the horizontal lines with opposing arrows. Clearly the lengths of these lines decrease with increasing gain. In this case the gain was varied by selection of the reflectivities of the end mirrors (chosen here to be equal for simplicity). The existence of a constant gain–bandwidth product often limits the usefulness of these devices in some high-speed FOCS.

Another consequence of feedback in optical amplifiers is manifested when the gain is plotted over a broader range of frequencies than was used in Fig. 4.30, namely, the appearance of a gain curve with multiple peaks. The spacing between these peaks is given approximately by $c/2nL$. This phenomenon is displayed in Fig. 4.31. To obtain this curve, we needed to select a reasonable single-pass gain curve $G_s(v)$, and the particular spectrum selected was a lorentzian spectral profile with a normalized 3-dB bandwidth of 12.0. Other broadly tuned spectra with the same bandwidth would yield the same key characteristics.such as the appearance of multiple peaks and nulls as well as a slow but steady dropoff in the height of these peaks as $|\xi|$ increases. It is tacitly assumed here that the amplifier is operating significantly below laser threshold and the input is such that saturation effects are avoided. We note here that the amplifier gain equation in terms of normalized frequency (defined in Fig. 4.31 legend) is given by

$$G_A(\xi) = \frac{(1 - R_1)(1 - R_2)G_s(\xi)}{\left[1 - G_s(\xi)\sqrt{R_1 R_2}\right]^2 + 4G_s(\xi)\sqrt{R_1 R_2}\,\sin^2\{\pi\xi\}} \qquad (4.22)$$

To determine this, let G_p be the gain at the peak of the center lobe. From Eq. (4.22), we have

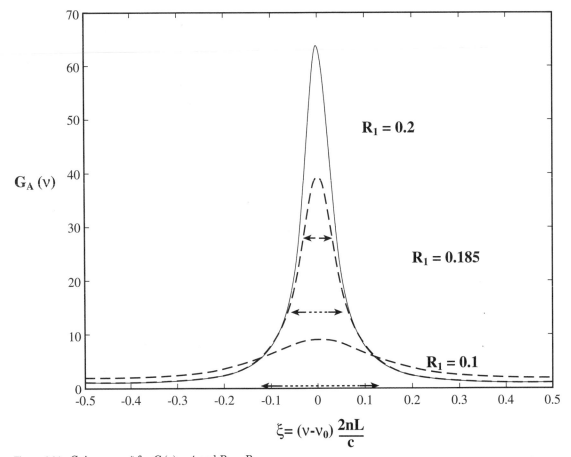

Figure 4.31   Gain versus ξ for $G_s(\nu) = 4$ and $R_1 = R_2$.

$$G_p = \frac{(1 - R_1)(1 - R_2)G_s}{\left(1 - G_s\sqrt{R_1 R_2}\right)^2} \qquad (4.23a)$$

or

$$\left(1 - G_s\sqrt{R_1 R_2}\right)^2 = \frac{(1 - R_1)(1 - R_2)G_s}{G_p} \qquad (4.23b)$$

where G_s stands for $G_s(0)$. Now, to determine the 3-dB bandwidth B of the central mode, we note that $\nu - \nu_0 = B/2$ at the point where the second term in the denominator of Eq. (4.22) is equal to the first term in the denominator. This results in

$$B = \frac{c}{\pi nL} \sin^{-1}\left\{ \sqrt{\frac{\left(1 - G_s\sqrt{R_1 R_2}\right)^2}{4 G_s\sqrt{R_1 R_2}}} \right\} \qquad (4.23c)$$

To obtain a large gain, it is necessary that $G_s\sqrt{R_1 R_2}$ be less than, but near to, unity. Under these conditions the small-angle approximation $\sin(\Delta) \approx \Delta$ can be used for the $\sin^{-1}$ function. This, along with Eq. (4.23b), leads to the following gain–bandwidth product:

$$A_p B = \frac{c}{\pi nL} \sqrt{\frac{(1 - R_1)(1 - R_2)}{4\sqrt{R_1 R_2}}} \qquad (4.23d)$$

where $A_p = \sqrt{G_p}$ is the peak amplitude gain. One method for adjusting the gain of a semiconductor optical amplifier is to control the drive current. The single-pass gain cancels out in the derivation of Eq. (4.23d), and therefore this type of gain control leads to a constant gain–bandwidth product.

4.6.3 Traveling-wave semiconductor OA

In most applications the multiple peaks displayed in Fig. 4.32 are undesirable. One way to avoid them is to eliminate the reflections at the ends (facets) of the cavity shown in Fig. 4.30. This will result in a semiconductor version of a single-pass TWOA. One important asset of TWOAs is that they generate far more favorable gain–bandwidth tradeoff characteristics than do cavity amplifiers. Let the basic exponential gain parameter $\alpha(\nu)$ due to population inversion have the following lorentzian spectral shape:

$$\alpha(\nu) = \frac{\alpha(\nu_0)}{1 + [(\nu - \nu_0)/W]^2} \qquad (4.24a)$$

where $2W$ is the full 3-dB bandwidth of the exponential gain coefficient α. Now the optical intensity grows exponentially with distance L (see Sec. 4.3.1), and this, in turn, leads to the following expression for the single-pass gain:

$$C_s(\nu) = \exp\left\{ \frac{g_0}{1 + [(\nu - \nu_0)/W]^2} \right\} \qquad (4.24b)$$

where $g_0 = C_s(\nu_0) \propto L$. It can be shown that the net 3-dB bandwidth of the gain curve for a TWOA represented by Eq. (4.24b) is given by [11].

Note how the net bandwidth depends on a function of $\ln(g_0)$; therefore, moderate increases in gain will have little effect on W_{net}. It can also be shown [11] that single-pass

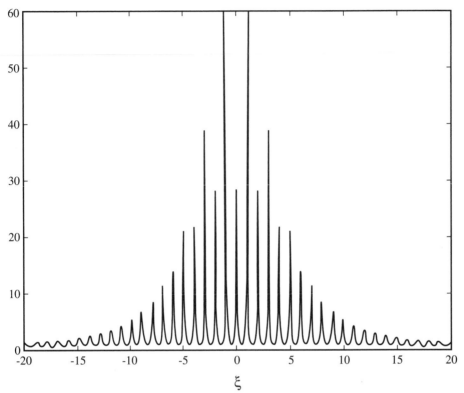

Figure 4.32 Gain versus normalized frequency plotted over a broad range for $G_s = 4$ and $R = 0.2$.

$$W_{net} = W \sqrt{\frac{\ln(2)}{\ln(g_0) - \ln(2)}} \tag{4.24c}$$

gain in decibels as a function of frequency is given by

$$G_{dB}(\nu) = \frac{G_{dB}(0)}{1 + [(\nu - \nu_0)/W]^2} \tag{4.24d}$$

In principle, a cavity-type optical amplifier can be converted to a TWOA by the use of antireflection (AR) multiple-layer dielectric coatings at the ends of the cavity, thereby setting $R = 0$. However, it turns out that even very small reflectivities can seriously affect the performance of a TWOA. To demonstrate this phenomenon, we shall employ Eq. (4.24d) to obtain $G_s(\xi)$ for substitution in Eq. (4.22) to arrive at the set of gain curves depicted in Fig. 4.33.

In all three figures the peak facet-to-facet gain was 20 db, and the 3-db half-bandwidth parameter W of Eq. (4.24d) was arbitrarily selected to be 10

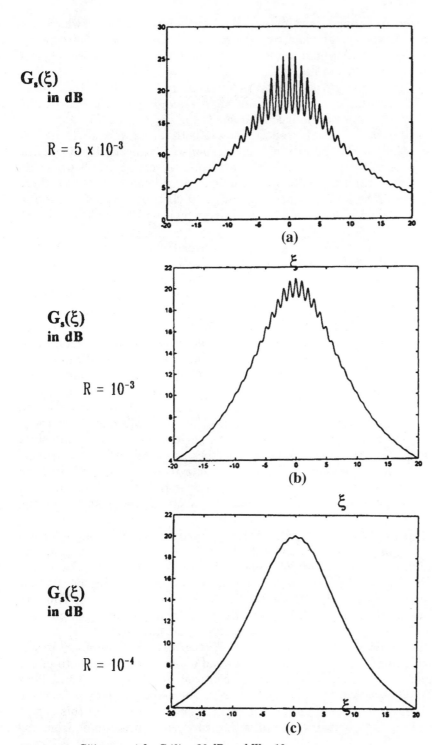

Figure 4.33 $G(\xi)$ versus ξ for $G_s(0) = 20$ dB, and $W = 10$.

(in normalized frequency coordinates). First consider Fig. 4.33c, which was obtained for the very small reflectivity of $R = 10^{-4}$. This curve is very nearly that of the ideal TWOA which would be obtained for $R = 0$. Next we consider the gain curve displayed in Fig. 4.33a.

Note that even with a reflectivity as small as $R = 5 \times 10^{-3}$, the overall response is more like a cavity amplifier than a single-pass TWOA. In Fig. 4.33b, $G(\xi)$ (where the normalized frequency ξ is defined in Fig. 4.31) was obtained for the case where the reflectivities of both facets were set to 10^{-3}. It can be seen that in the vicinity of peak response $G(\nu)$ oscillates about the ideal gain curve with an amplitude of ~1 dB and a frequency of ~1 normalized frequency unit (NFU). Unlike the gain curve depicted in Fig. 4.31, there is significant gain at all frequencies within $\pm$ 10 NFU of $\xi = 0$. As an example, consider a 0.5-mm-long optical amplifier which will have an NFU ~86 GHz.

Today most channels will have much narrower transmission spectra than that, so they will experience only a portion of one cycle of the variations in gain depicted in the graph. Nevertheless, even then there will be some distortion due to the nonflat gain curve. The fact that the gain is frequency-dependent also creates the need for more stringent frequency stability on transmitter laser outputs as well as sterner accuracy constraints on TWOA center frequencies. It is worthwhile to note here that it takes only a very small amount of distortion to prevent a communication system from attaining the 10^{-9} error rate criterion often employed in practical fiber-optic systems.

If we now set $R = 10^{-4}$ for both facets, the gain curve displayed in Fig. 4.33c accrues. Note that in this case with the lower reflectivity, the ripples in the gain have almost vanished. Therefore, the resulting distortion to an amplified signal would be much smaller. The various plots in Fig. 4.33 show that it is necessary to reduce the facet reflectivities to the 10^{-3} to 10^{-4} range. It turns out that these extremely low reflectivities are very difficult to achieve on a production basis by using only dielectric AR coatings and that some basic design changes are necessary. One such scheme is shown in Fig. 4.34.

By structuring the amplifier so that the facets are at a sufficient angle with the active region, the reflections will carum out of the active region and thereby avert the gain oscillations depicted in Fig. 4.33. However, the angled pigtail connection to each facet is also expensive to construct. Furthermore, it is not unusual to lose ~5 dB on entering and exiting the facets. Thus, in order to achieve a net gain of 20 dB, it will be necessary to have a single-pass gain of 30 dB. This now brings us to the topic of gain saturation.

The semiconductor optical amplifier (SOA) is *transversely pumped,* meaning that the pumping rate is uniformly spread along the entire length of the active region of the SOA. This stands in contrast to the erbium-doped fiber version of TWOA, which we shall soon discuss. There the pumping is from the ends and is said to be longitudinally pumped. Now, regardless of this aspect of the pumping, the primary amplification process drains atoms from the

Input

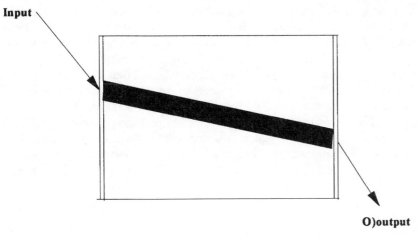

O)output

Figure 4.34   Traveling wave optical amplifier with angled facets.

upper state at a faster rate where the signal is strong than where it is weak. Now let $\alpha[I,z]$ denote the gain coefficient at a distance z from the input facet and where I represents the optical intensity at that point. The fact that à depends on the intensity stresses the fact that the stimulated emission at z affects the population inversion at that point. It can be shown [1,11,12] that the gain coefficient saturates according to

$$\alpha[I,z] = \frac{\alpha_0}{1 + (I(z)/I_{\text{sat}})} \qquad (4.25)$$

where α_0 is gain coefficient in the absence of an input signal and I_{sat} is clearly the intensity level at which the gain decreases by 3 dB. Now it is easy to see why we cannot increase the single-pass gain indefinitely by lengthening the distance between facets. An important population-inversion level is the point where the active region becomes transparent, specifically, where the net gain coefficient (including losses) is unity. The amplification process is generated by the portion of the population inversion that is in excess of the transparency level. Now Eq. (4.25) shows that as the optical intensity increases at a specified point, the gain will eventually saturate and thereby limit the facet-to-facet gain. However, as long as the mechanism is driven sufficiently hard to maintain the population inversion above the transparency level throughout the active region, there will be amplification everywhere in the active region.

At this point it is worthwhile to note that although the gain coefficient depends in a nonlinear manner on the optical intensity $I(z)$ [see Eq. (4.25)], the gain-saturation effect does not generate harmonics of the optical frequency carrier. There is, however, one very important and familiar type of signal deterioration generated by the gain-saturation nonlinearity, namely, interchannel crosstalk. This phenomenon may be manifested when we utilize a single TWOA to amplify a number of channels simultaneously. For example, suppose that there is one input channel sufficiently strong to drive the amplifier into the saturation region It is then conceivable that this single input channel can cause a reduction in population inversion and thereby reduce the amplification for all channels passing through the TWOA. This effect would most likely increase the error rate for all the channels. In this manner gain saturation creates a form of crosstalk since the signal in one channel affects the other channels. Also note that because of the absence of reflectors here, there is no question of the amplifier reaching lasing threshold.

4.7   Erbium-Doped Fiber Amplifier

4.7.1   Introduction

Optical amplifiers employing Nd-doped glass were developed for some of the laser fusion experiments. Although these devices provided large gains, they do not operate at the primary wavelengths employed in fiber-optic communications, namely, $\lambda_0 = 1.3$ and 1.55 μm. Other rare-earth elements were studied, and it was found that the Er^{3+} ion contained a metastable state with an extremely long lifetime of ~ 10 ms, and furthermore, this metastable level was separated from the ground state by photons with wavelengths ~ 1.53 μm, thereby placing this transition in the heart of the lowest loss portion of optical fibers. Some of the more important energy levels that are used in commercially available OAs are displayed in Fig. 4.35.

In Fig. 4.35 the energy levels are labeled by the free-space wavelength of the photon transition from that line to ground level. In the erbium-doped fiber amplifier (EDFA), amplification takes place for *free-space* wavelengths in the vicinity of $\lambda_0 = 1.53$ μm. Each level depicted in Fig. 4.35 is really a composite or band of many closely spaced levels. The pumping schemes in vogue today usually employ semiconductor lasers at either $\lambda_0 = 1.480$ μm or $\lambda_0 \sim 0.98$ μm; the latter is used more frequently. The pumping at 1.480 μm raises Er^{3+} ions from near the bottom of the ground state to near the top of the 1.53-μm band. From there it drops within the 1.53-μm band to a slightly lower level. In general, all the pumping schemes excite energy levels above the 1.53-μm band, and they decay from those levels rapidly to the upper state of the amplifier transition at ~ 1.53 μm.

It can be seen from Fig. 4.36 that there are other absorption lines in addition to those displayed there. All these absorption lines have energy levels that lie above that of the EDFA metastable state, and each absorption line can be used for EDFA pumping purposes. The earliest versions of the EDFA was pumped by argon gas lasers operating at wavelengths near 0.532 μm. For our purposes here, the most important aspect of the absorption spectrum

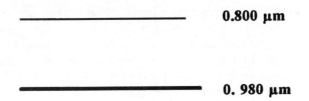

Figure 4.35   Some energy levels for Er^{3+} ions in glass.

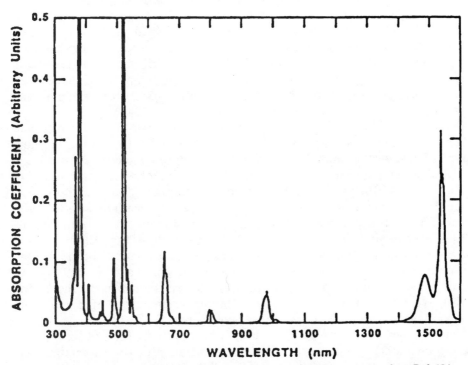

Figure 4.36   Absorption spectrum for Er^{3+} in glass. (*Reprinted with permission, from Ref. 13.*)

shown in Fig. 4.36 is to note how much more narrow these absorption lines are than the pumping-level linewidths in Nd-YAG shown in Fig. 4.7. Pumping by broad-spectrum optical sources such as flash lamps simply won't do the job for the EDFA. Lasers are required here to do the pumping; more specifically, diode lasers are best suited to this pumping task. Commercial interest has focused on pumping at the wavelengths at 0.980 and 1.480 μm because of the availability of high-power diode lasers at those wavelengths. The 0.800-μm line was of early interest because of the ready availability of inexpensive powerful diode lasers at this wavelength. Unfortunately there exists an energy level in Er^{3+} that is spaced at $h\lambda_0$ for $\lambda_0 \sim 0.790$ μm above the metastable state at ~ 1.539 μm. When the EDFA is pumped at $\lambda_0 \sim 0.790$ μm, photons will be absorbed, resulting loss of photons from the pump signal will absorb photons from the pump signal, and simultaneously reduce the population of the upper state of the 1.53-μm transition. This process reduces the population inversion density along with the gain coefficient. Although optical amplification has been achieved using a 800-μm pump, it is less efficient than other pumping wavelengths mentioned previously.

4.7.2 EDFA structures

Three possible arrangements often used for the EDFA are shown in Fig. 4.37.

In Fig. 4.37 the notation WSC stands for *wavelength-selective couplers.* Such specialized optical couplers are necessary for inserting the powerful pumping signal, which may be as large as 200 mW with minimum insertion loss while at the same time avoiding a coupling loss mechanism for the much weaker input signal. It is much easier to design a WSC for a 0.98-μm pump wavelength, because of its large separation from the signal wavelength, than for a 1.48-μm pump wavelength, and this is another reason why the pumping wavelength $\lambda_0 = 0.98$ μm is the more commonly used.

The notation EDF in Fig. 4.37 stands for *erbium-doped fiber,* which is usually a single-mode optical fiber which has been doped with erbium, in many cases codoped with aluminum to facilitate the erbium doping process. To get the most out of a particular length of fiber, as much of the erbium is placed in the core of the fiber as possible. This places the key atoms in the region where both the pump and the signal optical fields are most intense and thereby maximizes the interaction between these fields and the erbium atoms. Finally we note that the core radius of the EDF is smaller than the core of the external fiber, and this requires some unique interfacing techniques. One reason for the smaller core in the EDF is that the pump signal must be both amplified and made to propagate in a single transverse mode inside the EDFA. We note that the wavelength of the higher-energy pump photons are always shorter than those of the signal photons [see Eq. (3.13a)].

Note that an optical isolator has been included at EDFA exit for the forward-pumped case (Fig. 4.37a). This device attenuates the external reflections generated at various system surfaces. These spurious signals can gener-

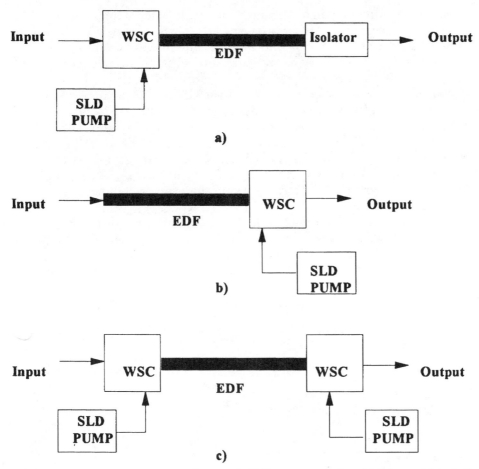

Figure 4.37 EDFA setups: (*a*) Forward-pumped; (*b*) Reverse pumped; and (*c*) pumped from both ends.

ate considerable amounts of unwanted fluctuations in the EDFA which behave very much like random noise. Although not shown in Fig. 4.37, the backward pumping setup of Fig. 4.37*b* often requires some sort of an optical filter to prevent the pump signal from propagating backward toward the transmitter end. An optical isolator at the transmitter will not block this stray pump signal since this signal is not due to reflection. The isolator requires two-way passage through the device. If, however, the fiber length between transmitter and EDFA is sufficient, the inherent attenuation of optical fibers may well be large enough to negate the need for the optical filter.

Since fiber amplifiers are pumped from the ends, the pump signal weakens as it propagates through the EDF. Consequently, the population inversion density and hence the gain coefficient decrease with distance from the pumping side. Take, for example, the forward-pumped EDFA, which will be characterized by maximum gain at the input side with reduced amplification coefficient as we progress toward the exit port. The opposite is true for the reverse-pumped EDFA of Fig. 4.37b, where the maximum gain coefficient occurs at the exit port and decreases as we move toward the input side of the EDFA. These two different pumping schemes generate different amounts of amplified spontaneous emission. We shall see the effects of this phenomenon when we discuss amplifier noise in Chap. 6. In Fig. 4.37c we depict an EDFA pumped from both ends. This provides a more even gain coefficient along the length of the EDF and can also be used to increase the overall gain of the amplifier. A variation of the two-sided pumping arrangement is to insert a second forward-directed WSC about midway down the EDF, and thereby allow for a longer EDFA with greater gain than can be obtained from the single pumped version.

Since the pump signal decays as it propagates through the EDF, it can become too weak to generate sufficient population inversion to produce amplification. The site where this occurs is known as the *transparency point*

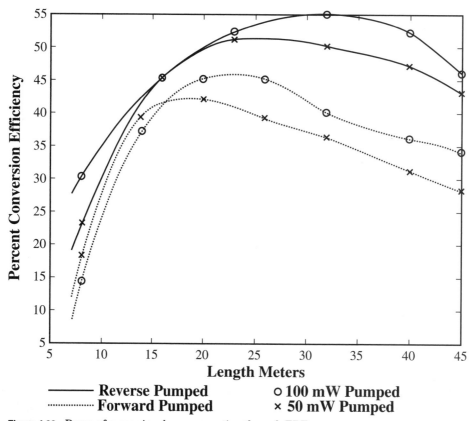

Figure 4.38 Decay of pump signal on propagation through EDF.

beyond which the signal will decay. This implies that there exists an optimum length for the EDF beyond which the performance diminishes. This trend can be seen in Fig. 4.38.

In Fig. 4.38 data are plotted for both forward and backward pumping at λ_0 = 0.978 μm, and for each pumping direction separate curves displayed for 50- and 100-mW pump powers. We note that the power conversion efficiency (PCE) has a maximum for each curve. The reduction in PCE versus EDFA length that occurs beyond these maxima indicates that the excess length is operating below transparency and therefore attenuating the signal in this portion of the amplifier. The following trends are apparent in Fig. 4.38:

1. Backward pumping demonstrates better PCE at all EDF lengths.

2. As expected, the optimum fiber length is longer for the higher pump level, but the trend is not dramatic.

3. The forward pumping maximum occurs at shorter lengths than in the reverse-pumped case.

To understand the spectral dependence of the EDFA gain curve, consider Fig. 4.39. The shape of the gain versus wavelength in the plots in Fig. 4.39 are fairly typical for all pumping wavelengths. Specifically, note the peak gain at $\lambda_0 \sim 1530$ nm and the substantial gain that can be achieved at $\lambda_0 \sim 1550$ nm. In this case the gain is highly sensitive to pump wavelength, which is caused by

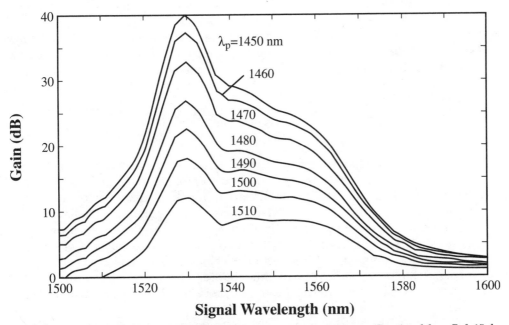

Figure 4.39 Gain versus wavelength for several pumps near $\lambda_0 = 1450$ nm. (*Reprinted from Ref. 15, by permission.*)

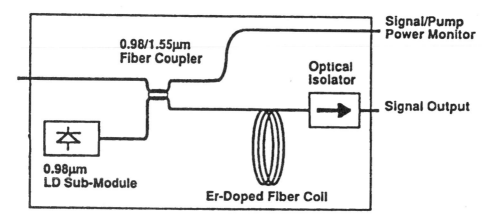

Schematic diagram of the EDFA.

Figure 4.40   Schematic diagram of the EDFA. (*Reprinted from Ref. 16, by permission.*)

the proximity of the pump wavelength to the amplifier wavelength. We note that the sensitivity to pump wavelength is much smaller for EDFA pumped at 980 nm.

Figure 4.40 presents a diagram which demonstrates how the EDFA may be compactly packaged. Note how the erbium-doped fiber is wound in small loops and stored as spool within the package. In designing this amplifier, one must consider the bend losses incurred in such an arrangement.

The wavelength-sensitive coupler for the device shown in Fig. 4.40 was such that ~98 percent of the pump was coupled into the EDF while only 1 percent of the signal at $\lambda_0 = 1.54$ μm was coupled out of the system [15]. The pump laser diode was driven at 80 mA to achieve a maximum net gain of 33 dB at $\lambda_0 = 1.536$ μm.

4.8   Summary

In this chapter we have described two devices that are very important components of modern FOCS, namely, the optical amplifier (OA) and the semiconductor laser diode (SLD). The amplification in both of these devices is provided by the combination of stimulated emission and population inversion. We also note that spontaneous emission is a source of noise in these devices, but as we shall see in Chaps. 6 to 8, these are not the only sources of noise in FOCS. In the remainder of this section we shall indicate how some of the phenomena that were described in this chapter affect the design and performance of FOCS, and where in the text more detailed descriptions of these effects can be found.

We begin with the SLD and note that single-frequency SLDs are employed as sources in almost all high-speed FOCS. This is particularly true for wavelength-division-multiplexed systems (a broad form of frequency-division multiplexing), which seem to have become the backbone of large-scale regional networks (see Chap. 9 Sec. 9.2). The usual source of multiple-output frequencies encountered in practice is the exciting of undesired axial modes. Equation (4.10) presents a formula for axial mode frequency intervals which results in frequency spacings on the order of 200 GHz for typical SLD dimensions . Although these undesired SLD outputs are usually much weaker than the desired output, the large spacing between axial modes frequencies along with the large number of channels in large systems with many wavelength multiplexed channels can result in significant interference between separate channels. In addition large scale dispersion distortion may be generated from these undesired axial modes. Single-frequency SLDs virtually eliminate interchannel interference and greatly reduce dispersion effects.

Another important attribute of the SLD is that it can be tuned by simply varying the bias current. This process suffers from mode hopping for simple SLD structures and the continuous tuning range (CTR) is limited. For simpler SLDs, the CTR may be limited to tens of gigahertz (see Fig. 4.28); however, CTRs on the order of one terahertz have been obtained with more complex SLD structures such as those depicted in Fig. 4.29. This tuning capability plays a very important role in optical heterodyne detection systems (see Chap. 8), where tunable single-frequency lasers are often used as local oscillators as well as signal sources. In particular, SLDs of the type shown in Fig. 4.29 can be tuned in microseconds, enabling rapid reconfiguration of nodes in large-scale FOCS networks (see Sec. 9.2). Although ambient temperature can also be used to adjust SLD output frequency, the usual method for tuning SLDs is to control the bias current while maintaining as constant a temperature as possible. The reason for this is that temperature control is both coarser and reacts much more slowly than does current control. We note here that the bias current affects the SLD output frequency because it controls the free charge carrier density within the active region of the SLD, and the charge carrier density affects the index of refraction along the full length of the active region. The index of refraction, in turn, affects the resonant frequency of the optical cavity within the SLD. This same phenomenon can also be employed to frequency-modulate the SLD output. In particular, this type of direct frequency modulation is relatively easy to implement, and it is often employed in high-speed FSK transmission systems. An example of such a system is described in Sec. 8.4.

We begin the OA portion of this summary section by comparing the semiconductor version of the single-pass TWOA with the erbium-doped optical fiber version. We note that one important limitation that exists in the state of the art in commercially available EDFAs today is the fact that they operate only in the general vicinity of $\lambda_0 = 1.53$ μm. On the other hand, by suitable selection of heterojunction layers, semiconductor optical amplifiers (SOAs) can be constructed to operate at almost all wavelengths of interest in fiber-optic communication systems (FOCS). Another advantage of SOA over EDFA is the simpler pumping mechanism for SOAs. Nevertheless, at this point in time the EDFA has assumed a dominant position in FOCS being deployed today.

There are two primary reasons why EDFAs are preferred:

1. *The polarization properties of these optical amplifiers.* In particular, we note that, because of the cylindrical symmetry of the EDF, the performance of these optical amplifiers is virtually independent of the polarization of the input field. This symmetry does not exist in the active region of the SOA, and consequently the performance of the SOA strongly favors one linear component of polarization over the other. One proposed solution to this problem is to use various structures that employ two separate and complete SOAs; one for each orthogonal polarization, acting in concert as a single amplifier. Nevertheless, polarization considerations are much easier to deal with in the cylindrically symmetric EDFAs.

2. *Crosstalk.* The second major advantage of the EDFA over the SOA is concerned with this phenomenon. We first encountered this topic in Sec. 4.4.3, where crosstalk was generated by simultaneous amplification of several strong channels. Our focus here is on the problem generated by amplification of a single channel of a binary ON/OFF-keyed (OOK) digital channel. The crosstalk in this case is between adjacent bits and is caused by an imbalance in ON-state pulses depleting the population inversion for the next pulse interval and so on. The magnitude of this effect is dependent on the natural decay time of the upper-state atoms. If it is long enough to average over a sufficient number of bits, this type of crosstalk becomes negligible. Therefore, because the decay time of Er^{3+} is ~ 10 ms while the decay time in SOA is on the order of nanoseconds, the FDFA is much less susceptible to crosstalk than is an SOA. On the other hand, when the data rate is sufficient, say, on the order of 10 Gbits/s, then both types of amplifiers will be fairly immune to this form of crosstalk. Actually the long lifetime of the metastable state of erbium will also mitigate the effects of interchannel crosstalk.

Both the SLD and the OA generate noise in FOCS. Spontaneous emission in SLDs produces a form of noise known as *phase noise,* which is particularly troublesome in optical heterodyne systems. The discussion of that particular form of noise is presented in Chap. 8, which is devoted to optical heterodyne detection. In Chap. 6 we discuss the signal-to-noise aspects of OAs. There we shall see that prime source of noise is amplified spontaneous emission. There is also randomness in the stimulated absorption and emission process, which adds to the noise landscape, as does the amplified noise component of the amplifier input. All of these topics are treated in Chap. 6.

Problems

1 Derive
  (**a**) Equation (4.14a) from Eq. (4.13)

(**b**)   Equation (4.14b)

(**c**)   Describe the nature of the output radiation from a laser as the pump power is raised in steps from 0 to twice the threshold pump level.

Diode lasers

2   A particular GaAs diode laser has the following parameters: length = 400 μm and width of the active region = 1 μm. The threshold current for the device is observed to be 20 mA when the ends are cleaved (no coatings). After this measurement, one end is coated so that it is 100 percent reflecting, and the threshold current is now found to be 12.5 mA. Find

(**a**) $\bar{\alpha}$ in cm^{-1} (reciprocal centimeter) units

(**b**) $\bar{\beta}$ in cm/A units

Answers: $\bar{\alpha} = 9.5$ cm^{-1}; $\bar{\beta} = 7.6 \times 10^{-3}$ cm/A.

3   Answer the following questions about the diode laser with 100 percent reflection on the nonoutput side in the preceding problem:

(**a**)   Could $\bar{\alpha}$ have been measured directly by passing light through an undoped block of the same GaAs material that was used to manufacture the laser? What might differ between the two schemes? Note that the doping process itself does not affect the optical transmission properties of the block.

(**b**)   What would be the threshold current if the output-side reflection were increased to 60 percent?

(**c**)   Suppose that you had coated the output side but did not know its percentage of reflection. You find, however, that the threshold current is 11 mA. Find the percentage reflection of the output side for this device.

4   A particular diode laser that has been used by the author of this text has the following specifications: $\lambda_0 = 0.83$ μm, $\Delta f = 1.5 \times 10^7$ Hz, and power = 30 mW. Find the temperature that a blackbody source would require to match these parameters.

Optical amplifiers

5   Consider an erbium-doped silica fiber traveling-wave amplifier which yields a power gain of 1000 over 50-m with a 5-mW laser pump.

(**a**)   What is the gain exponential coefficient α in nepers per meter?

(**b**)   How much longer would this amplifier need to be to achieve a gain of 2000?

(**c**)   What would be the frequency spacing between axial modes if the original amplifier were made into a laser by placing reflectors at the end of the fibers?

6   Consider a high-speed FOCS operating at 1.55 μm and using pulses of 1 ns duration.

(**a**)   Let this signal be passed through an 0.5-mm semiconductor TWOA. Assume the index of refraction to be $n = 3.6$, and estimate how many round-trip reflections will occur during the time it takes one pulse to pass through the TWOA.

(*b*)   Repeat part **a** for a 20-m-long EDFA with $n = 1.5$.

References

1. B. E. A. Saleh and M. C. Teich, *Fundamentals of Photonics,* Wiley, New York, 1991: (*a*) Chap. 11, (*b*) Chap. 12, (*c*) Chap. 13, (*d*) Chap. 15.

2. A. E. Siegman, *An Introduction to Lasers and Masers,* McGraw-Hill, New York, 1971, in par-

ticular: (*a*) Chaps. 5 and 10, (*b*) Sec. 10.7.

3. K. Nakagawa and S. Shimada, "Optical amplifiers in future optical communication systems," *IEEE Mag. Lightwave Commun. Syst.* 57–62 (Nov. 1990).

4. G. P. Agrawal, *Fiber-Optic Transmission Systems,* Wiley, New York, 1992.

5. P. Bhattacharya, *Semiconductor Optoelectronic Devices,* Prentice-Hall, Englewood Cliffs, N.J., 1994, p. 294.

6. J. Wilson and J. F. B. Hawkes, *Optoelectronics: An Introduction,* 2d ed., Prentice-Hall, Englewood Cliffs, N.J. 1989, pp. 187–201.

7. G. Keiser, *Optical Communications,* 2d ed., McGraw-Hill, New York, 1991, Chap. 4.

8. K. Kobayashi and I. Mito, *J. Lightwave Technol.* **6,** 1623–1633 (1988).

9. Y. Yamamoto, "Characteristics of AlGaAs Febry-Perot cavity type laser amplifiers," *IEEE J. Quant. Electron.* **QE-16,** 1047–1052 (1980).

10. G. P. Agrawal, *Fiber-Optic Transmission Systems,* Wiley, New York, 1992, Chap. 8.

11. A. E. Seiegman, *An Introduction to Lasers and Masers,* McGraw-Hill, New York, 1971, pp. 210–222.

12. P. E. Green, Jr., *Fiber-Optic Networks,* Prentice-Hall, Englewood Cliffs, N.J., 1993.

13. W. J. Miniscalo, "Erbium-doped glasses for fiber amplifiers at 1500 nm," *J. Lightwave Technol.* **9,** 234–250 (1991).

14. R. I. Lamming, J. E. Townsend, D. N. Paynr, F. Meli, G. Rrasso, and E. J. Tarbox, "High-power erbium-doped-fiber amplifiers operating in the saturated regime," *IEEE Photonics Technol. Lett.* 285– (1991).

15. C. R. Giles and E. Desurvire, "Propagation of signal and noise in concatenated erbium doped fiber optical amplifiers," *J. Lightwave Technol.* **9,** pp. 147–154 (1991).

16. M. Shimizu, M. Horoguci, M. Yamada, I. Nishi, J. Noda, T. Takeshita, M. Oayasu, S. Uehara, and E. Sugita, "Compact and highly efficient fiber amplifier modules pumped by a 0.98 laser diode," *J. Lightwave Technol.* **9,** 291–296 (1991).

17. J. Wilson and J. F. B. Hawkes, *Optoelectronics—an Introduction,* 2d ed., Prentice-Hall, Englewood Cliffs, N.J., 1989.

Optical Detectors

5.1 Detection of Light

5.1.1 The detection process

The primary detection process for electromagnetic radiation involves the absorption of photons from the incident field. However, because of the low energies of photons at radio and microwave frequencies, the effects of quantization of the electromagnetic field are not readily apparent at those frequencies. The detection of light differs greatly from rf detection[1] primarily because of the enormous difference in the photon energy for these two vastly different regions of the spectrum At rf the detection process is typically imitated with an antenna made out of good conductors.[2] Then, after this primary detection, the current induced in the antenna is amplified and processed electronically to obtain a signal suitable for demodulation and final detection. Therefore, the basic interaction in rf detection involves the response of conduction band electrons to the incident electromagnetic field. Now good conductors contain an enormous number of conduction band electrons distributed over a virtual continuum of energy levels. Consequently, there is no observable lower bound on photon energy for absorption at rf. This differs markedly from the photoelectric effect as well as optical detection in general.

In contrast to rf detection, the primary interaction in optical frequency detection takes place between the incident light radiation and valence band electrons which are bound to atoms within the detector. For example, the light absorbed by atoms lying in the depletion region of a photodiode is of primary interest in fiber-optic communication systems. The detection process is initiated by the absorption of a photon and the concurrent elevation of an

---

[1]We shall denote the combination of both rf and microwave frequency ranges simply as rf in this chapter.

[2]Some very sensitive microwave receivers employ a maser in a preamplifier arrangement similar to the optical preamplifier described in Chap. 4. This type of device does employ quantum-mechanical effects such as stimulated emission and is an exception to the discussion in this paragraph.

electron in the valance band up to the conduction band. In order for this event to occur, the energy of the incident photons must have sufficient energy to bridge the bandgap. Unlike rf detection, this condition places a lower limit on the frequency that can be detected by a particular photodiode. One manifestation of this phenomenon is the existence of a low-end cutoff frequency observed in the photoelectric effect. While the responses of rf antenna systems are frequency-dependent, there is no clearly observable cutoff frequency below which the antenna will not respond at all. We now introduce two types of light detectors that have been used in optical frequency communications and lidar (optical radar) systems: the photomultiplier and the photodiode. Fiber-optic communication systems (FOCS) use photodiodes almost exclusively. In this chapter, we focus on several variations of these devices. First, we describe the photomultiplier tube (PMT) because photon statistics will play an important role in our coverage of noise in FOCS; photon statistics are usually measured using PMTs.

5.1.2 Photomultipliers

In 1905 Einstein demonstrated that the photoelectric effect was a clear manifestation of the quantization of light into discrete units of energy called *photons*. He explained experimental observations of the cutoff frequency in the photoelectric effect by noting that the photon energy required to observe the photoelectric effect must be such that

$$hf \geq E_g + \text{surface work function}$$

The work function in this expression is defined as the kinetic energy required to overcome the potential barrier that always exists at surface boundaries. Not every absorption of photons that satisfy the cutoff condition leads to an emitted electron; yet an average over a large number of such events produces a current response that is proportional to the incident optical intensity. If the frequency of the incident radiation is less than cutoff condition, virtually no current will be observed regardless of the intensity of the incident light. Thermal agitation and some stray high-energy particles from cosmic ray showers will generate a very small "dark" current that will be generated even when there is no light incident on the detector. Thermally generated dark current can be reduced significantly by cooling the photoelectric detector while shielding the PMT with a commercially available "mu" metal covering shield that greatly reduce the effects caused by high-energy particles. A representative diagram for a simple PMT is given in Fig. 5.1.

The detection process is initiated when an incident photon kicks an electron out of the cathode and that electron is then accelerated toward the first dynode. When the electron strikes the dynode, it generates a number of secondary electrons (which are, in turn, accelerated into the next dynode, etc.) At each dynode, an incident electron generates a random number of secondary electrons. This process continues until the anode is reached; at the

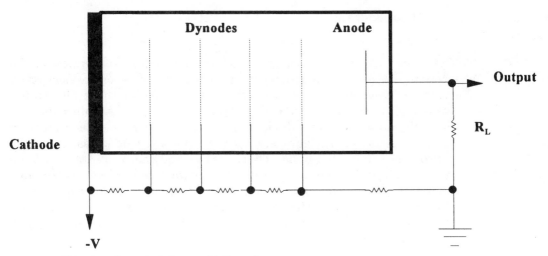

Figure 5.1 Side view of a typical photomultiplier tube.

anode, a short burst-of-current pulse is generated in the output circuitry. In a typical PMT detection system, this current pulse, initiated by a single photo-electric absorption event, now contains a large number of electrons. With the aid of further amplification, the event can be registered in a high-speed counter. Figure 5.1 displays only four dynodes, but a practical PMT designed for photon-counting experiments often contains as many as 10 or 14 dynodes and has a net dynode amplification greater than 10^6. This magnification process is random at each dynode; the net response to each photon is itself a random variable. What results is that statistics of the first few dynodes dominate statistics of the final output [1]. Using a larger average gain at the first dynode and smaller average gains at the subsequent dynodes is one way to reduce final output variance. When properly cooled and shielded, the PMT is one of the most sensitive detectors of low-level light in existence. This is true even though less than 10 percent of the incident photons are detected by these devices. Furthermore, this detection process has the flavor of matched-filter detection; the high sensitivity accrues from the ability of PMT systems to observe macroscopic electronic responses to individual photon absorptions events. This detection process differs considerably from the macroscopic currents generated by the detection of a very large number of photons, as is common in photodiode detection (as we shall soon see). Although we shall frequently refer to photons per bit as a measure of system sensitivity, no practical or planned FOCS actually uses photon counting as the primary detection mechanism. Although we will discuss reasons for this later on in the text, for now we offer the explanation (without proof) that the photon-counting requirement for very sharp pulse pair resolution requires enormously fast detection circuitry when used in high-speed FOCS.

5.2 Semiconductor Photodiodes

5.2.1 General characteristics

The primary interaction in semiconductor photodiode detection is the absorption of a photon, and the simultaneous excitation of an electron in the valance band up to the conduction band. Each of these absorption events results in the production of one hole–electron pair. At minimum, the incident photon must have sufficient energy to support the transition of an electron across the energy gap between the top of the valence band and the bottom of the conduction band. This energy gap, known as the *bandgap,* and is denoted by E_g. However, unlike the photoelectric effect, the conduction band electrons remain within the semiconductor; there is no surface work function to contend with. Such behavior can be expressed more compactly by noting that the free-space wavelength of the photon must satisfy $\lambda_0 < \lambda_c$ where

$$\lambda_c \ (\text{in } \mu\text{m}) \approx \frac{1.24}{E_g \ (\text{in eV})} \qquad (5.1)$$

Although Eq. (5.1) is a necessary condition for the detection of a photon in a photodiode, it is not a sufficient condition. The optical field can be detected by a photodiode only if these photon absorption events lead to observable currents in external circuits. If the effect of a hole–electron pair that was generated by a specific absorption is canceled by a recombination event before it has been observed in the external current, it simply will be as if that event never occurred. Not all the photons absorbed in the photodiode are observed in the diode response. Further, we also note here that photodiodes, unlike laser diodes, are designed to operate over a broad range of wavelengths. Note that the bandgap is a key parameter in photodiode design.

In Chap. 4, we pointed out that although silicon is not used for diode lasers because of the indirect bandgap, silicon photodiodes can be and are used for detection in a number of commercial CD and CD-ROM (compact disk–read-only memory) systems. However λ_c for silicon is 1.06 μm, which rules out silicon as detector for modern fiber-optic systems employing transmission at 1.3 or 1.55 μm. Another important difference between diode lasers and photodiodes is that the latter is reverse-biased. Since the received optical signal is usually weak, very little heat is generated in these detectors. This property alone makes it much easier to embed photodiodes, as opposed to lasers, into optoelectronic integrated circuit (OEIC) devices. The existence of considerable heat generation and dissipation problems greatly limits the options available to the designer of OEIC-containing lasers, and it is one of the main reasons for the great interest in small quantum-well lasers. Next, we examine the three most important optical detectors employed in FOCS in the following order: *p–n, p–i–n,* and avalanche photodiodes (which also happens to be the order of complexity of these devices).

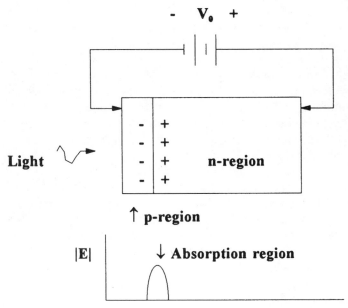

Figure 5.2   End-illuminated p–n photodiode.

5.2.2 *p–n photodiodes*

A typical photodiode (PD) setup is shown in Fig. 5.2. In this end-illuminated version, light enters the end of the PD through a very thin p region, passes through the depletion region of the p–n junction, and continues propagating through the n region of the photodiode. Photons are being absorbed all along this path; the strength of the light beam decreases exponentially as it propagates. This device could have been illuminated from the side; however, this mode of operation is seldom employed, for reasons that we will discuss shortly.

The depletion region (DR) plays a major role in the operation of a photodiode. In the DR there is a strong electric field directed from the immobile[3] positive charges to the immobile negative charges shown in Fig. 5.2. Much like the situation in a parallel capacitor, the field inside the DR is far stronger than the fields outside. This strong field inside the DR creates a strong potential barrier which impedes majority carrier flow across the DR. For example, an electron in the n region that happens to be near, but not in, the DR region will be attracted only weakly to enter the DR. Conversely, an electron inside the DR will be driven very forcibly toward the n region. A parallel phenomenon occurs for holes and the p region. Furthermore, as the term suggests, the

---

[3]These immobile charges are the residual charge remaining with the doping atoms. A donor is short one electron; hence a positive charge and an acceptor atom has an extra electron. These atoms are fixed in the lattice of the semiconductor and therefore are immobile.

DR is almost completely devoid of holes and conduction band electrons. For this reason, the impedance of the DR is much larger than it is in the other regions of the PD; virtually the entire applied bias voltage appears across the very thin DR. This results in a very large potential barrier which prevents excess majority charge carriers from crossing the DR. Charge neutrality is maintained by excess electrons from the n region flowing through the external circuitry and recombining with excess holes in the p region. Except for a very small contribution from the reverse-bias current, the current response of the photodiode is directly proportional to the rate that photons are absorbed in the DR.

So far in this section we have concentrated on photon absorption and how this process generates hole–electron pairs. To fully describe the operation of the photodiode, we need to understand the properties of the inverse process, namely, recombination. As we noted in Chap. 4, recombination occurs when a conduction band electron drops down to the valance band, thereby eliminating one electron and one hole in the process. Although this may seem to be an oversimplification, one could view this process as a collision between an electron and a hole. We shall consider such a view in the following discussion. Next, we note that while the photon absorption process is virtually independent of the doping levels, the effect of recombination on photodiode operation is very strongly dependent on whether the region of the semiconductor is n-type or p-type. In the next three paragraphs we examine this topic more closely.

When a photon is absorbed in the depletion region, the resulting hole and electron are quickly swept out of the DR by the intense electric field in that region. The hole is forced out of the DR into the p region of the photodiode, and the electron is driven into the n region; they become excess majority carriers in their respective regions. In both cases, the excess carrier density is largest just outside the DR; these excess carriers generate a diffusion current directed away from the depletion region and toward the electrical contacts at the ends of the bulk[4] regions. Now, the large potential barrier created by the strong electric field in the DR prevents both of these types of majority carriers from recrossing the depletion region. Therefore, there now exists an excess positive charge in the p region, and an excess negative charge in the n region, each trapped in their respective regions. These excess charges can be neutralized only by the external electric current as described in the previous paragraph. Specifically, when a conduction band electron in the n region reaches the electric contact at the right edge of the device (see Fig. 5.2), the electron is impelled by the power supply to flow through the external circuitry (including the power supply) all the way around to the conductor at the p region, where it recombines and annihilates a hole. Important to note here that neither the electron nor the hole involved in this recombination is the original charge carrier generated by the photon absorption.

---

[4]The term *bulk region* refers to the n- and p-type regions outside the depletion region.

When a photon is absorbed in either of the bulk regions that are external to the DR, electrical neutrality is not disturbed: an equal amount of positive and negative charge is created. These charges drift relatively slowly within these bulk regions and collide with other charge carriers as they meander along. By far, the most frequent type of collision is between majority carriers, an interaction which does not cause recombination. Therefore, the lifetime of majority carriers is relatively long. Conversely, a minority carrier will almost surely collide with a majority carrier within a short time and thereby generate a recombination event. Since one charge carrier of each polarity is annihilated, the overall charge neutrality of the bulk region is not affected by this process. As a consequence, the minority carriers created by photon absorption outside the DR are very short-lived; in most instances these events are not observed externally. In the next paragraph we describe one important exception to the scenario described above.

Some hole–electron pairs generated by photon absorptions that occur within the bulk region, but very near to the DR (say, within a diffusion length), diffuse into the depletion region, separate in the DR, and are subsequently detected. While this process does not add much to the overall response of the diode, it does add a delayed response, thereby generating undesirable trailing edge effects. This phenomenon is displayed in Fig. 5.3 for an optical input consisting of rectangular digital pulses. Although the "trailing edge" effect is quite small, it can generate sufficient intersymblol interference to prevent the attainment of the low-error-rate specifications commonly employed in modern high-speed FOCS. Later on in this chapter we shall describe a number of techniques (some rather elaborate) for avoiding photon absorption regions that cause this effect.

We now examine the detection characteristics of the simple p–n junction photodiode. The basic semiconductor absorbs light as it passes through the PD, which leads to

$$S(z) = S_0 e^{-\alpha z} \tag{5.2}$$

where $S(z)$ is the power density of the optical field inside the photodiode at a distance z from the front surface, S_0 represents the power density just inside the input end of the photodiode, and α is the same absorption coefficient

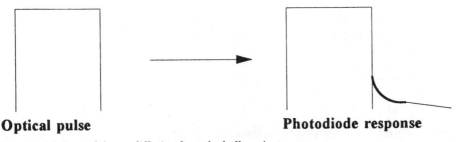

Optical pulse   **Photodiode response**

Figure 5.3 Effects of charge diffusion from the bulk region.

parameter that we encountered in Chap. 4. In photodiodes the only important excitation mechanisms for raising electrons to the conduction band are absorption of photons from relatively weak input and thermal agitation. In the absence of light and with the PD at room temperature, virtually all atoms in the PD are in the lower state. Although Eq. (5.2) is essentially the same as Eq. (4.9b), in Eq. (5.2) we use S to represent the power density (intensity) and reserve the variable I to denote current.[5] Note also that S, as it is used here, is a scalar quantity, and does not represent the Poynting vector.

Plots of the absorption coefficient versus wavelength for several semiconductor materials that are often used in practical photodiodes are presented in Fig. 5.4.

---

[5]The symbol I will be used at various places in the text to represent either optical intensity or electric current. The distinction will be based on the subject matter or context wherever this symbol appears.

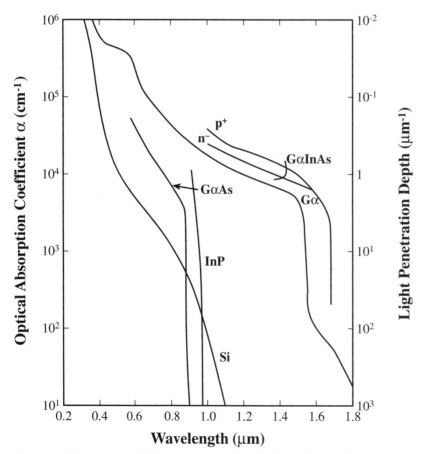

Figure 5.4 Absorption coefficient versus wavelength. (*Reproduced with permission from Ref. 2.*)

Some salient features are apparent from these curves. First, the absorption coefficient is wavelength-dependent [i.e., $\alpha = \alpha(\lambda)$], which implies that the detector sensitivity will vary according to the wavelength of the incident light. Figure 5.4 also clearly demonstrates the existence of the cutoff wavelength λ_c discussed earlier in this chapter. Some particular cutoff wavelengths that will be of interest to us are Si $\lambda_c = 1.06$ μm, Ge $\lambda_c = 1.6$ μm, and InGaAs $\lambda_c = 1.7$ μm.

Data plotted in Fig. 5.4 also demonstrate that semiconductor photodiodes based on GaAs or pure InP substrates will not be able to detect light at 1.3 and 1.55 μm. Such photodiodes are not useful in modern FOCS. However, an InGaAs variant of these semiconductor materials (as well as some other mixtures of the type employed in heterojunction diode lasers), is sensitive at these two wavelength regions. Although the absorption coefficient is not a function of doping concentration, it is affected (in a static sense) by the selection of the composition of the layers in a heterojunction version of the photodiode.

Now let $P(z)$ be the total power absorbed *across a uniform cross section of area A* over a length z beginning at the input end of the photodiode. Then

$$P(z) = A[S_0 e^{-0} - S_0 e^{-\alpha z}] = AS_0[1 - e^{-\alpha z})]$$

$$= P_0[1 - e^{-\alpha z}] \quad \text{(where } P_0 = AS_0 = \text{incident power)}$$

Clearly, $P(z)$ is a function of the absorption coefficient $\alpha = \alpha(\lambda)$. Almost all of the light that enters the semiconductor will be absorbed within the photodiode if the path is long enough. However, only those photon-induced absorptions that occur in (and a few that occur within a diffusion length from) the DR region are actually observed.

We next examine some measures of the sensitivity and efficiency of photodiodes. Let us reexamine the end-illuminated photodiode depicted in Fig. 5.2; it is reasonable to assume that the photon absorption rate in the ultrathin p region is negligible, which leads to the following relation:

$$I_\text{p} = \left(\frac{P_0}{hf}\right)\left(1 - r\right)\left(1 - e^{-\alpha d}\right)e \qquad (5.3)$$

where I_p = photon-generated external current
  d = width of depletion region
  r = reflection coefficient expressed as fraction of photons reflected at input surface of photodiode
  e = charge of an electron

The structure of Eq. (5.3) is easily explained by noting that each of the three factors in that equation may be described in more visual terms as follows:

$$\left\{\frac{P_0}{hf}\right\} = \text{number of photons per second incident on the photodiode sur-}$$
face.

$(1-r)$ = fraction of photons entering the diode. The reflection coefficient is given by the same formula that we encountered in Chap. 4:

$$r = \left\{\frac{n_1 - n_2}{n_1 + n_2}\right\}^2$$

$(1 - e^{-\alpha d})$ = fraction of photons absorbed in the depletion region that generate externally observed hole–electron pairs.

We have seen that the product of the three bracketed terms in Eq. (5.3) is equal to the number of electrons per second that flow in the external circuit which were generated by the absorbed light. The overall quantum efficiency of the detector η is defined as

$$\eta = \frac{\text{number of photon-induced external electrons}}{\text{number of incident photons}}$$

Next we note that I_p/e is the number of photon absorptions observed as electrons flowing in the external circuitry of the photodiode (in the absence of dark current). Then, from Eq. (5.3), we have

$$\eta = \frac{I_p/e}{P_0/hf} = (1 - r)\,(1 - e^{-\alpha d}) \tag{5.4}$$

One major consideration in designing a photodiode-based receiver is to maximize the quantum efficiency of the photodiode detector itself. An integral part of this task is to eliminate optical reflections at the entry port of the photodiode. As we noted in our discussion of optical amplifiers in the previous chapter, we can make $r \approx 0$ by employing appropriate dielectric coatings known as *antireflection* (AR) coatings. These AR coatings are made up of alternate layers of two dielectrics with different indices of refraction. The thickness of these layers must be very precise on a scale of fractions of wavelength, which is why these AR coatings yield only $r \approx 0$ (i.e., less-than-perfect AR coatings). Moreover, AR coatings are designed to operate at a specific wavelength and lose efficiency when operated at another wavelength. Consequently, photodiodes designed for high quantum efficiency will have far tighter restrictions on input wavelength than is specified by the much broader condition expressed by the combination of Eq. (5.1), specifically, $\lambda_0 < \lambda_c$. For example, consider a photodiode designed for maximum efficiency at 1.55 μm. This device would respond to light at 1.3 μm, but at a reduced quantum efficiency caused by the AR coating problem. A photodiode designed for 1.55 μm probably would not be used in long-range 1.3 μm FOCS. On the other hand, if a general-purpose photodiode is desired for use with a broad range of input wavelengths, the AR coatings can be omitted entirely. This would result in a reduction in the quantum efficiency by the $(1 - r)$ factor in Eq. (5.4), but the effect of this factor almost independent of incident wavelength.

Another important consideration in the design of efficient photodiodes is the size and shape of the depletion region. In particular, we must make the optical path through the depletion region efficiently long that most of the light is absorbed there. For example, let the length of the path be in the DR d = $3/\alpha$, then find that ~95 percent of photons that enter the DR will be absorbed within that region. The thickness of the DR (denoted by d) depends somewhat on the bias voltage and the doping concentrations. A typical value for d is on the order of ~4 μm.

Example 5.1 Let the incident light on an InGaAs photodiode be centered at λ_0 = 1.55 μm. Determine the percentage of photons absorbed in the depletion region. We employ Fig. 5.4 to find that $\alpha \sim 10^4$ cm^{-1} = 10^6 μm^{-1} for InGaAs at ~1.55. Then $(1 - e^{-\alpha d}) = 1 - e^{-4} \approx 0.982 \Rightarrow 98.2$ percent absorbed in DR.

Next, let's suppose that there is no AR coating at the input port. Let the indices of refraction at this port be 1.45 for the glass fiber and, say, n = 3.2 for the PD. Then we have

$$r = \left\{ \frac{3.2 - 1.45}{3.2 + 1.45} \right\}^2 = 0.142 \qquad \text{and} \qquad 1 - r = 0.858$$

Thus 85.8 of the incident photons enter the photodiode.

Overall quantum efficiency η = $(0.982)(0.858)$ = 0.842.

In this example the DR is deep enough to absorb almost all light that enters. Finally, we note that not every photon absorbed in the DR is observed externally because a few charge pairs recombine in the DR; they don't reach the external circuitry. However, this constitutes a very small fraction of the hole–electron pairs generated in the DR.

Photodiodes can also be illuminated from the side and thereby increase the length of the light path within the depletion region.[6] When the depletion region is viewed from the side, however, the DR presents a much smaller target (recall d~4 μm) for the incident light than is the case with end illumination. Hence in most cases the quantum efficiency of the side-illuminated p–n junction PD would be reduced considerably from comparable end-illuminated devices.

5.2.3 p–i–n photodiodes

The p–i–n photodiode alters the basic p–n junction photodiode structure by inserting a nondoped (intrinsic) region between the n-doped and p-doped regions as shown in Fig. 5.5.

As indicated in Fig. 5.5, the depletion region in the p–i–n photodiode includes the entire intrinsic region. Just as in the p–n junction, there exists a potential barrier that opposes majority carrier flows into the extended DR that is formed between the two interfaces with the intrinsic region. This electric field is generated by the residual stationary charges at these two inter-

---

[6]This is valid only for simple p–n junction photodiodes. It will not be valid for p–i–n photodiodes, described in the next subsection.

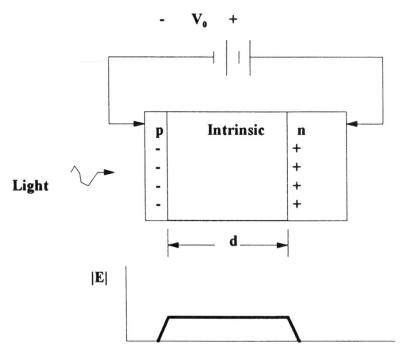

Figure 5.5   End-illuminated *p–i–n* photodiode.

faces. Holes and free elections inside the DR are expelled in opposite directions into regions where they are majority carriers. As was the case in the simple *p–n* junction photodiode, these excess carriers lead to externally observed electric currents.

One major advantage of the *p–i–n* photodiode is the improved frequency response characteristics that accrue from the reduced capacitance of the DR (C_{DR}). The effect of the additional length on C_{DR} is demonstrated by the well-known formula for a parallel-plate capacitors: $C = \epsilon(A/d)$. Conversely, frequency response also depends on the transit times of charges through the depletion region; the extended depletion region of the *p–i–n* structure has a slowing influence on transit time. In addition to the longer transit-time effect, the larger DR results in a weaker electric field than is found in the *p–n* junction devices for the same bias voltage. Nevertheless, the *p–i–n* structure yields some of the fastest photodiode responses ever observed.

There are several other advantages of the *p–i–n* device over the simple *p–n* junction. First, virtually all the photons that enter the DR are absorbed there and virtually none pass through to the next region (*n* region in Fig. 5.5). While there is also some improvement in the $(1 - e^{-\alpha d})$ factor in Eq. (5.4), such is not a major consideration here. Perhaps the most important advantage of the *p–i–n* structure is that it greatly reduces the number of photons absorbed in the bulk region, which, in turn, virtually eliminates the trailing-edge effect described earlier. Additionally, this property enhances the photo-diode frequency response.

Current flows in the bulk regions mainly by diffusion, while electric-field-driven drift current is the principal current mode in the extended depletion region. What results is another form of improvement of response time of *p–i–n* diodes over *p–n* photodiodes; most of the current flow is in the form of the faster drift current. To obtain frequency responses on the order of 50 GHz, the intrinsic region and the cross section must be made as small as possible. Because of the narrow cross section, there is difficulty in coupling light into the photodiode; some of the fastest reported *p–i–n* detectors have significantly smaller quantum efficiencies. Although there are pros and cons in comparisons between *p–n* and *p–i–n* photodiodes, *p–i–n* photodiodes are preferred over *p–n* photodiodes because of the reduction of the trailing-edge effect and the accompanying improvement in frequency response.

Another very useful variation of the *p–i–n* photodiode is to employ a hetero-junction structure exclusively for the intrinsic region of the photodiode. Unlike heterojunction lasers, these devices are designed to operate over a large range of wavelengths; matching the heterojunction *p–i–n* photodiode to a single operating frequency is not necessary. In particular, the energy gap in the intrinsic region is designed so that λ_c is compatible with reception of the λ_0 of the transmitted signal, say, 1.3 or 1.55 μm of radiation. Note that the latter case will work for both of the preceding wavelength ranges. On the other hand, the doped bulk regions are constructed out of the host semiconductor, either GaAs or InP, without heterojunction layers. This makes the bandgaps in those regions too large to absorb radiation at the wavelengths of the received optical signal (see Fig. 5.4). In this way, there is no photon absorption (hence no absorption loss) or photon-induced charge carrier generation in the bulk region. In turn, this prevents the delayed diffusion effect and thereby eliminates the trailing-edge effect. This compound design results in high-speed photodiodes with very large 3-dB cutoff frequencies. One possible variation of such a setup is displayed in Fig. 5.6.

5.2.4 Responsivity

The overall sensitivity of a photodiode is most simply described as the ratio of photodiode response current (I_p) to the incident optical power (P_0). This ratio is known as the *responsivity* and is denoted by R. Thus

$$R = \frac{I_p}{P_0} = \frac{\eta e}{hf} = \frac{\eta \lambda_0 e}{hc} \tag{5.5}$$

The responsivity may be expressed in any of the following equivalent units: A/W (amperes per watt), μA/μW, or nA/nW, and so on. There are several distinct reasons for the dependence of the responsivity of a photodiode on the wavelength of the incident light. First, the response of the host semiconductor for the photodiode is in itself a function of wavelength, as can be seen from Fig. 5.4; at the very

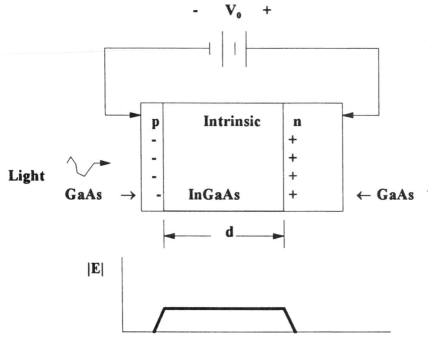

Figure 5.6 End-illuminated p–i–n photodiode with heterojunction intrinsic region.

least, the bandgap of the semiconductor determines the cutoff wavelength. The wavelength-dependent photon energy affects responsivity in a somewhat surprising way. This effect may be explained by noting that photon energy is inversely proportional to wavelength. Thus, for example, 1 mW at 1.3 μm contains less photons per second than does 1 mW at 1.55 μm. Now, since each detected photon generates one external electron, a larger current will be observed for the 1.55-μm case (provided other aspects such as quantum efficiency are equal). Responsivity is often graphically displayed for several commonly used semiconductors. These curves are of limited value, however, because responsivity also depends on device dimensions, antireflection coatings, and other factors. Succinctly, responsivity depends on photodiode structure as well as the properties of the semiconductors used to construct the photodiode.

At *low input levels* such as those encountered at the receiving end of fiber-optic communication systems, we find that I_p is linearly proportional to P. Thus, both R and η are functions of λ only and not dependent on optical power level. One application for calibrated photodiodes is for measuring the intensity of laser beams. For strong lasers, there may be saturation effects which lead to nonlinear intensity dependence. In such cases the responsivity will also depend on power level. A typical R/wavelength plot is shown in Fig. 5.7.

In the ideal case the responsivity rises linearly with wavelength because of the reduction in energy per photon. It then drops sharply to zero at $\lambda = \lambda_c$.

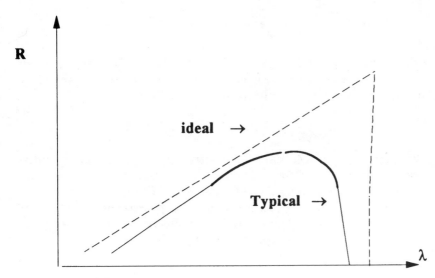

Figure 5.7 Typical R–λ curves.

Responsivity actually observed differs from the ideal primarily because the absorption coefficient is itself a function of wavelength (see Fig. 5.4). The upper wavelength cutoff is determined by the bandgap, while a low-end wavelength cutoff can be observed when the absorption coefficient becomes so large that very little light reaches the depletion region.

Example 5.2 Let 2 μW at $\lambda_0 = 1.55$ μm be the input to a photodiode *without AR coatings*. The observed response to this input is 1.8 μA. For this PD, find (*a*) responsivity and (*b*) quantum efficiency.

$$R = \frac{1.8}{2.0} = 0.9 \text{ from Eq. (5.5)} = \eta = \frac{Rhc}{\lambda_0 e} = 0.721$$

Now, what is the response if the same PD is illuminated with 5 μW at 1.3 μm? Here, because of the absence of AR coatings, we still have $\eta = 0.721$. But

$$R = \frac{\eta \lambda_0 e}{hc} = 0.754 \qquad I_p = RP_0 = 3.77 \text{ μA}$$

5.2.5 Avalanche photodiodes (APDs)

When the reverse bias on a semiconductor diode is set close, but not quite up to, the zener breakdown level, there is strong acceleration of free electrons and holes by intense electric fields in the depletion region. Collisions between these highly energized charges and semiconductor atoms can generate secondary hole–electron pairs through a process known as *impact ionization*. These secondary charge carriers will also be accelerated; when they achieve enough kinetic energy, they can generate more hole–electron pairs, and so on. Aptly termed the "avalanche effect," this process increases the external current by a factor of $\overline{G}$ defined as

$$\overline{G} = \frac{I}{I_{\mathrm{p}}} \tag{5.6}$$

where I is the actual current response of the APD and I_{p} is the external current that would have been observed had there been no avalanche gain at all. The responsivity of the APD is simply stated as

$$R_{\mathrm{APD}} = \overline{G}R$$

Actually, the gain G for each initial hole–electron pair is a random quantity and the parameter G represents average gain. Clearly, this randomness adds noise to the detection process, and, as we shall see in Chap. 7, it turns out to be a very significant amount of noise. Nevertheless, the APD performs a valuable service when used properly by negating the effects of receiver generated electrical noise. A typical APD is displayed in Fig. 5.8.

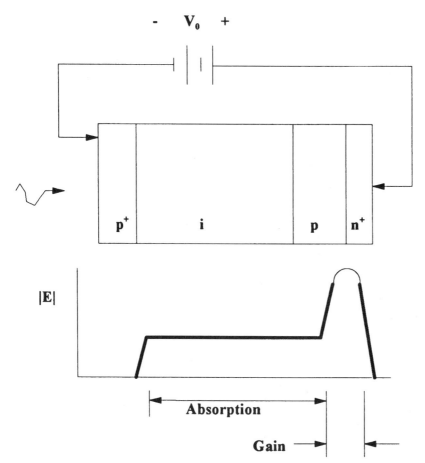

Figure 5.8 End-illuminated APD.

Light enters the APD through the thin p^+ region on the left-hand side of the figure. Virtually all photon absorption takes place in the intrinsic region, which in this case is a lightly doped p region. As was the case in the $p-i-n$ photodiode, the electric field in the intrinsic region drives the holes and electrons in opposite directions. Specifically, holes are drawn toward the p^+ region by the negative power supply contact while the electrons are drawn toward the positive power supply contact on the n^+ region. Along the way the electrons enter the avalanche region and initiate the amplification process there. A strong reverse-bias voltage is applied to this region which is very close to the zener breakdown level. There it generates an *intense field in the depletion region associated with the $p-n^+$ junction*. The $p-n^+$ depletion layer is a high-impedance region; thus there is a strong **E** field in this region as shown in Fig. 5.8; the avalanche gain takes place entirely within this very small region. If the reverse bias exceeds the zener breakdown level, the output current is independent of the incident light. Thus, as one can see from the avalanche gain versus reverse bias shown in Fig. 5.9, it is necessary to very

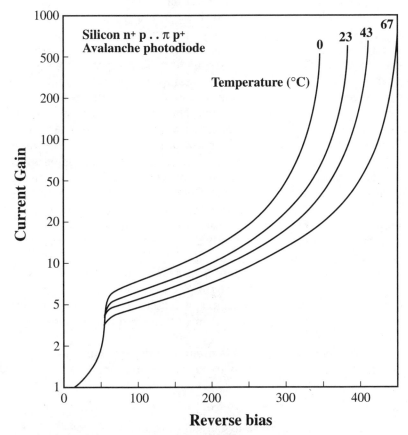

Figure 5.9 Current gain versus bias for several temperatures. (*From Ref. 3; Copyright © 1978, AT&T. All rights reserved. Reprinted with permission.*)

carefully control this voltage to avoid entering the breakdown region. From Fig. 5.9, we can also discern that the ambient temperature must be carefully controlled to maintain steady gain.

The probabilities per unit length of a collision between an accelerated charge carrier and a semiconductor atom creating an additional hole–electron pair are denoted by the impact ionization coefficients α_e and α_h for electrons and holes, respectively. These parameters increase as the electric field increases and vice versa. A parameter of particular interest for understanding the noise characteristics of APDs is the ratio of these two parameters κ, which is defined as $\kappa = \alpha_h/\alpha_e$.

When $\alpha_h \ll \alpha_e$ ($\kappa \gg 1$), electron impact ionization dominates the avalanche process and proceeds from the p side to n^+ side of the avalanche region of the APD. Average gain may be maximized by using materials with large α values and $\kappa \sim 1$. However, such devices will have relatively slow frequency response because the avalanche chains reverberate (do not die out) inside the avalanche gain region. In practice, the APD is usually designed to satisfy frequency response criteria at the price of using moderate current gain. This can be accomplished by having either holes or electrons dominate the avalanche process. Thus, we desire $\kappa \approx 0$ (or possibly $\kappa \gg 1$). As we shall see in the next chapter, $\kappa \approx 0$ yields better noise characteristics.

As can be seen from Fig. 5.9, there are stringent requirements on both the power supply voltage stability and the ambient temperature control of APDs. One might now be tempted to ask why not simply electronically amplify the p–i–n output and avoid these complications. The avalanche amplification mechanism produces predetection gain; such is the reason for using APDs in a number of applications. To clarify this point, we note that the primary sources of noise in most receivers are the thermal and shot noises generated at the receiver. The avalanche gain process does not amplify these noises, although randomness in the avalanche process does add some noise of its own (we shall discuss this subject in the next chapter). On the other hand, a p–i–n photodiode followed by electronic amplification will amplify the receiver generated noises as well as the photodiode output. The APD output also requires electronic amplification to achieve suitable signal levels, but the signal portion of the APD output will already have been amplified by the avalanche process, thereby diminishing the effect of the receiver-generated electronic noise. In Chap. 6 we shall see that the avalanche detection process can totally mask the effects of receiver noise and result in unimproved overall signal-to-noise ratio.

The key to this noise reduction process is predetection gain. However, there are other detection systems that also provide predetection gain:

1. Optical preamplifiers placed directly in front of a p–i–n photodiode

2. Heterodyne detection systems (to be discussed in Chap. 8)

3. Photomultipliers (too bulky for fiber optics)

Of these, only the optical preamplifier actually amplifies the optical signal before detection; the others achieve predetection gain during the detection process. However, none of them amplify the noise generated by the receiver electronics. All these detection processes add noise to the desired signal in one way or another, but sufficient predetection gain often is such that we can totally ignore the locally generated electronic noise. The most important point here is that the resulting S/N can be much larger than would be obtained if the dominant noise were the electronically amplified thermal noise.

Problems

1 A particular photodiode generates on the average one-hole electron pair for three incident photons. Let $\lambda_0 = 0.8$ μm and find
  (a) The quantum efficiency of the device
  (b) The mean photocurrent when the received optical power is 0.1 mW

2 A particular photodiode has a quantum efficiency of 50 percent at $\lambda_0 = 0.9$ μm. Find
  (a) Its responsivity at this wavelength
  (b) The received optical power if $I_p = 1$ μA
  (c) The corresponding number of photons received per second

3 When 800 photons/s are incident on a photodiode operating at $\lambda_0 = 1.3$ μm, they generate 550 electrons per second. Calculate the responsivity of this device.

4 Consider the p–i–n receiver configuration shown in Fig. P5.4. The incident radiation is at $\lambda_0 = 1.5$ μm, and the detector diode is a p–i–n device. Also assume that absorption in the thin p^+ region can be ignored. Find the width of the intrinsic region d that is required to obtain a 50 percent overall quantum efficiency for this receiver setup.

5 A particular silicon APD with a quantum efficiency of $\eta = 0.88$ is biased just below the avalanche region ($\overline{G} = 1$). There it is found to have a responsivity of 0.58 A/W. We next bias this same device into the avalanche gain region and illuminate it with 5 nW of $\lambda_0 = 0.83$ μm of radiation. The APD current is then measured and found to be 60 nA. For these conditions, find
  (a) Avalanche gain $\overline{G}$.

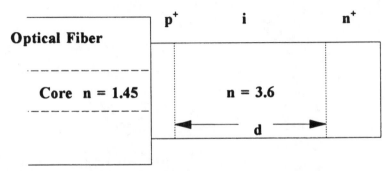

Figure P5.4 GaInAs p–i–n photodiode

(**b**)   Number of photons incident per second.

(**c**)   Number of electrons that flow through the APD per second.

(**d**)   Calculate the ratio of your answers to parts **b** and **c** above and compare it with your answer to part **a.** Use the attached figures to explain why they differ.

6   The responsivity of a particular photodiode operating at $\lambda_0 = 1.55$ μm is 0.7 A/W where an appropriate antireflection coating has been attached to the input port.

(**a**)   Find the overall quantum efficiency for this setup.

(**b**)   Find the overall quantum efficiency for the preceding setup with the coating removed, and you have an air to GaAs interface ($n = 3.6$).

(**c**)   Now suppose that the reverse bias on the photodiode is increased into the avalanche region where $M = 20$. Assume that except for the gain everything else remains the same as in part **a,** *including the quantum efficiency,* and find the output current when the device is illuminated by 0.2 mW at $\lambda_0 = 1.3$ μm.

References

(*Note:* All the texts listed in Refs. 1 to 7 of Chap. 4 contain significant material on photodiodes.)

1. G. Lachs, "The statistics for the detection of light by nonideal photomulipliers," *J. Quantum Electron.* **QE-10,** 591–596 (1974).
2. J. E. Bowers and C. A. Burrus, Jr., "Ultrawide band long wavelength p-i-n photodetectors," *J. Lightwave Technol.* **10,** 1336–1350 (1987) (see Fig. 6).
3. H. Melchior, A. R. Hartman, D. P. Sshinke, and T. E. Seidel, "Planar epitixial silicon avalanche photodiode," *Bell Syst. Tech. J.* **57,** 1791–1807 (July–Aug. 1978).

6

Receiver Noise Considerations

6.1 Introduction

We begin this chapter by discussing some general aspects of the noise encountered in fiber-optic communication systems (FOCS). The performance of FOCS as determined by important system characteristics such as maximum data rate, digital error rate, maximum distance between repeaters, and receiver sensitivity are affected by a variety of noise types that exist in FOCS transmission. These noises may be subdivided into three principal categories: predetection, during-detection, and postdetection noise. In this chapter we introduce only the various species of noise that we shall encounter in the remaining chapters of this text, where we will discuss them in more detail.

6.1.1 Predetection noise

Predetection noise is usually created in optical frequency communication systems (especially open-space-type systems) by stray background light incident on the optical detector. However, because of the enclosed nature of fiber-optic systems, this type of disturbance is seldom encountered in FOCS. Another predetection source of randomness in a FOCS is generated by the unpredictable fluctuations observed in the polarization alignment of the optical signal as it progresses down the optical fiber. The physical source of this random polarization is the birefringence of the fiber itself coupled with random pressure variations in the surrounding environment of the fiber. We also note that there exists a form of noise unique to propagation in multimode FOCS known as *mode-partitioning noise*. This type of randomness in the propagating signal is caused by unpredictable fluctuations in the distribution of energy among the modes, which, in turn, causes fluctuations at the receiving end of the fiber. A similar effect also results when multimode lasers are employed as signal sources. We shall disregard these last two forms of mode partitioning noise in this text because we will concentrate entirely on single-mode fiber propagation and single-frequency laser diode sources.

The principal source of predetection noise encountered in high-speed single-mode FOCS are the semiconductor laser diodes (SLDs) themselves. It turns out that SLDs *generate considerably more noise than do their macroscopic counterparts.* The basic source of this random behavior is the spontaneous emission process. It generates random fluctuations in both the intensity and the phase of the SLD output, which are known, respectively, as *relative intensity-noise* (RIN) and *phase noise.* The latter noise source is particularly troublesome in optical heterodyne detection systems. A salient feature common to both of these types of noise is that, unlike additive noise, the effects of these noises can't be negated by merely increasing laser output power. Phase noise in particular often sets upper bounds on achievable receiver S/N ratios, which, in turn, sets lower bounds on achievable error rates. We note that a bit error rate criterion of 10^{-9} is often used in comparative tests of FOCS systems, and it is conceivable that under certain conditions the phase noise generated by the laser source may make this very low error rate unattainable. We shall study this topic in greater detail in the text two chapters when we examine the performance of complete FOCS and optical heterodyne detection systems. A reader interested in more detail on this subject will find an excellent description of these phenomena in the treatise by Agrawal [11] listed at the end of this chapter.

6.1.2 Detection-process noise

Each predetection gain process is characterized by a specific set of random effects. The APD, for example, generates noise in the avalanche process, while the photomultiplier generates noise in the dynode amplification process. Coherent detection results in a high level of shot noise, and the optical amplifier contributes noise due to amplified spontaneous emission as well as some inherent quantum mechanical effects. The basic optical detection process also generates what is known as *quantum noise,* which sets a fundamental lower limit on noise power. It is a consequence of the basic physics of the detection process. We shall discuss this process in some depth here in this chapter to prepare us for the coverage of coherent detection systems in Chap. 8.

6.1.3 Postdetection noise

Postdetection noise is generated in the electronic circuitry associated with the photodiode and the first stage of the electronic amplifier that immediately follows the photodiode. This consists primarily of thermal noise and shot noise (including dark current noise). Each of these noises may be modeled as zero-mean white gaussian noise. The white noise is subsequently shaped by the frequency response of the receiver circuitry. Let X_1 and X_2 represent two separate statistically independent sources of noise. Now let then [1,2]

$$Y = a_1 X_1 + a_2 X_2$$

$$\sigma_Y^2 = a_1^2 \sigma_{X,1}^2 + a_2^2 \sigma_{X,2}^2$$

Furthermore, if $X_1(t)$ and $X_2(t)$ are gaussian random processes, then $Y(t)$ is also a gaussian random process. It will remain gaussian noise until it encounters a nonlinear operation such as envelope detection or an FM limiter.

There is also a class of nongaussian noise distortion that is generated in the receiver circuits by the phase and RIN noises of the laser diodes. For example, the random phase in SLD outputs known as *phase noise* is a gaussian random process. However, because it appears inside a nonlinear cosine function, the net result is a nongaussian random process. Furthermore, phase noise generates amplitude fluctuations when it is passed through a band-limited linear filter. Thus the predetection noise represented by phase noise and RIN give rise to noise in the electronic portion of the receiver that will have statistical properties that differ from the input noise. We shall now take a closer look at the various noises that affect reception in FOCS.

6.2   Quantum Noise (Statistics)

6.2.1   General formulas

In this section we introduce some basic formulas for computing the probability distribution for the number of photons observed in a T-second interval. This will be done initially from the viewpoint of the photoelectric effect in which the absorption of a photon by an atom is manifested by the ionization of that atom with the emission of an electron. We shall soon, however, apply the statistical techniques to find the distribution of the number of electrons optically excited to the conduction band by the absorption of light. We begin with the formula for the probability of a single atom being ionized.

Let p_a be the probability that a particular atom illuminated by incident light emits an electron in T seconds. It can be shown that

$$p_a = K_0 \int_0^T |E(t)|^2 \, dt = \beta_1 \int_0^T I(t) \, dt \tag{6.1}$$

where $I(t) \propto |E(t'')|^2$ = optical intensity. The terms K_0 and β_1 depend on the detector material and are functions of λ. They also depend on units used, such as photons per second per square meter or watts per square meter. The subscript on β_1 denotes that this parameter applies to a single atom.

In the following discussion we assume that each atom in the detector acts independently of all the other atoms in the detector. Then for each atom we have two possible results in T seconds. That atom either emits an electron or it doesn't (we assume that the probability that an atom emits an electron absorbs an electron and again emits in T seconds is negligibly small). This statistical model leads to the binomial distribution for the number m of primary electrons emitted in T seconds:

$$p(m) = \binom{N}{m} p_{\mathrm{a}}{}^{m} (1 - p_{\mathrm{a}})^{N-m} \tag{6.2}$$

where N is the number of atoms in the detector, typically $>10^{20}$, and the "binomial coefficient" is

$$\binom{N}{m} \doteq \frac{N!}{m!\,(N-m)!}$$

Now, for typical digital data rates and the physical structure of photodiodes normally used in FOCS receivers, the number of atoms in the photon absorption region of the detector is much larger than the average number of photons counted in the T digital pulse duration (baud time). Hence the probability that a particular atom is ionized during a single detection interval is extremely small (i.e., $p_{\mathrm{a}} \ll 1$). We now employ this condition to the formula for the mean of the binomial distribution to obtain

$$a = <m> = Np_{\mathrm{a}} <<< N$$

It is well known from probability theory that these inequalities imply that the binomial distribution is accurately approximated by the Poisson distribution:

$$p(m) = \frac{a^m e^{-a}}{m!} \tag{6.3a}$$

In general, we have

$$a = \beta \int_0^T I(t)\,dt$$

where the parameter β (without a subscript) includes implicitly the number of atoms N in the detection portion of the photodiode. Now, when $I(t) = I_0 =$ constant, we can express the mean as

$$a = \beta I_0 T = \text{constant}$$

and the photoelectron count distribution is as given by Eq. (6.3a). This is called a *homogeneous Poisson distribution,* and the random process represented by the photon stream is called a *Poisson process* (PP). When the incident field is time-varying, that is, $I(t) = f(t)$, it is a *nonhomogeneous Poisson process.* The photoelectron count distribution for the detection of such a field is given by

$$p(m) = \frac{\left\{\beta \int_0^T I(t)\,dt\right\}^m \exp\left\{-\beta \int_0^T I(t)\,dt\right\}}{m!} \tag{6.3b}$$

Another important variation of this equation is the situation in which the incident field itself is random. Two examples of this phenomenon are the electromagnetic fields emitted in blackbody radiation, and the output of a SLD perturbed by relative intensity noise. There are now two degrees of randomness involved in the detection of this type of optical frequency radiation, and it is known as a *doubly stochastic Poisson process* (DSPP). One degree of randomness comes from the field itself, and the other comes from the inherent randomness of the detection process. The photoelectron count distribution is obtained by averaging the right-hand side of Eq. (6.3b) over the statistics of the incident field. Thus, for a DSPP, we have

$$p(m) = \mathbf{E} \frac{\left\{ \beta \int_0^T I(t)dt \right\}^m \exp\left\{ -\beta \int_0^T I(t)dt \right\}}{m!} \tag{6.3c}$$

Note that Eq. (6.3c) also applies to a random time-varying optical field such as one would encounter with in the output of a modulated SLD perturbed by RIN.

One salient property of both the homogeneous and nonhomogeneous Poisson processes is that the photon counts for nonoverlapping intervals are statistically independent. For these processes, the time of occurrence of each photoelectron is independent of any of the other occurrences. Let x be the waiting time between events in a nonhomogeneous Poisson process. Then the probability density for the waiting time between events can be shown to be

$$f(x) = \beta I(x) \exp\left\{ -\beta \int_0^x I(t)dt \right\} u(x) \tag{6.3d}$$

A very useful tool for working with the probability distributions of integer random variables (sometimes called *lattice distributions*) is the generating function of the probability distribution. The usual definition of the generating function is

$$G(s) = \mathbf{E}\,[s^m] = \sum_{m=0}^{\infty} p(m)s^m \tag{6.4a}$$

As an example, we consider the generating function for the homogeneous Poisson distribution:

$$G(s) = \sum_{m} p(m)s^m = \sum_{m} = 0 \,\frac{(as)^m e^{-a}}{m!} = e^{-a}\,e^{-as} = e^{-a(1-s)}$$

An alternative form of the generating function that is more convenient for various Poisson-related processes is defined by

$$Q(s) = G(1-s) = \sum_{m} P(m)(1-s)^m \tag{6.4b}$$

We then obtain the following $Q(s)$-type generating functions for the various versions of the Poisson process:

$$Q(s) = e^{-as} \qquad \text{homogeneous} \tag{6.4c}$$

$$Q(s) = \exp\left\{-s\beta \int_0^T I(t)dt\right\} \qquad \text{nonhomogeneous} \tag{6.4d}$$

$$Q(s) = \mathbf{E}\left[\exp\left\{-s\beta \int_0^T I(t)dt\right\}\right] \qquad \text{DSPP} \tag{6.4e}$$

Although both forms of the generating functions provide the same information about photon statistics, it turns out that the $Q(s)$ format yields a somewhat simpler generating function than does the $G(s)$ format.

The generating function plays a very similar role for distributions of integer random variables, as does the characteristic function for continuous random variables. For example, the generating function for the sum of two statistically independent integer random variables is the product of the generating functions of each individual generating function. Let M and N be statistically independent integer random variables, and let $R = M + N$. Then, as in characteristic functions, we can show that $Q_R(s) = Q_M(s) Q_N(s)$. This can be verified by employing Eq. (6.4b) for each of the preceding generating functions and matching like exponents of $(1 - s)^k$ on both sides of the preceding equation; we then obtain the discrete form of convolution:

$$p_R(r) = p_M(m) * p_N(n) = \sum_{\ell=0}^{r} p_M(\ell)p_N(m - \ell) \tag{6.4f}$$

Another aspect of generating functions that parallels characteristic functions is the fact that the moments of integer probability distributions can be obtained by differentiating the corresponding generating function. Beginning with Eq. (6.4b)

$$Q(s) = p(0) + p(1)(1 - s) + p(2)(1 - s)^2 + p(3)(1 - s)^3 + p(4)(1 - s)^4 + \cdots$$

it can be shown [6] that the rth factorial moment C_r is given by

$$Cr \doteq \mathbf{E}\left[\frac{m!}{(m-r)!}\right] = (-1)^r \left.\frac{d^r Q(s)}{ds^r}\right|_{s=0} = (-1)^r Q^r(0) \tag{6.5a}$$

These factorial moments can be combined to obtain the ordinary moments $m_k = E[X^k]$. Thus, for example, $C_1 = E[m] = m_1 = $ mean and $C_2 = E[m(m-1)] = E[m^2] - E[m] = m_2 - m_1$

Also the variance may be expressed in terms of the factorial moments as

$$\sigma^2 = C_2 + C_1 - C_1^2 \tag{6.5b}$$

The probability distribution can also be obtained from the generating function, but in this case the process is very different from the Fourier transform method used with characteristic functions. Once again, beginning with Eq. (6.4b), one can show that

$$(p(m) = \frac{(-1)^m}{m!}\, Q^{(m)}(1) \tag{6.6}$$

Students familiar with digital signal processing may discern that generating functions are really a form of Z transforms. As such, the expansion of the generating function in a power series is an attractive alternative to repeated differentiation for obtaining $p(m)$.

6.2.2 Narrowband radiation

In this subsection we assume that the time–bandwidth product of the optical radiation is $\ll 1$, i.e., specifically, $TB_T \ll 1$, where B_T is the bandwidth of the incident radiation. Then we can assume that the intensity I is approximately constant, over the detection interval T, but we allow the intensity I to vary randomly from interval to interval. In words, I in this model is a random variable rather than either a constant or a random process. The narrowband assumption enables us to simplify the following integral which appears in Eqs. (6.4d) and (6.4e) to

$$\beta \int_0^T I(t)dt = \beta I T$$

Thus, if I is random (DSPP), and the probability density $f(I)$ governs the statistics, we get

$$Q(s) = \mathbf{E}_I \left[e^{-s\beta IT} \right] = \int_0^\infty f(I)\, e^{-s\beta IT} dI \tag{6.7a}$$

$$p(m) = \int_0^\infty \frac{f(I)\,(\beta IT)^m e^{-\beta IT}}{m!}\, dI \tag{6.7b}$$

We will next investigate some specific examples of narrowband radiation. We begin with an ideal laser in the absence of RIN and phase noise. Such a laser is characterized by a constant intensity output and $f(I) = \delta(I - I_0)$. Substituting in Eqs. (6.7a) and (6.7b), we obtain

$$Q(s) = e^{-s\beta I_0 T} = e^{-sa} \Rightarrow \text{homogeneous poisson}$$

and

$$p(m) = \frac{a^m e^{-a}}{m!} \qquad (\text{where } a = \beta I_0 T)$$

Now let $a = \langle m_C \rangle = $ average number of photon counts in T seconds (subscript on $\langle m_C \rangle$ stands for coherent). This results in

$$p(m) = \frac{\langle m_C \rangle^m\, e^{-\langle mC \rangle}}{m!} \tag{6.8a}$$

$$Q(s) = e^{-s\langle m_C \rangle} \tag{6.8b}$$

which leads to $C_1 = <m_c>$ and $C_2 = <m_c>^2$:

$$E[m] = <m_c> \tag{6.8c}$$

$$\sigma_C^2 = C_2 + C_1 - C_1^2 = <m_c> \tag{6.8d}$$

Thus the notation $<m_c>$ in Eqs. (6.8a) and (6.8b) is justified by Eq. (6.8c). We note here that $< >$ is commonly used in the physics literature to denote *ensemble average,* which differs from its frequent use in the electrical engineering literature to denote *time average.* Since Eq. (6.8a) is a homogeneous Poisson distribution, it is no surprise that the variance = mean.

For our next example, we consider narrowband natural (nonlaser) light. This type of light can be obtained by passing blackbody radiation through a narrowband filter. The electric field for narrowband thermal light can be mathematically modeled by

$$E(t) = n_c(t) \cos(\omega_0 t) - n_s(t) \sin(\omega_0 t)$$

where $n_c(t)$ and $n_s(t)$ are slowly varying with respect to the center frequency ω_0. By central-limit-theorem arguments (superposition of a large number of uncorrelated or weakly correlated contributions), n_c and n_s are jointly gaussian zero-mean equal variance, gaussian random variables. Furthermore, if optical spectrum is symmetric about the central frequency ω_0, then $n_c(t)$ and $n_s(t)$ are statistically independent random processes [17]; the joint density of n_c and n_s is given by

$$f(n_c, n_s) = \frac{1}{2\pi\sigma^2} \exp\left[-\frac{n_c^2 + n_s^2}{2\sigma^2}\right] \tag{6.9a}$$

Now let $I \Rightarrow |E|^2 = K_1 [n_c^2 + n_s^2] > 0$. Then, using standard change of variable techniques [1] for probability densities, we obtain

$$f(I) = \frac{1}{2K_1\sigma^2} \exp\left\{-\frac{I}{2K_1\sigma^2}\right\} = \frac{1}{\alpha} \exp\left\{\frac{-I}{\alpha}\right\} u(I)$$

$$E\{I\} = \int_0^\infty I f(I) \, dI = \alpha \doteq <I> = 2K_1\sigma^2$$

$$f(I) = \frac{1}{<I>} \exp\left\{\frac{-I}{<I>}\right\} u(I) \tag{6.9b}$$

and substituting Eq. (6.9b) in Eq. (6.7b), we find

$$p(m) = \int_0^\infty \frac{(\beta I T)^m \exp\left\{-I\left(\beta T + \frac{1}{<I>}\right)\right\}}{<I> m!} \, dI$$

Now let $U = \beta IT$ and $a = \beta <I> T$ to arrive at

$$p(m) = \int_0^\infty \frac{U^m}{am!} \exp\left\{-U\left(1 + \frac{1}{a}\right)\right\} dU \qquad \text{and}$$

$$p(m) = \frac{a^m}{(1 + a)^{m+1}} = \frac{<m_T>^m}{(1 + <m_T>)^{m+1}} \qquad (6.9c)$$

where $<m_T> = a = \beta <I> T$ = average number of *thermal* photons in T seconds due to narrowband natural radiation.

Equation (6.9c) is the Bose–Einstein probability distribution that we encountered earlier in Chap. 4. The generating function $Q(s)$ for this distribution is obtained by using Eq. (6.9b) in Eq. (6.7a) to arrive at

$$Q(s) = \int_0^\infty \frac{\exp\left\{-I\left(s\beta T + \frac{1}{<I>}\right)\right\}}{<I>} dI$$

$$Q(s) = \frac{1}{1 + \beta <I> Ts} = \frac{1}{1 + s<m_T>} \qquad (6.9d)$$

from which we find $C_1 = <m_T>$ and $C_2 = 2<m_T>^2$. Thus the mean and variance for a single mode of thermal light are

$$E[m] = <m_T> \qquad (6.9e)$$

$$\sigma_T^2 = <m_T> + <m_T>^2 \qquad (6.9f)$$

Equation (6.9f) shows that the variance for the number of photons in a specified time interval T for narrowband thermal light is larger than the corresponding variance for pure coherent light (ideal laser) with the same average number of photons. The $<m_T>^2$ term is the excess variance, and this leads to a phenomenon called *photon bunching*, which means that there is a greater tendency of a second photon to follow sooner after the occurrence of a preceding photon than would be the case if each photon event were statistically independent of the others. Note that there is no bunching effect for the output of an ideal laser. One consequence of photon bunching is that photon counts in nonoverlapping intervals are no longer statistically independent.

Next we consider *optical radiation*, which consists of a superposition of narrowband thermal radiation and the output of an ideal laser whose frequency falls in the center of the thermal light spectrum. The net electric field in this case is a sum of narrowband gaussian noise and a pure sinusoid. Thus

$$E(t) = n(t) + A \cos \omega_0 t = (n_c + A)\cos \omega_0 t - n_s \sin \omega_0 t$$

which we restate as $E(t) = X_1 \cos \omega_0 t - X_2 \sin \omega_0 t$ where X_1 and X_2 are statistically independent gaussian random variables and $E[X_1] = A$ while $E[X_2] = 0$. Then the joint density of X_1 and X_2 may be expressed as

$$f(X_1, X_2) = \frac{1}{2\pi\sigma^2} \exp\left\{ \frac{-(X_1 - A)^2 + X_2^2}{2\pi\sigma^2} \right\} \qquad (6.10a)$$

Now, if we let the envelope of $|E(t)| = r = \sqrt{X_1^2 + X_2^2}$, then we obtain the well-known Rice density:

$$f(r) = \frac{r}{\sigma^2} \exp\left[-\frac{r^2 + A^2}{2\sigma^2} \right] I_0 \left(\frac{rA}{\sigma^2} \right) u(r) \qquad (6.10b)$$

We then find the probability density of the optical intensity I by using the following change of variable:

$$I = Kr^2 \Rightarrow r = \sqrt{\frac{I}{K}}$$

we find:

$$f(I) = \frac{1}{2K\sigma^2} \exp\left\{ -\frac{I + KA^2}{2K\sigma^2} \right\} \boldsymbol{I}_0 \left(\frac{A}{\sigma} \sqrt{\frac{I}{K}} \right) u(I) \qquad (6.10c)$$

This version is known as the *Rice–Nakagami density*. We now employ Eq. (6.10c) in Eq. (6.7b) to obtain

$$p(m) = \frac{a^m}{(1 + a)^{m+1}} exp\left\{ -\frac{b}{1 + a} \right\} L_m\left(-\frac{b}{a(1 + a)} \right) \qquad (6.10d)$$

where

$$L_m(x) = \sum_{\ell=0}^{m} \binom{m}{\ell} \frac{(-x)^\ell}{\ell!} = \text{Laguerre polynomial}$$

and where $a = \beta(2K\sigma^2)T$ and $b = \beta KA^2 T$.

Clearly the parameter a depends only on σ^2; therefore, it is generated by the thermal part of the input. The parameter b, on the other hand, depends only on the amplitude squared of the coherent portion of the input radiation, and therefore it clearly represents the contribution of the coherent portion of the input. This interpretation is fortified by noting that, with the aid of l'Hôpital's rule, one can show that

and

$$\lim_{a \to 0} p(m) = \frac{b^m e^{-b}}{m!} = \text{poisson} = \text{pure coherent}$$

$$\lim_{b \to 0} p(m) = \frac{a^m}{(1 + a)^{m+1}} = \text{Bose–Einstein} = \text{thermal}$$

We can interpret the parameters as follows:

$a = \langle m_T \rangle$ = average number of photons due to thermal part alone

$b = \langle m_C \rangle$ = average number of photons due to coherent part alone

Then we have

$$p(m) = \frac{<m_T>^m}{(1 + <m_T>)^{m+1}} \exp\left\{-\frac{<m_C>}{1 + <m_T>}\right\} L_m\left(-\frac{<m_C>}{<m_T>(1 + <m_T>)}\right) \quad (6.10e)$$

Substituting Eq. (6.10c) in Eq. (6.7a), we obtain

$$Q(s) = \frac{1}{1 + s<m_T>} \exp\left\{\frac{-s<m_C>}{1 + s<m_T>}\right\} \quad (6.10f)$$

After considerable bookkeeping, one finds $C_1 = <m_C> + <m_T>$ and

$$C_2 = <m_C>^2 + 2 <m_T>^2 + 4 <m_T> <m_C> \quad \text{which result in}$$

or

$$\sigma_S^2 = \sigma_c^2 + \sigma_T^2 + \text{mixing effects}$$

$$E[m] = <m_C> + <m_T> \quad (6.10g)$$

$$\sigma_S^2 = <m_C> + <m_T> + <m_T>^2 + 2<m_C><m_T> \quad (6.10h)$$

where the subscript S stands for the superposed field, the subscript T stands for the thermal field, and the subscript C stands for the coherent field.

Figure 6.1 displays several examples of the photon-count distributions for superposed coherent and thermal radiation. Note that these curves are defined only at integer values of the argument but are graphed as a continuous curves for convenience. Although the coherent field and the thermal field are independent sources, we note the following peculiar characteristics of the superposed field. By comparing Eqs. (6.8d) and (6.9f) with Eq. (6.10h), we see that

$$\sigma_S^2 \neq \sigma_C^2 + \sigma_T^2 \quad (6.10i)$$

We also see from Eqs. (6.8b), (6.9d), and (6.10f) that

$$Q_S(s) \neq Q_C(s)Q_T(s) \quad (6.10j)$$

The anomalies represented by Eqs. (6.10i) and (6.10j) may be explained by noting the following:

1. The superposition of two electric fields in an isotropic homogeneous region of space (including vacuum) are governed by Maxwell's equations.

2. These equations are linear partial differential equations with constant coefficients for the electromagnetic field in free space.

3. In particular, the superposition of a coherent and a thermal field may be expressed as

$$E_S(t) = E_C(t) + E_T(t)$$

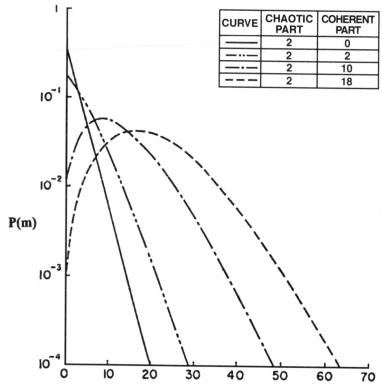

CURVE	CHAOTIC PART	COHERENT PART
——	2	0
—··—	2	2
—·—	2	10
– – –	2	18

Figure 6.1 Some photon-count distributions for superposed coherent and thermal radiation.

4. Note the absence of any nonlinear function in this relation. However, we note that the response of a photodiode to this composite field is expressed by the nonlinear relation

$$i_S(t) = K|E_C + E_T|^2$$

The fact that the variance of the photocounts for the sum field does not equal the sum of the corresponding variances of the constituent fields clearly demonstrates that it is incorrect to view photons simply as pellets. We see that photon number statistics display interference effects as evidenced by the last term on the right-hand side of Eq. (6.10h). These interference effects are responsible for such observable phenomena as

1. Optical interference in general, as demonstrated, for example, by Young's double-slit experiment

2. Coherent (optical heterodyne) detection

3. Holography

4. Interferometry

We note here that the typical SLD sources used in FOCS are particularly noisy so that $<m_T>$ may not be trivial. Nevertheless, we do find that the SLD output is such that

$$<m_C> >> <m_T>$$

and

$$\sigma_S^2 \approx <m_C> + 2 <m_C><m_T>$$

Thus the variance of the number of photons in a T-second interval for a SLD output (denoted here as σ_S^2) could be such that $\sigma_S^2 >> <m_C>$.

6.2.3 Wideband radiation

Recall that

$$Q(s) = \mathbf{E}\left[\exp\left\{-s\beta \int_0^T I(t)dt\right\}\right] \tag{6.11a}$$

We now assume that the optical field is an ergodic random process which implies the following (and more!):

$$<E^2(t)> = \lim_{T\to\infty} \frac{1}{T}\int_0^T E^2(t)\, dt \propto \lim_{T\to\infty} \frac{1}{T}\int_0^T I(t)\, dt \tag{6.11b}$$

Note: The left-hand term in Eq. (6.11b) is meant to be ensemble average here even though in this case it agrees with common IEEE usage. We also note that the $E[\cdot]$ notation for ensemble average is awkward when the variable is itself denoted as E. Ergodicity is the principal situation where the two interpretations are equal. Now when $TB >> 1$, T is in effect $\sim\infty$, which implies that

$$\int_0^T I(t'')dt'' \approx <I>T \qquad (\text{where } <I> = E[I])$$

Therefore

$$Q(s) = \exp[-s\beta<I>T] \tag{6.11c}$$

$$P(m) = \frac{(\beta<I>T)^m\, e^{-(\beta<I>T)}}{m!} \tag{6.11d}$$

It can be seen from Eqs. (6.11c) and (6.11d) that wideband thermal light generates a simple Poisson photon-count distribution. For example, for a wideband thermal source such as an LED where output bandwidth is $\sim 10^{12}$, the photon-count statistics will be Poisson and not Bose–Einstein. To show this, we note that an appropriate value for T in digital communications is the baud duration which in LED systems is $\sim 10^{-7}$. Thus $BT \sim 10^5$ and simple homogeneous (or nonhomogeneous if modulated) Poisson statistics apply.

6.2.4 Arbitrary time–bandwidth product case

In most high-speed FOCS using good laser diodes as sources the time–bandwidth product of the optical signal is determined primarily by modulation. This usually results in time–bandwidth products in the range of 0.5 to 2, and neither narrowband nor the wideband approximations apply. The determination of $p(m)$ in these cases is considerably more complicated than in the preceding analysis. We next look into the photon-count statistics for arbitrary time–bandwidth products [4,5].

Let there exist a complete set of functions $\{\Psi_k(t)\}$ defined on the interval $(0,T)$ with the following property:

$$\int_0^T \Psi_j(t)\Psi_k(t)dt = \delta_{jk} = \begin{cases} 0 & if \quad j \neq k \\ 1 & if \quad j = k \end{cases} \tag{6.12}$$

An example is the ordinary Fourier series where

$$\Psi_0(t) = \frac{1}{\sqrt{T}}$$

$$\Psi_{2k}(t) = \sqrt{\frac{2}{T}} \cos\left\{\frac{2\pi kt}{T}\right\}$$

$$\Psi_{2k-1}(t) = \sqrt{\frac{2}{T}} \sin\left\{\frac{2\pi kt}{T}\right\}$$

then

$$x(t) = \sum_{\text{all } k} a_k \Psi_k(t)$$

$$a_j = \int_0^T x(t)\Psi_j(t)dt \tag{6.13}$$

Equation (6.13) can be verified by replacing $x(t)$ by $\sum$all k $a_k \Psi_k(t)$ here [in Eq. (6.13)] and then using Eq. (6.12) to eliminate all but one of the terms. Since $x(t)$ is a random waveform while the $\Psi_k(t)$ are not random, the coefficients a_k must be random variables. Thus the random process may be represented by joint density of the a_k. Furthermore, if $x(t)$ is gaussian, then the a_k are jointly gaussian because the process for determining the a_k is itself a linear process. However, the orthogonality of the Ψ_k does not by itself imply that the various a_j are statistically independent.

Recall that the electric field for thermal radiation is a zero-mean gaussian process. By taking the average of Eq. (6.13), we see that the a_j are zero-mean gaussian random variables. To examine the question of statistical independence of the a_k, let $x(t)$ represent the electric field of thermal radiation. Then

$$E[a_j a_k] = \boldsymbol{E}\left\{\int_0^T \int_0^T x(t)x(v)\Psi_j(t)\Psi_k(v)dt\,dv\right\} \tag{6.14a}$$

where $R_x(\tau)$ is the autocorrelation function of the incident field (shifted down to baseband) or, in other words, with the optical frequency carrier removed. Our task of determining the

$$E[a_j a_k] = \int_0^T \int_0^T R_x(t-v)\Psi_j(t)\Psi_k(v)dt\,dv \qquad (6.14\text{b})$$

photon-count distribution will be greatly simplified if we can find an orthonormal set of $\Psi_k(t)$ that diagonalizes the covariance matrix represented by $E[a_j a_k]$. This can be accomplished for random processes by using the $\Psi_k(t)$ defined by the eigensolutions of the following (Karhunen–Loeve) integral equation:

$$\int_0^T = R_x(t-v)\Psi_k(v)\,dv = \lambda_k\Psi_k(t) \qquad (6.14\text{c})$$

When Eq. (6.14c) is substituted into Eq. (4.14b), we find that

$$E[a_j a_k] = \lambda_k\partial_{jk} \qquad (6.14\text{d})$$

The symmetry properties of the autocorrelation function are such that they cause the eigenvalues λ_k to be real and nonnegative: $0 < \lambda_k$. The joint density of the a_k for thermal radiation is then given by

$$f(\vec{a}) = \Pi_k\left\{\frac{1}{\sqrt{2\pi\lambda_k}}\,\exp\left[-\frac{a_k^2}{2\lambda_k}\right]\right\} \qquad (6.15\text{a})$$

It can also be demonstrated that the joint density of the a coefficients for superposed coherent and thermal radiation are

$$f(\vec{a}) = \Pi_k\left\{\frac{1}{\sqrt{2\pi\lambda_k}}\,\exp\left[-\frac{(a_k-\phi_k)^2}{2\lambda_k}\right]\right\} \qquad (6.15\text{b})$$

where

$$\Phi_k = \int_0^T S(t)\,\Psi_k(t)dt$$

Equations (6.15a) and (6.15b) clearly demonstrate that the eigenvalues λ_k are the variances of the corresponding a_k coefficients. Now let

$$\sqrt{K_0}\,E(t) = \sum_{\text{all }k} a_k\Psi_k(t)$$

Then

$$K_0\int_0^T |E(t')|^2\,dt' = \beta\int_0^T I(t')\,dt' = \sum_{\text{all }k} a_k^2$$

which then yields

$$Q(s) = E\left[\exp\left\{-s\sum_k a_k^2\right\}\right] = \Pi_k E\left[\exp\left\{-sa_k^2\right\}\right]$$

where the last step is due to the statistical independence of the individual a_k. Thus

$$Q(s) = \prod_k Q_k(s) \quad \text{where} \quad Q_k(s) = E[\exp(-sa_k^2)] \quad (6.15c)$$

We now put this in a form that more clearly indicates the physical source of the various rems. Let $\gamma(t)$ be the normalized autocorrelation function for the incident field:

$$\gamma(t) = \frac{R(\tau)}{R(0)} \quad (\text{where } |\gamma(t)| \le 1)$$

$$\int_0^T \gamma(t-v)\, \psi_k(v)\, dv = \chi_k\, \psi_k(t) \quad (6.16a)$$

This does not affect the $\Psi_k(t)$, but it causes

$$\sum_k \chi_k = 1 \quad (6.16b)$$

We also normalize the signal $S(t)$ such that $\int^T S^2(t')\, dt' = 1$, which results in

$$\sum_k \phi_k^2 = 1 \quad (6.16c)$$

Combining Eqs. (6.15b), (6.15c), and (6.16), we obtain the formats for $Q_k(s)$ and $Q(s)$:

$$Q_k(s) = \frac{1}{1 + s\chi_k <m_T>^{m+1}} \exp\left\{\frac{-s\phi_k^2<m_C>}{1 + s\chi_k<m_T>}\right\} \quad (6.17a)$$

$$Q(s) = \prod_k \left\{\frac{1}{1 + s\chi_k<m_T>} \exp\left\{\frac{-s\phi_k^2<m_C>}{1 + s\chi_k<m_T>}\right\}\right\} \quad (6.17b)$$

where $<m_C>$ and $<m_T>$ have the same meaning as before. By differentiating Eq. (6.17b), we obtain

$$E[m] = \sum_k \{\chi_k<m_T> + \Phi_k^2<m_C>\} = <m_T> + <m_C>$$

Now from the structure of the generating function $Q_k(s)$ given by Eq. (6.17a), we see that this is the generating function for narrowband superposed thermal and coherent radiation, and we can write

$$p_k(m) = \frac{(\chi_k<m_T>)^m}{1 + \chi_k<m_T>)^{m+1}} \exp\left\{\frac{-\phi_k^2<m_C>}{1 + \chi_k<m_T>}\right\} L_m\left(\frac{-\phi_k^2<m_C>}{\chi_k<m_T>(1 + \chi_k<m_T>)}\right) \quad (6.17c)$$

Since

$$Q(s) = \prod_k Q_k(s)$$

We can find the overall photon-count distribution by repeated discrete convolution [see Eq. (6.4d)] of Eq. (6.17c) over the index k:

$$p(m) = p_1(m) * p_2(m) * p_3(m) * \cdots$$

Equation (6.17c) was derived using a normalized autocorrelation function, which, in turn, normalizes the sum of the eigenvalues [see Eq. (6.16b)]. Thus, if we arrange the eigenvalues χ_k in descending order ($\chi_1 \geq \chi_2 \geq \chi_3$, etc.), the eigenvalues must become very small as k becomes large. Although the size of the signal components ϕ_k^2 does not coincide with the size of the eigenvalues χ_k, the constraint on the sum of the ϕ_k^2 represented by Eq. (6.16c) implies that there must be a value of k beyond which the ϕ_k^2 become negligibly small. When $\chi_k \approx 0$ and $\phi_k^2 \approx 0$, then

$$p_k(m) \approx \delta_{k0} = \begin{cases} 1 & \text{if } k = 0 \\ 0 & \text{otherwise} \end{cases}$$

This type of distribution has no effect on the discrete convolution process. Therefore an approximate photon-count distribution can be generated on a computer with a modest number of convolutions. For example, the approximate photon-count distributions (within ~1% [5]) shown in Fig. 6.2 were obtained by using 50 discrete convolutions along with a simple truncation correction.

These specific results were obtained for the superposition of coherent radiation with lorentzian-shaped thermal radiation. For the notation in the Fig. 6.2, we have

$$S(\omega) = \frac{A}{\Delta^2 + (\omega - \omega_0)^2} \rightarrow \frac{A}{\Delta^2 + \omega^2} \quad \text{(low-pass version)}$$

and the normalized autocorrelation function is given by

$$\gamma(\tau) = \exp\{-\Delta|\tau|\}$$

All the photon-count distributions depicted in Fig. 6.2 were computed for $\langle m_C \rangle = 40$ and $\langle m_T \rangle = 4$. The time–bandwidth product, however, is unique for each curve and is as denoted in the figure. Note that the narrowband situation is represented by the $\Delta T = 0$ curve and the homogeneous Poisson photon-count distribution (with mean = 44) is represented in Fig. 6.2 by the $\Delta T = \infty$ curve. As ΔT increases, the following trends are apparent:

1. The variance decreases, as indicated by the spread of the distribution from a maximum for the $\Delta T = 0$ case to a minimum as $\Delta T \rightarrow \infty$ (Poisson) case.

2. The photon-count distributions approach the Poisson distribution as $\Delta T \rightarrow \infty$.

Item 2 can be verified by noting that the sum of two homogeneous Poisson random variables is also a Poisson random variable, and that the photon-count distribution for wideband thermal radiation is Poisson-distributed.

Since there is no background radiation to worry about in FOCS, the ther-

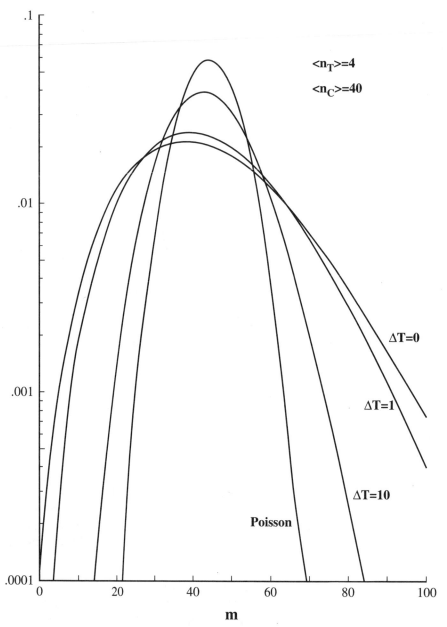

Figure 6.2 Some $p(m)$ for several time–bandwidth products ΔT. (*Reproduced with permission from G. Lachs J. Appl., 1971 ref 5*).

mal portion of the received radiation as represented by $<m_T>$ must come from the transmitter laser diodes. An unmodulated, good single-frequency SLD will have a lorentzian-shaped output spectrum with bandwidth $\Delta \leq 20$ MHz. Now, as an example, let $\Delta = 20$ MHz and assume that this particular SLD is ON/OFF-keyed (OOK) to generate 1-Gbit/s OOK transmission. The pertinent counting time T (baud duration) would be $\sim 10^{-9}$; then we would have $\Delta T = 0.02$ as the time–bandwidth product for the photon-count statistics. Thus the photon statistics for the received radiation would behave like narrowband superposed coherent and thermal radiation. It can be seen from Fig. 6.2 that the photon-count distribution for an intermediate time–bandwidth product such as $\Delta T = 0.5$ is better fit by the photon-count distribution of the narrowband superposition of coherent and chaotic radiation $\Delta T \sim 0$ than by a simple homogeneous Poisson distribution. We note, however, that this narrowband model doesn't incorporate the effects of phase noise because, as we shall see in Chap. 8, phase noise is not additive noise. It does turn out that, because of the absence of background light in FOCS and the simple nature of OOK detection, phase noise does not seriously affect OOK detection.

All the optical fields that we have considered in this section have classic *equivalents,* meaning that the averaging process required to compute $p(m)$ in Eq. (6.3c) can be carried out using ordinary probability densities. One very important characteristic of these fields is that they all have photon-count distributions [see Eqs. (6.8c), (6.9f), and (6.10h)] in which

$$\sigma^2 \geq E[m] \tag{6.17d}$$

These photon-number distributions were obtained by first evaluating the generating function represented by Eq. (6.4e). This was then followed by conversion of the generating function into $p(m)$. The averaging process implied by Eq. (6.4e) employs a probability density for either the optical intensity or the electric field. The averaging in these cases is carried out in a quantum-mechanical format that uses a quantum-mechanically based weighing function known as the *P representation* [6] that violates some conditions that a classic probability density must satisfy. Now all the fields that we discussed earlier in this chapter have P representations that have definite equivalent classic probability densities. It is possible, however, to generate optical fields whose photon-number distributions have a variance that is less than the mean. These quantum-mechanically based fields have no classic equivalents. One particular version of nonclassic fields where Eq. (6.17d) is not satisfied is known as "squeezed state" optical fields [7], in which the variance of the photon-number distribution is less than the mean. In principle, the detection of squeezed-state optical fields generates lower levels of shot noise than does the detection of the classic optical fields that we examined earlier in this section. However, these fields are not of interest for FOCS because the attenuation process incurred during transmission as well as the effects of optical amplification (see Sec. 6.4) convert the states of these fields into states in which the photon-number statistics do satisfy the inequality represented by Eq. (6.17d).

6.2.5 Probability of error (photon counting)

In this section we shall examine the probability of error based on a photon-counting mode of detection. Such a detection scheme is not used in practice because it requires a detection bandwidth ~ 100 times greater than that required for current-based detection. Nevertheless, the following simplified analysis provides a direct example of how a photons-per-bit (ppb) criterion is associated with a specific error rate. Let us begin by considering an ON/OFF-keyed (OOK) binary communication system with the following properties:

1. $p(0) = p(1) = \frac{1}{2}$

2. $p(m \mid 0) = p_0(m) =$ probability of m photoelectron detected when "0" (OFF state) is transmitted

3. $p(m \mid 1) = p_1(m) =$ probability of m photoelectron detected when "1" (ON state) is transmitted

Figure 6.3 displays two typical photocount distributions as well as a possible detection threshold. These photon-count distributions are defined only for integer values of the argument of $p(m)$; as in Fig. 6.2, they are often displayed as continuous curves. On the basis of Fig. 6.3, a logical decision rule for selecting which binary signal was transmitted would be

$$\text{Assume} \begin{cases} 0 & \text{if} \quad m < m_{\text{th}} \\ 1 & \text{if} \quad m \geq m_{\text{th}} \end{cases} \tag{6.18}$$

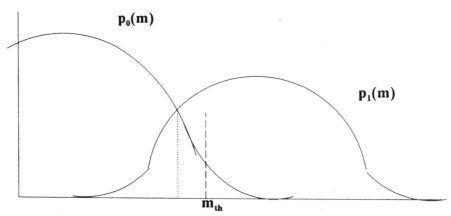

Figure 6.3 Threshold for binary ON/OFF keying.

Detection based on Eq. (6.18) is known as *threshold detection,* and m_{th} is known as the *threshold level.* This integer is usually selected to minimize the probability of error as expressed by

$$P_e = P(1)P(e\,|\,1) + P(0)P(e\,|\,0) \tag{6.19a}$$

where $P(e\,|\,1)$ is the conditional probability of error given a 1 was sent and $P(e\,|\,0)$ is the conditional probability of error when a 0 is sent. Probabilities $P(0)$ and $P(1)$ are the a priori probabilities of sending 0 or 1 at the transmitter. In almost all binary communication systems, these probabilities are equal, which leads to

$$P_e = \tfrac{1}{2}P(e\,|\,1) + \tfrac{1}{2}P(e\,|\,0) \tag{6.19b}$$

The quantities $P(e\,|\,1)$ and $P(e\,|\,0)$ are determined using the decision algorithm defined by Eq. (6.18), and since the equal sign is associated with the 1, we have

$$P(e\,|\,1) = \sum_{m=0}^{m_{th}-1} p_1(m) \tag{6.20a}$$

$$P(e\,|\,0) = \sum_{m=m_t}^{\infty} p_0(m) = 1 - \sum_{m=0}^{m_{th}-1} p_0(m) \tag{6.20b}$$

where the last version of Eq. (6.20b) is used for actual computations (finite summation).

Now, for $P(0) = P(1) = \tfrac{1}{2}$, it turns out that the optimum threshold is the intersection of $p_1(m)$ with $p_0(m)$ as indicated by the dotted line in Fig. 6.3. However, in general, this intersection will not occur at an integer, and to agree with the decision rule expressed by Eq. (6.18), we select m_{th} as the first integer above the intersection of the conditional distributions. Since photon counting is not used for detection, we shall not pursue this line further except for the ideal (but unrealistic) situation where

$$p_0(m) = \delta_{m0} = 1 \qquad \text{only for } m = 0$$

and

$$p_1(m) = \frac{<m_C>^m\, e^{-<m_C>}}{m!}$$

Clearly $m_{th} = 1$ is the logical choice here, and the only way in which an error can occur is for $m = 0$ when a 1 is sent. Thus

$$P_e = \tfrac{1}{2}p_1(0) = \tfrac{1}{2}e^{-<m_C>}$$

Now a commonly used criterion for the error rate in FOCS is

$$P_e = 10^{-9} = \tfrac{1}{2}e^{-<m_C>}$$

which results in $<m_C> = -\ln(2 \times 10^{-9}) = 20.03$.

To obtain the average photons per bit, we note that zero photons are sent for a 0. Thus

$$<\text{Photons/bit}> = \frac{20.03}{2} \approx 10$$

Ten photons per bit is often quoted as the minimum energy requirement to achieve $P_e = 10^{-9}$. However, it is not a true lower bound on the ppb quantity because this so-called ideal detection setup neither allows for the use of modern coding techniques nor considers the possibility of squeezed-state sources. Nevertheless, existing OOK-type FOCS are not yet close to achieving this "optimum" performance level. We shall soon discuss examples of optimum ppb parameters for current-based detection as opposed to photon-counting-based detection.

6.3   Receiver Noise

6.3.1   Introduction

The preceding analysis employed photon-count statistics to obtain such parameters as probability of error and photons per bit. These results are directly applicable only to systems using photon-counting detection systems. However, the pulse-pair resolution required for photon-counting detection in ultra-high-speed FOCS is beyond the state of the art. In virtually every existing FOCS the primary detection variable is obtained from current response to the optical input. In particular, the processing of this current is the basis for both analog and digital detection. We must therefore examine the various noise sources that affect the response current in order to determine such FOCS performance characteristics such as error rate and photons per bit. The following noise sources will be now discussed.

1. Shot noise

2. Dark current noise

3. Thermal noise

4. Avalanche effect gain noise

5. Other forms predetection noise covered later in this book

It is useful to note at this point that the variance of the sum for statistically independent random variables is such that

$$\sigma_{\text{TOT}}^2 = \sum_i \sigma_i^2 \quad \text{and} \quad \sigma_{\text{TOT}} = \sqrt{\sum_i \sigma_i^2} \neq \sum_i \sigma_i$$

6.3.2   Shot noise

Electric current is composed of the inherently random flow of discrete units of charge whose magnitude is equal to the charge of an electron (1.6×10^{-19} C).

Shot noise arises in electronic circuits as a result of the random nature of the number of charges that pass a particular point in a circuit. In optical detection the hole–electron pair generated by photon absorption is the primary source of external current This type of shot noise may be treated as either a homogeneous or nonhomogeneous Poisson process depending on the modulation format employed in the FOCS. We begin this section by examining shot noise for situations in which DSPPs are present, as is the case with actual laser diodes.

The response current for a photodiode can be written as [8]

$$I(t) = \sum_k a_k h(t - t_k) \tag{6.21a}$$

where t_k is the instant that the kth photon is absorbed, a_k is the magnitude of that response (this allows for random gain effects such as those found in APD detectors), and $h(_t)$ is the impulse response of the photodiode to a single detected photon. Photons are indistinguishable from one another, so we cannot pin an identifier on any photon, but we can label the instant that an unidentified photon is observed as t_k [see Eq. (6.21)]. Note again that the subscript on t_k refers only to the observation instant and the subscript k is not a label on any particular photon. These observation instants are random variables, and for most of the fields that we employ in FOCS, they are almost statistically independent. Not all photons incident on the receiver photodiode are observed because the quantum efficiency of a real photodiode is such that $\eta < 1$. The effect of each photoelectron on the detection circuitry is extended in the time-delay direction, (*hereby maintaining causality*) by the impulse response of the receiver electronics. A typical impulse response for a photodiode is

$$h(t) = \exp[-\alpha t]u(t) \tag{6.21b}$$

where $u(t)$ is the unit step function. It has been also shown by Rice [8] that the variance of the output current may be written in the following form:

$$\sigma_I^2 = \frac{\nu <a^2>}{2\alpha} + \left\{ \frac{\nu <a>}{\alpha} \right\}^2 \left[\frac{\alpha q}{\nu(1 - q)} - 1 \right] \tag{6.22a}$$

where ν is the average number of detected photons per second. Then, noting that the charge of the electron is embedded within the magnitude of the response a_k. We can express the average current as

$$<I(t)> = \nu <a> \int_{-\infty}^{\infty} h(t)dt = \nu <a> \int_{0}^{\infty} e^{-\alpha t}\, dt = \frac{\nu <a>}{\alpha} \tag{6.22b}$$

The parameter q that appears in Eq. (6.22a) is defined as

$$q = \int_{0}^{\infty} p(x) \exp[-\alpha x]dx \tag{6.22c}$$

where $p(x)$ is the probability density of the time elapsed between adjacent photon detections. This density can be obtained by taking the second derivative of the generating function $Q(s)$ with respect to the counting time T [9]. For an ideal laser input, the time-interval distribution and the corresponding q are

$$p(x) = v\,e^{-vx}\,u(x) \quad \text{and} \quad q = \frac{v}{v + \alpha}$$

This specific value of q causes the entire second term on the right-hand side (RHS) of Eq. (6.22a) to vanish. Thus the variance of the current response of a photodiode to an ideal laser input is

$$\sigma_{\mathrm{I}}^2 = \frac{v<a^2>}{2\,\alpha} \tag{6.23}$$

The exponentially decaying probability density $p(x)$ for the ideal laser output is also the well-known time-interval density for a homogeneous Poisson process. It represents the physical and statistical situation where the time of occurrence for each photon absorption is statistically independent of the occurrence of all other absorptions. Thus there is no photon bunching effect for the photodiode detection of the output of an ideal laser. A salient characteristic of the Poisson process is that the variance equals the mean. For optical inputs in general, however, the second term on the RHS of Eq. (6.22a) does not vanish. In these cases there is an increased variance that represents the excess shot noise. This excess noise is attributed to what is known as the photon bunching effect, and the effect of this term is called "bunching noise."

While the calculations for $p(x)$ and q are very simple for the ideal laser; the calculations of $p(x)$ and q for superposed thermal and coherent radiation are very complicated, indeed [10]. To begin with, we need to make the following substitutions in Eq. (6.10f) (the generating function for narrowband superposed thermal and coherent radiation):

$$<m_{\mathrm{T}}> = r_{\mathrm{th}}T \quad \text{and} \quad <m_{\mathrm{C}}> = r_{\mathrm{C}}T$$

where r_{th} is the average number of thermal photons per second and r_{C} is the average number of coherent photons per second. The differentiations required to obtain $p(x)$ are straightforward but lengthy. Similarly, the evaluation of q leads to a long, complicated equation. We shall omit these equations and summarize some of the key results with the aid of Fig. 6.4.

In Fig. 6.4 the abscissa represents the average number of thermal photons detected within the decay-time constant of the detector impulse response. Now, this quantity will be fairly large in heterodyne detection because of the strength of the local oscillator. The ordinate is the ratio of the bunching noise to the noise level that would be observed if there were no bunching noise. Thus, when this ratio = 1, there is a 3-dB degradation in system performance due to the bunching effect. The B/G parameter specifying the various curves

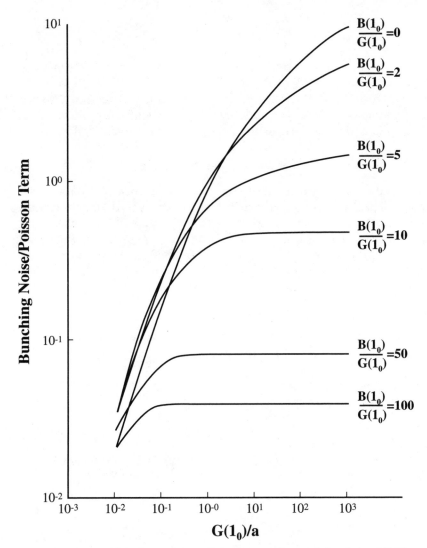

Figure 6.4 Ratio of bunching noise to Poisson term for several values of $B/G = r_C/r_{th} = $ S/N. (*From Ref. 10.*)

is equivalent to $<m_C>/<m_T>$. Depending on this value, the excess bunching noise will be either negligible or significant.[1]

Let us now consider the shot noise generated by the detection of the radiation from an ideal laser with an output power level P_0. Then the average current $\bar{i}$ is

---

[1]The author cannot help wondering whether this effect is partly responsible for RIN in SLD outputs.

$$\bar{i} = I_p = \eta_d e \Phi = \frac{\eta_d P_0 e}{hf}$$

where we denote η_d as the quantum efficiency *of the detector in order to avoid confusion with the noise power spectral density.*

Now, in order to examine noise fluctuations, we define $i_s = i - \bar{i}$, where i_s is a zero-mean random process. It is well known that for shot noise the variance of the fluctuations is linearly proportional to the average current flow in the photodiode:

$$\bar{i}_s = 0 \quad \text{and} \quad \overline{i_s^2} \propto I_p$$

Thus, for a steady-state optical input, the shot noise in a photodiode is a homogeneous Poisson process. Furthermore, the current is the sum of many nearly independent photoelectron, so the total current will (because of the central-limit theorem) be a gaussian random process. One can show that for frequencies within the useable bandwidth of photodiodes, the power spectral density (PSD) of the shot noise is well approximated by a constant v_N (white noise), where

$v_N = 2e(I_p + I_d) = $ noise power per hertz

$I_p = e<m> = e \times$ number of detected photoelectrons per second

$I_d = $ dark current due to current that flows with no light input

Now let the photodiode output current drive a linear filter of bandwidth B. Then the variance of the output current of that filter is

$$\sigma_{i_s^2} = \overline{i_s^2} = 2eB(I_p + I_d) \tag{6.24a}$$

Typically the dark current is very small in an operating FOCS, and for the most part we shall ignore it. Specifically, Eq. (6.24a) represents the shot noise contribution to the post-filter-detection circuitry.

6.3.3 Thermal noise

Thermal noise, which is also known as *Johnson noise* or *Nyquist noise,* is generated by the random motion of conduction band electrons as they pass through a resistor. The number of electrons involved is extremely large, and therefore thermal noise is also a gaussian random process. It has been verified experimentally that the spectral density of thermal noise is flat up to frequencies in the far-infrared (far below visible) region. For a resistor of R ohms (Ω at temperature T kelvins, the current-based noise power density v_N is given by

$$v_N = \frac{2kT}{R}$$

where $k = $ Boltzman's constant (1.38×10^{-23} J/K). This spectral density differs completely from the well-known blackbody spectrum given by

$$S(\omega) = \frac{8\pi h\omega^3}{c^3} \frac{1}{\exp\{hf/kT\} - 1} \tag{6.24b}$$

This equation was obtained by Planck for a three-dimensional blackbody. Thermal noise, however, is generated by only one component of motion of the charges. Specifically, only the velocity component that contributes to the thermal noise points in the same direction as the electric field generating the drift current flow through the resistor. It has been shown that a one-dimensional derivation of blackbody radiation yields a flat spectrum with the v_N given above (see Saleh and Teich [7], p. 682). Now let B be the bandwidth of the receiver electronics. Then, from the equation for v_N, we have

$$\overline{i_s^2} = \frac{4kTB}{R} \tag{6.24c}$$

In systems without predetection gain, thermal noise usually dominates over electronic noise. The principal reason for this is that shot noise is proportional to $(I_p + I_d)$, which is usually very small in the absence of predetection gain since:

1. The dark current I_d tends to be quite small since the diode is reverse-biased. An exception is APD biased near breakdown.

2. The detection current I_p is also very small since the received signal is likely to be weak. If it isn't, then the noise is of no consequence, anyway.

To demonstrate the magnitude of the various receiver-generated noises, consider a p–i–n diode in which $v_d = 0.55$, $\lambda_0 = 1.3$ μm, $I_d = \ln A$, $B = 10^7$, $R = 4.7$ kΩ, $P_0 = 100$ nW, and $T = 300$ K. First we compute the shot noise contribution:

$$I_p = \left\{\frac{\eta_d P_0}{hf}\right\} e = \left\{\frac{\eta_d P_0 \lambda_0}{hc}\right\} e \tag{6.24d}$$

$$= \frac{(1.6 \times 10^{-19})(0.55)(1 \times 10^{-7})(1.3 \times 10^{-6})}{(6.626 \times 10^{-34})(3 \times 10^8)} = 57.6 \text{ nA}$$

$$\overline{i_s^2} = 2eB(I_p + I_d) = 1.87 \times 10^{-19} \text{ A}^2$$

of which less than 2 percent comes from the dark current. We next compute the thermal noise contribution:

$$\overline{i_s^2} = \frac{4(1.38 \times 10^{-23})(300)(10^7)}{4700} = 3.5 \times 10^{-17} \text{ A}^2$$

(*Note:* $\overline{i_s^2} >> \overline{i_s^2}$ by $= 22.7$ dB.) Finally $\overline{i_{TOT}^2} = \overline{i_s^2} + \overline{i_s^2} = 3.5187 \times 10^{-17} \approx \overline{i_s^2}$. Note that

$$\sigma_{\text{TOT}} = 5.93 \text{ nA} \qquad not \qquad \sigma_t + \sigma_s = 6.35 \text{ nA}$$

This last result is not a very significant difference for the numbers in this example, but it could make a greater difference for other parameter values. If we ignore the shot noise entirely, we get $\sigma_t = 5.916$ nA, which is almost equal to the total noise.

A simplified diagram of a p–i–n detector system (including a linear amplifier) is shown in Fig. 6.5. The circuit elements with subscripts A are amplifier input parameters, while C_D is diode capacitance (due primarily to the depletion region). Actually, what we label as R_L is the combined resistance of the load resistance and the diode impedance parameters. Also, we note that usually the amplifier will be a high-impedance MOSFET (metal oxide semiconductor field-effect transistor), so $R_A \gg R_L$. Thus the parallel combination of these two resistances simplifies to

$$R_L \mid\mid R_A \approx R_L$$

Now, although usually $C_D \gg C_A$, we will include the parallel combination as $C_T = C_D + C_A$. Thus the frequency response of a pin photodiode will be that of a low-pass filter with a 3-dB cutoff frequency given by

$$B_{3 \text{ dB}} = \frac{1}{2\pi R_L C_T}$$

In high-speed data applications one of the primary tasks of the amplifier is to match the photodiode output impedance to the transmission line traveling-wave circuitry that is often used in detection circuits for ultra-high-speed digital communication systems.

We note that the electronic amplifier itself also adds noise to the detection process. A parameter that is sometimes employed to characterize this effect is the "noise figure." This is a somewhat vague quantity whose definition sometimes depends on the application. This particular parameter is of most value when similar devices are being compared for a specific particular application, and often runs into trouble when vastly different types of devices and/or systems are being compared. We shall employ the following definition taken from Schwartz [12, p. 227]:

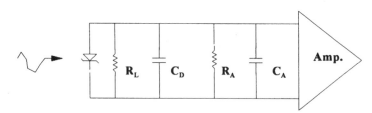

Figure 6.5 Simplified p–i–n detection circuit.

The noise figure Γ [the usual symbol for this parameter is F, but to avoid confusing this with usage for the excess noise figure in this text, we adopt Γ here] *is defined as the ratio of the noise power appearing at the system output to that would appear if the system generated no noise of its own, the only noise power then being due to thermal noise power at the input only.* A noise figure of 2 (3 dB) thus means that the system noise contribution equals that introduced at the input to the system.

In general, this means that

$$\overline{i^2_{\text{TOT}}} = \overline{i_{\text{t}}^2} + \overline{i^2_{\text{amp}}} = \left(\frac{4kTB}{R_{\text{L}}}\right)\Gamma$$

We shall soon see that predetection gain adds significant amounts of shot noise to the mix. In preparation for our presentation on the avalanche photodiode and heterodyne detection systems, we add the shot noise back into the total noise. The overall signal-to-noise ratio (SNR) for a p–i–n is then given by

$$\text{SNR} \doteq \frac{I_{\text{p}}^2}{i^2_{\text{TOT}}} = \frac{I_{\text{p}}^2}{2eB(I_{\text{p}} + I_{\text{d}}) + (4kTB\Gamma/R_{\text{L}})} \tag{6.25}$$

where we ignore the electronic amplifier gain since it affects equally both the numerator and the denominator. We assume that the 3-dB cutoff frequency of the amplifier electronics is larger than the channel bandwidth of the received signal Then the parameter B in Eq. (6.25) is the channel bandwidth and not the detection circuit 3-dB bandwidth where

$$B < B_{3\,\text{dB}} = \frac{1}{2\pi R_{\text{L}} C_{\text{T}}}$$

This places a maximum on the value that R_{L} can take on and still pass the signal with tolerable distortion. A further reduction in bandwidth is frequently implemented in the digital detection and decision circuitry that follows the amplifier. A simplified view of this is presented in Fig. 6.6.

Thus we have that max $R_{\text{L}} = 1/2\pi C_{T}B$ for a baseband-type signal after detection by a photodiode. This sets an upper bound on R_{L}, which, in turn, sets a lower limit on the thermal noise component of the total noise. To obtain ultrafast responses at the receiver, it is necessary to design p–i–n photodiodes with very small values of C_{d}. There are commercial photodiodes available

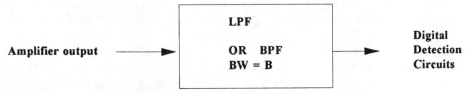

Figure 6.6 Flowchart of digital detection and decision circuitry.

today with baseband responses at 50 GHz. These photodiodes often employ small termination impedances, say, $R = 50\ \Omega$, to match the characteristic impedances of the coaxial cables employed to interconnect these devices with the remainder of the receiver. One consequence of this low impedance is that it yields a relatively large thermal noise power. It turns out, however, that these particular photodiodes are often used in detection systems with predetection gain in which the effects of thermal noise are frequently negated.

6.3.4 Avalanche photodiode noise considerations

The avalanche effect amplifies the photon-induced current and the internal dark current, but it does not amplify the thermal noise. There is also a surface component of the dark current that travels along the outer surface of the diode structure that is not amplified. It is usually very small in well-designed devices, and we shall ignore it.

The optically induced output current for an APD is given by $i = \eta e\Phi\overline{G} = I_p\,\overline{G}$, and, including dark current, we get $i = (I_p + I_d)\overline{G}$. Recall from probability theory that if $Y = aX$, then $\sigma_Y^2 = a^2\sigma_x^2$. This implies that the amplified shot noise would be given by

$$\overline{i_{SA}^2} = 2\,eB(I_p + I_d)\overline{G}^2$$

However, this equation ignores the randomness of the avalanche gain process itself. With random gain, we get

$$\overline{i_{SA}^2} = 2eB(I_p + I_d)\,\overline{G^2}F$$

where the excess noise factor F is defined as

$$F = \frac{E[\overline{G^2}]}{(\overline{G})^2} = 1 + \frac{\sigma_G^2}{(\overline{G})^2} \tag{6.26a}$$

where $\sigma_G^2 = \text{variance}(G)$ and $E[\overline{G^2}] = \sigma_G^2 + (\overline{G})^2$. Therefore, we find that $F \geq 1$. Note that when the amplification process is not noisy, we find that $\sigma_G^2 = 0$ and $F = 1$. The excess noise arises from two factors. One is the randomness of the location of an initial hole–electron pair within the avalanche region, and the other has to do with excess randomness incurred by the presence of two types of charge carriers. The avalanche chain tends to reverberate when both electrons and holes contribute significantly to the avalanche process. If, on the other hand, only one type of charge carrier is responsible for the avalanche gain, then this second form of randomness does not occur.

Let α_e and α_h be the impact ionization coefficients for electrons and holes, respectively. Now we define the ionization ratio for these quantities as $\kappa = (\alpha_h/\alpha_e)$. Then we can show that for the case where only electrons are injected initially at the boundary of the avalanche gain region, the excess noise factor is

$$F = \kappa \overline{G} + (1 - \kappa)\left[2 - \frac{1}{\overline{G}} \right] \qquad (6.26b)$$

Plots of Eq. (6.26b) for various values of κ are shown in Fig. 6.7 for the same case that led to Eq. (6.26b), specifically, the case where only electrons injected initially at the boundary of the avalanche region give rise to the avalanche mechanism. Note also that the APD depicted in Fig. 5.8 is such a device.

Now Eq. (6.26b) is somewhat complicated, and we shall use a simplified approximation to the excess noise factor. A simpler relation is suggested by noting the nearly linear variation of $\log(F)$ versus $\log(\overline{G})$ displayed in Fig. 6.7 for large $\overline{G}$:

$$\log(F) \approx x \log(\overline{G})$$

Thus

$$F \approx \exp\left[\log(\overline{G}^x)\right] \Rightarrow F \approx \overline{G}^x$$

where x is a parameter that lies in between 0.7 and 1.0 for GaAs devices and between 0.3 and 0.5 for Si APDs). This factor results in the following approximation:

$$\overline{i^2_{\text{SA}}} \approx 2eB(I_p + I_d)\,\overline{G}^{2+x} \qquad (6.27)$$

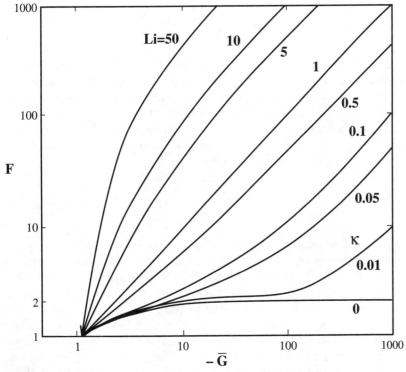

Figure 6.7 Excess noise factor F versus average current gain. (*Reproduced with permission from John Wiley & Sons, New York. See Ref. 7.*)

Then, using Eq. (6.25) as a guide, we find that the overall SNR for an APD then becomes

$$\text{SNR} \approx \frac{\overline{G}^2 I_p^2}{2eB(I_p + I_d)\overline{G}^{2+x} + 4kTB\Gamma/R_L} \tag{6.28}$$

Equation (6.28) shows why thermal noise may be neglected in avalanche-diode-based detection systems when $\overline{G} \gg 1$. It can be seen from this equation that a maximum SNR exists by noting that the denominator exponent of $\overline{G}$ is larger than the numerator power of $\overline{G}$. Thus as $\overline{G} \to \infty$ SNR $\to 0$. Clearly SNR $= 0$ when $\overline{G} = 0$. The optimum $\overline{G}$ can be found by finding the value of $\overline{G}$, which causes

$$\frac{d(\text{SNR})}{d\overline{G}} = 0 \quad \text{which yields} \quad G_{\text{opt}}^{2+x} = \frac{4kT\Gamma}{xeR_L(I_p + I_d)}$$

Just a comment here on coherent detection (CD) systems. In these systems there also exists an equivalent to the $\overline{G}^2$ factor due to the conversion gain. When $\overline{G}$ is large enough, the thermal noise can be ignored. It turns out, however, that the excess factor $\overline{G}^x$ does not arise in CD systems, and this can lead to a significant advantage of CD over APD detection. This is especially so since

$$\lim_{\overline{G} \to \infty} \text{SNR} \neq 0 \quad \text{when} \quad x \to 0$$

The parameter κ which appears in Eq. (6.26b) for GaAs lies between 0.3 and 1.0 and leads to a value of $x \approx 1$. For silicon APDs, κ is much smaller and leads to a much lower value for the excess noise, thus making silicon a superior material for APD. Unfortunately, silicon does not respond at either 1.3 or 1.55 μm and therefore cannot be used at those wavelengths. These properties do explain however why Si APD detectors are often employed where the transmission wavelength is appropriate.

6.4 Noise in Optical Amplifiers

6.4.1 Introduction

In Chap. 4 we described some of the operational aspects of optical amplifiers. The focus in this section is on the noise characteristics of these devices with a major interest in the effects of optical amplifiers on the signal-to-noise ratio (SNR). The material presented in this section is a relatively brief treatment of a very complex subject. Students interested in pursuing this topic in greater depth will find an extremely comprehensive treatment in Desurvire's book [13].

The three types of predetection gain used in fiber-optic communication systems are avalanche photodiode receivers, conversion gain in optical heterodyne detection (OHD), and the optical amplifier. Typically only one of these detection systems would be used in a particular system, and each of these processes contributes noise to the final detection statistics and affects the performance of that system. The noise contributed by the APD is reflected in the excess noise factor F discussed in the previous section. Phase noise introduced by the local SLD at the receiver in OHD as well as the phase noise from the transmitter laser can significantly limit OHD performance. This topic is discussed at length in Chap. 8. We next consider the noise process in optical amplifiers.

Photon emission and absorption events occur at random locations and random times within a traveling-wave optical amplifier (TWOA) and are responsible for the basic random nature of these processes. Simulated emission, absorption and, spontaneous emission can be modeled as birth, death, and immigration processes, respectively, in terms used in mathematical models for the ecology. We begin our coverage of noise in optical amplifiers by noting the primary sources of randomness in the TWOA:

1. Stimulated emission events are birth processes in which photons in the TWOA give rise to more photons. One important aspect of the randomness here is that it affects the amplification initiated by an incident photon. Furthermore, this process may be complicated in high-gain amplifiers by the photon flux density affecting the population inversion near the output end of the TWOA.

2. Clearly absorption is a death process to photons. Absorption can also affect the population inversion in a random manner.

3. Spontaneous emission is an immigration process and is not affected directly by the photon flux and thus has no connection with "parents"; hence the term *immigration*. Spontaneous emission, however, is indirectly dependent on the photon flux by the effects of stimulated emission and absorption on the populations of both the upper and lower states within the TWOA.

6.4.2   Derivation of master equation for quantum version of OA noise

We begin this study by considering the variation of the optical intensity as denoted by the photon flux density, as it passes through a thin slab of width Δz (see Fig. 4.4) at a distance z from the entrance to the amplifier. Now, because the slab is so thin, we assume that there is at most a change of one photon in the beam as it passes through this infinitesimal slab. We are going to derive formulas for the mean and variance of the photon density at the point z, which we shall employ to determine the signal-to-noise ratio. The primary variable here is P_n, which we define as the probability of n photons within the Δz slab located at an arbitrary point within the amplifier.[2]

---

[2]Note that for this portion of the text we have changed our notation to P_n in place of $p(n)$.

We first derive the primary equation for P_n that we shall employ to determine the statistical properties of the optical amplifier output. Since only one photon transition occurs in Δz, we need consider only the n, $n + 1$, and $n - 1$ states of the optical field in the slab. There are two transitions that decrease the probability of finding n photons inside in the slab (and hence decrease P_n), and two transitions that will increase P_n:

1. P_n is decreased by either an emission event for a field in a state with n photons (hence changing to field state with $n + 1$ photons), or an absorption event which changes the field to an $n - 1$ photon abate.

2. P_n is increased by either an emission event for a field in a state with $n-1$ photons (hence changing to field state with n photons) or an absorption event for a field in the $n + 1$ state which drops the field to a state with n photons.

Recognition of these processes leads to the following "master" equation for the derivative of P_n with respect to the distance z from the input end:

$$\frac{dP_n}{dz} = a(z)[nP_{n-1} - (n + 1)P_n] + b(z)[(n + 1)P_{n + 1} - nP_n] \qquad (6.29a)$$

where $a(z) = \alpha_e N_2(z)$ and $b(z) = \alpha_a N_1(z)$, and where α_e and α_a are the emission and absorption coefficients (cross sections), respectively.[3]

Note that $a(z)$ is proportional to the upper-state population density while $b(z)$ is proportional to the lower-state population density. It turns out that the cross sections for absorption (α_a) and emission (α_e) for Er^{3+} are not equal. We note that both terms in the bracket multiplied by $a(z)$ are emission terms while those that in the brackets, multiplied by $b(z)$ are absorption terms. Recall that emission for a field in the n photon state is proportional to $n + 1$ as a result of the sum of the contributions from both stimulated and spontaneous emission while the absorption terms are proportional simply to n. This system is an example of a Markov random process, and Eq. (6.29a) is a type of equation known in the literature as a *forward Kolmogorov equation* [13,14]. We shall soon see that we do not need to solve the master equation [Eq. (6.29a)] for P_n explicitly in order to determine the basic quantities required to determine the SNR, namely, $<n(z)>$ and $\sigma^2(z)$.

6.4.3   Solution to master equation

We now compute the rate of change of the mean of the photon-number distribution $<n>$ as a function of distance from the entrance z. This may be obtained by evaluating

$$\sum_n n \frac{dP_n}{dz} = \frac{d<n>}{dz} =$$

---

[3]The z dependence in Eq. (6.29a) is contained in the $a(z)$ and $b(z)$ parameters.

$$\sum_n \left\{ a(z)[n^2 P_{n-1} - n(n+1)P_n] + b(z)[n(n+1)P_{n+1} - n^2 P_n] \right\} \qquad (6.29b)$$

It is shown in App. 6A that this equation leads to

$$\frac{d<n>}{dz} = a(z)[<n> + 1] - b(z) <n> \qquad (6.29c)$$

whose solution is given by

$$<n(z)> = G(z)<n(0)> + N(z) \qquad (6.29d)$$

where

$$G(z) = \exp\left\{ \int_0^z [a(z') - b(z')]dz' \right\} \qquad (6.29e)$$

and

$$N(z) = G(z) \int_0^z \frac{a(z')}{G(z')} \, dz' \qquad (6.29f)$$

One very important conclusion that can be drawn from Eq. (6.29d) is that when there is no input signal at all (i.e., $<n(0)> = 0$), then $<n(z)> = N(z)$. Now this result, along with the structure of Eq. (6.29f), demonstrates that the quantity $N(z)$ represents amplified spontaneous emission (ASE). Now for the special case where both a and b are independent of z, we find that

$$G(z) \doteq G = \exp\{(a - b)z\} \qquad \text{and} \qquad N(z) = a \exp\{(a - b)z\} \int_0^z \exp\{-(a - b)z\}$$

which yields

$$N(z) = \frac{a}{a - b} G\left[1 - \frac{1}{G} \right] = \frac{a}{a - b} (G - 1)$$

Now let $N_{ASE} \doteq$ noise total power in a bandwidth B due to the ASE. We may express N_{ASE} in the following form:

$$N_{ASE} = n_{sp} B \, h\nu \, (G - 1) \qquad (6.29g)$$

where $n_{sp} = a/(a - b)$.

For the general case where both a and b depend on z, namely, $a(z)$ and $b(z)$, then n_{sp} can be expressed in the following form [13]:

$$n_{sp} = \frac{N(z)}{G(z) - 1} = \frac{G(z)}{G(z) - 1} \int_0^z \frac{a(z')}{G(z')} \, dz' \qquad (6.29h)$$

When $G \gg 1$, the ASE power $N(z)$ can be expressed simply as the amplification of n_{sp} photons at the input to the TWOA: $N(z) \approx n_{sp} G(z)$. However, the parameter n_{sp} turns out to be awkward for small-gain amplifiers. In particular, it will become negative below the transparency condition where $G < 1$. Another parameter that equals n_{sp} when $G \gg 1$, but remains positive for small-gain amplifiers, is the term n_{eq}, defined as follows:

$$n_{eq}(z) \doteq \frac{N(z)}{G(z)} = \frac{n_{sp}(z)(G(z) - 1)}{G(z)} = \int_0^z \frac{a(z')}{G(z')} dz' \qquad (6.29i)$$

This parameter can be used to define an equivalent input noise power $2 n_{eq}$. Some values for the equivalent input power are plotted in Fig. 6.8.

6.4.4 Evaluation of s2(z) for TWOA

We shall compute the variance indirectly by first finding C_2 and then using it in Eq. (6.5a) along with the solution for the mean represented by Eq. (6.29d). It turns out that in this case it is easier to determine C_2 using master equation Eq. (6.29a) to determine the second factorial moment than $<n^2(z)>$. These calculations are carried out in App. 6A to yield

$$C_2(z) = G^2(z)[<n^2(0)> - <n(0)>] + 4<n(0)> G(z)N(z) + 2N^2(z) \quad (6.30a)$$

Then, from Eqs. (6.5a), (6.29d), and (6.30a), we obtain

$$\sigma^2(z) = G^2(z)[\sigma_{in}^2(0) - <n(0)>]$$

$$+ 2<n(0)>G(z)N(z) + N^2(z) \qquad (6.30b)$$

$$+ G(z)<n(0)> + N(z)$$

where $\sigma_{in}^2(0)$ is the variance of the photon distribution for the input to the TWOA.

The last two terms on the RHS of Eq. (6.30b) are equivalent to a single mode (or, equivalently, narrowband radiation) of superposed coherent and chaotic radiation. This correspondence may be observed in the term-for-term comparison between Eqs. (6.10h) and (6.30b) as shown in Eq. (6.30c):

$$G(z)<n(0)> + N(z) + N^2(z) + 2 G(z)<n(0)>N(z)$$

$$<m_C> + <m_T> + <m_T>^2 + 2 <m_C><m_T> \qquad (6.30c)$$

The direct comparisons here are based on $<m_C> \doteq G(z)<n(0)>$ and the amplified mean of the input signal, and $<m_T> \doteq N(z)$ (the amplified spontaneous emission). We note that Eq. (6.29f) shows that the amplification $G(z)$ is embedded within the definition of $N(z)$.

The first term on the RHS of Eq. (6.30b) represents the contribution of the amplified "excess" photon fluctuations of the optical input to the TWOA. We found in Sec. 6.2 that excess variance leads to a phenomenon called "photon bunching." This excess noise will disappear from $\sigma^2(z)$ when the optical input to the TWOA has a simple homogeneous Poisson photon distribution in which $\sigma_n^2 = <n>$. Then the amplifier output may be considered as a single mode of superposed coherent and chaotic radiation. If there is excess variance at the

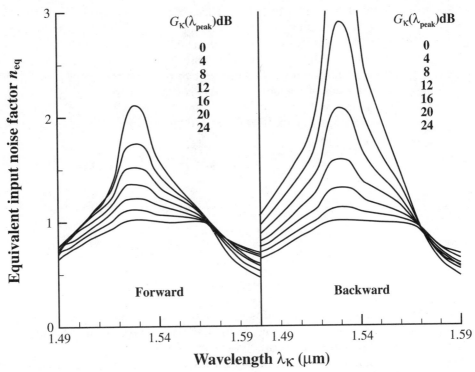

Figure 6.8 Optical amplifier equivalent noise factor spectrum. EDFA pumped at $\lambda_0 = 0.980$ μm. [*Reproduced by permission by John Wiley & Sons, from Ref. 13, Fig. 2.10 (p. 107).*]

input to the TWOA, the output statistics will differ somewhat from that of a single mode of coherent and chaotic radiation. If, on the other hand, the input signal has a variance less than the mean (squeezed state), then the net variance will be less than would be obtained with the semiclassic minimum variance of a Poisson distributed input. However, even when the TWOA input is in a squeezed state, the amplifier output will definitely not be in a squeezed state because of the added variance represented by the terms on the second line of Eq. (6.30b).

Another reason why squeezed-state signal sources are not especially useful for FOCS is the fact that the process of attenuation encountered by the optical field during propagation through an optical fiber channel will raise the variance very quickly toward the Poisson-based semiclassic minimum. This effect may be visualized by considering the situation in which a stream of photons consists of equally spaced photons. The variance of this artificial setup is zero. Now let photon stream pass through a screen wherein each incident photon has a probability p of passing through and a probability $q = 1-p$ that it is absorbed by the screen. If this event is statistically independent for each photon, then the photon-count statistics on the far side of the screen will be described by a binomial distribution and $\sigma^2 = Npq > 0$. If the number of incident photons per second is very large (as is the case in FOCS), then the binomial is accurately approximated by a simple Poisson distribution and thus the attenuated optical field will longer be in a squeezed state.

The preceding results were obtained for a single mode, but even a single-mode erbium-doped TWOA has two possible modes of propagation due to the two degrees of freedom for polarization. The signal might propagate as a single mode, but the spontaneous emission will go equally into both of these polarization modes. Equations (6.29d) and (6.30b) can be generalized for l modes to [14]

$$<n(z)> = G(z)<n(0)> + lN(z) \qquad (6.30d)$$

$$\sigma^2(z) = G^2(z)[\sigma_{in}^2(0)-<n(0)>]$$

$$+ 2<n(0)>G(z)N(z) + lN^2(z) \qquad (6.30e)$$

$$+ G(z)<n(0)> + lN(z)$$

The number of modes parameter l multiplies the $N(z)$ contribution to both the mean and variance. It also affects the $N^2(z)$ contribution to the variance. In the beat noise terms on the second line of Eq. (6.30e), l does not affect the beating between the ASE and the amplified signal at all. The reason for this is that there is no beating between orthogonal modes such as the two polarization modes and the only contribution to this term is beating between the signal and noise in the mode containing the signal.

In a similar vein, we note the mode number l in the $N^2(z)$ term rather than l^2. The $lN^2(z)$ term represents the sum of the ASE contributions from l noninterfering modes.

6.4.5 Optical noise figure

One way to measure the performance of a TWOA is to use the noise figure of the device to compare the signal-to-noise ratios of the output and input of the TWOA. In line with the standard definition of the signal-to-noise ratio as used in Eq. (6.25), that is, that the SNR is the signal portion of the mean-squared, divided by the variance, we write

$$\text{SNR}\,(z) \doteq \frac{(G(z)<n(0)>)^2}{\sigma^2(z)} \qquad (6.31a)$$

Then, by utilizing Eq. (6.30e), we arrive at

$$\frac{1}{SNR(z)} = \frac{1}{<n(0)>}\left[\frac{\sigma^2(0)}{<n(0)>} -1 + \frac{1 + 2N(z)}{G(z)} + \frac{lN(z)[N(z) + 1]}{G^2(z)<n(0)>} \right] \qquad (6.31b)$$

The optical noise figure at a distance z from the entrance to the TWOA is defined as the ratio of the SNR at the input to the TWOA to the SNR at z. Thus we have

$$F_o(z) \doteq \frac{\text{SNR}(0)}{\text{SNR}(z)} \qquad (6.31c)$$

and from the preceding two equations, we get

$$F_o(z) = \frac{\text{SNR}(0)}{<n(0)>}\left[\frac{\sigma^2(0)}{<n(0)>} - 1 + \frac{1 + 2N(z)}{G(z)} + \frac{\ell N(z)[N(z) + 1]}{G^2(z)<n(0)>}\right] \quad (6.31d)$$

The subscript on the noise figure $F_o(z)$ signifies that $F_o(z)$ is the noise figure for the optical field. It does not as yet include the electronic noise aspects of the detection process. We note here that the noise figure is always ≥ 1 since amplification by itself cannot improve the SNR of the optical field. Soon we shall demonstrate that a large-gain TWOA has a minimum noise figure of 2 or equivalently 3 dB. Let us now simplify Eq. (6.31d) by assuming that the photon statistics of the input signal are described by a homogeneous Poisson distribution which leads to $<n(0)> = \sigma^2(0)$.

1. The first term inside the main brackets of the equation is unity and cancels out the second term inside the brackets.

2. $\text{SNR}(0) = \dfrac{<n(0)>^2}{\sigma^2(0)} = <n(0)>.$

Thus, for a homogeneous Poisson input to the TWOA, we have

$$F_o(z) = \frac{1 + 2N(z)}{G(z)} + \frac{\ell N^2(z)}{G^2(z)<n(0)>} + \frac{\ell N(z)}{G^2(z)<n(0)>} \qquad (6.31e)$$

Now, if the gain $G(z)$ is very large, namely, $G(z) \gg 1$, and $<n(0)>$ is comparable to, or larger than l, then Eq. (6.31e), along with a comparison between Eqs. (6.29e) and (6.29f), leads to the following simple approximation to the optical noise figure:

$$F_o(z) \approx \frac{1 + 2N(z)}{G(z)} \approx \frac{1 + 2n_{sp}(z)[G(z) - 1]}{G(z)} \approx 2n_{sp} \qquad (6.31f)$$

Equation (6.29h) provided n_{sp} to Eq. (6.31f). Now we note that when $G(z) \gg 1$ and we also have full population inversion (i.e., all erbium atoms in the upper state), we can show that $n_{sp} \approx 1$ [13, p. 77]. Thus the minimum. noise figure for a high-gain TWOA is 2 (or 3 dB). However, we note that noise figures of $<$ 2 can be obtained at low amplification levels, but these smaller noise figures are obtained at the expense of incurring smaller amplification. Optical amplifiers commonly employed in FOCS are high-gain amplifiers (typically ~40 dB

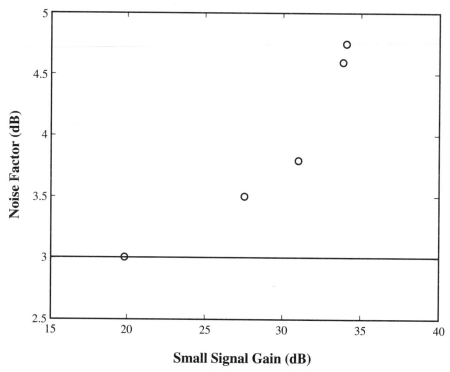

Figure 6.9   Noise figure for small-signal input with copropagating pump (54 mW) at 0.98 μm. Each point corresponds to an EDFA of 7m, 12m, 17m, 25m, and 30m in order of increasing gain. (*Adapted from Ref. 15.*)

or more), and the noise figure for real TWOAs can be significantly larger than 3 dB. This can be seen from Fig. 6.9, where some experimental data are presented demonstrating these basic trends for the noise figure.

All the data depicted in Fig. 6.9 were obtained for the same pumping power (54 mW), but each EDFA that was tested had a different length and therefore a different overall gain. The shortest EDFA amplifier tested produced the 3-dB noise figure, but it also had the smallest gain (20 dB). The reason for this is the strong pumping applied to this particular EDFA-achieved total population inversion, and as we have already seen, this leads to a 3-dB noise figure. As the length of the EDFA increases, the gain increases, but the fixed pump power gets absorbed as it moves through the TWOA. Eventually, as the amplifier length increases, a point is reached where the fixed pump no longer provides sufficient pump energy to achieve total population inversion. This, in turn, causes $n_{sp} > 1$ along with a larger noise figure. All these trends are clearly manifested by the data in Fig. 6.9.

Some experimental data for noise figure versus pump power is displayed in Fig. 6.10. All the data were obtained for a single EDFA. For this particular

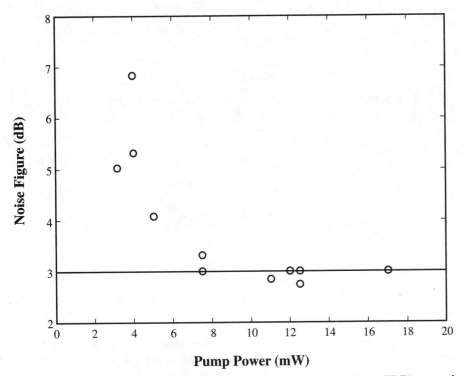

Figure 6.10 Some measured values of noise figure versus pump power for an EDFA pumped at 0.980 μm. (*Adopted from Ref. 16.*)

EDFA, it is clear that as pump power increases, the noise figure approaches 3 dB while as lower pump power levels, the noise figure is considerably greater than 3 dB. In both Figs. 6.9 and 6.10 the EDFA was pumped at 0.980 μm. Note that both figures produced data points with noise figure considerably higher than 3 dB. It is also important to note that even a high-gain-environment EDFA can generate noise figures larger than 3 dB. To arrive at a 3-dB noise figure, it is necessary to have both large gain and total population inversion for the entire length of the amplifier.

6.4.6 Signal-to-noise ratio for EDFA preamplifier-photodiode combination

In this subsection we examine the SNR for detection by a system consisting of a preamplifier TWOA followed by a *p–i–n* photodiode. In a manner similar to what we have done in previous sections of this chapter, we define

$$\text{SNR}_e \doteq \frac{I_{\text{sig}_e}^2}{\sigma_e^2}$$

where the subscript e indicates that these are post-electronic-detection quantities. In particular, if the input is 100 percent modulated, as in the case of ON/OFF-keyed digital transmission, then

$$I_{sig_e} = \eta G I_s$$

where I_s is optical power of signal portion of the input to the TWOA and η is the quantum efficiency of the photodiode. It can be shown [App. K of Ref. 13] that

$$\text{SNR}_e = \qquad (6.32a)$$

$$\frac{[\eta G I_s]^2}{2\eta B_e(G I_s + \ell I_N) + 2\eta^2 G I_s I_N\left(\dfrac{2B_e}{B_o}\right) + \ell\eta^2 I_N^2\left(\dfrac{2B_e}{B_o^2}\right)\left[B_o - \left(\dfrac{B_e}{2}\right)\right] + \left(\dfrac{4K_B T B_e}{R}\right)}$$

where B_e is the bandwidth of the photodiode and B_o is the bandwidth of the optical input.

Let us now consider the four primary terms in the denominator of Eq. (6.32a):

1. The first term generates the usual shot noise, where $G I_s$ is the bias current generated by the light signal and the contribution of the l orthogonal modes of amplified spontaneous emission (ASE).

2. The second term represents the beat noise contribution of the amplified signal mixing with the ASE.

3. The third term represents the self-mixing of l modes of ASE.

4. The last term is the usual thermal noise term.

Now, dividing the numerator and denominator of Eq. (6.32b) by $(\eta G)^2$, we arrive at

$$\text{SNR}_e = \qquad (6.33)$$

$$\frac{P_s^2}{\left(\dfrac{h\nu}{\eta G}\right)2B_e\left(P_s + \ell P_o\right) + 4P_s P_o\left(\dfrac{B_e}{B_o}\right) + \text{l}P_o^2\left(\dfrac{2B_e}{B_o^2}\right)\left[\dfrac{B_o - B_e}{2}\right] + \left(\dfrac{4k_B T B_e}{R}\right)\left[\dfrac{h\nu}{e\eta G}\right]^2}$$

where $I_s = eP_s/h\nu$, $I_N = eN/h\nu = eGP_o/h\nu$, and $P_o = n_{eq}h\nu B_o$.

We again associate the first term in the denominator with shot noise; the second term is an excess noise term generated by the beating of signal with noise, the third term is also an excess noise generated by the self-beating of the ASE, and the last term represents receiver-generated thermal noise. As we noted in Sec. 6.3, this excess noise leads to photon bunching, and a concomitant increase is the total shot noise. A typical value for R is 50 Ω, which, in the absence of predetection gain, generates enough thermal noise at room temperature for the thermal noise term to dominate the receiver noise. However, thermal noise is the only term in the denominator of Eq. (6.32) that is not amplified by the optical preamplifier (OP). Thus, for sufficiently large OP gain (i.e., $G \gg 1$), the thermal noise term becomes insignificant and we shall not consider it further. In addition, the first and last terms of the denominator become negligible because of the G factors in the denominators of those terms. We now further assume that $P_s \gg P_o$. This assumption is supported by the definition of Ps_o *given above, and by Eq. (6.29i), which shows that* n_{eq} is a relatively small term because the $G(z)$ factor has been removed. Note also that P_s is the amplified output signal power from the TWOA and retains the gain factor. Now, by employing these assumptions, we are able to obtain the following simpler formulation for SNR_d for the electronic response of p–i–n photodiode preceded by a large-gain optical preamplifier:

$$\text{SNR}_d^{\text{opt}} \approx \frac{P_s}{4P_o(B_e/B_o)} = \frac{P_s}{4\,n_{eq}h\nu B_e} \tag{6.34}$$

The superscript denotes that this is the maximum SNR_d obtainable from the OP/p–i–n diode combination. In this large-gain configuration, the dominant noise term is generated by the beating between the signal and the ASE. We shall use this equation in Chap. 7 to drive a value for the minimum member of photons per bit required to satisfy a system error rate specification.

6.4.7 Summary

The optical preamplifier is one of three predetection gain systems commonly used in modern fiber-optic communication systems (FOCS). The other two are the amplification obtained with avalanche photodiodes (described in Sec. 6.3) and conversion gain, encountered in optical heterodyne detection (to be discussed in Chap. 8). The primary advantage of predetection amplification for all of these detection schemes is that they essentially negate the effects of the receiver-generated electronic noise. This is accomplished by overwhelming the receiver-generated electronic noise with predetection-amplified noise terms. In particular, they override the effects of receiver thermal noise which would otherwise dominate the noise portion of the detection process. The reader will note that this was precisely what transpired in the preceding development in traveling-wave preamplifiers (TWOA). We also encountered this process when we studied the noise in avalanche photodiode (APD) detection systems. One feature that is common to all predetection amplification receivers is a very large gain: $G \gg 1$. Although this process resulted in a much larger amount of detection noise than we would find from the resident thermal noise, the accompanying increase in signal power due to signal amplification more than compensated for the increased noise. There is, however, a decrease in the signal-to-noise ratio (SNR) for all predetection systems. In the optical preamplifier (OP) case the SNR was diminished by at least 3 dB as stated in Eq. (6.31f), where the minimum noise figure was achieved for large gain *and complete population inversion*. We have seen that this noise figure may be considerably larger than 3 dB for nonideal conditions. In any case, the OP noise figure is usually much lower than the APD excess noise factor $F = G^x$ encountered earlier for APDs. It is worthwhile to note here that while the excess noise factor is proportional to amplification factor for APD detection, this is not the case for OP. The data exhibited in Fig. 6.9 would seem to contradict this statement, but we must remember that the increase in gain for that figure was obtained by lengthening the amplifier while maintaining the pump power. If there had been a concomitant increase in pump power so as to retain total population inversion throughout the amplifier, the 3-dB noise factor would have been observed. We shall see in Chap. 8 that there is a minimum loss of 3 dB in the optical heterodyne detection receiver.

Problems

1 In class we came up with the following photocount distribution:

$$p(m) = \frac{<m_\mathrm{T}>^m}{(1 + <m_\mathrm{T}>)^{m+1}} \exp\left\{-\frac{<m_\mathrm{C}>}{1 + <m_\mathrm{T}>}\right\} L_m\left(\frac{-<m_\mathrm{C}>}{<m_\mathrm{T}>(1 + <m_\mathrm{T}>)}\right)$$

(**a**) How would you measure $<m_\mathrm{T}>$ and $<m_\mathrm{C}>$ in a non-fiber-optic open-space communication system?

(**b**) What difficulties do you encounter in determining these parameters when semiconductor diodes are employed in fiber-optic communication systems?

(**c**) Under what conditions is the preceding $p(m)$ value applicable?

2 Show that the mean of narrowband thermal radiation is expressed by the formulas given in the text.

3 In our study of the photon-count distribution for narrowband radiation, we encountered a clear example that photons do not behave like pellets—that is to say, there are interference effects.

(**a**) Identify this example.

(**b**) Indicate two ways in which this effect is manifested.

4 (**a**) Find the receiver signal-to-noise ratio for a p–i–n receiver operating at 300 K where received power = 100 nW, $\lambda_0 = 1.55$ μm, $\eta = 0.6$, $I_d = 1$ nA, $R = 4.7$ kΩ, $B = 1$ GHz, and noise figure = 2.

Answer: SNR = 0.78.

(**b**) Note the relative contributions of dark current, shot noise, and thermal noise.

5 Repeat problem **4** with a 30-dB optical amplifier preceding the p–i–n photodiode.

Answer: SNR = 1.78×10^5.

Also determine how many decibels stronger the SNR is in this case.

6 Consider Figs. 6.9 and 6.10 and explain the following trends:

(**a**) Both figures show minimum values of approximately 3 dB.

(**b**) What specific aspect of the EDFA is demonstrated by Fig. 6.9?

(**c**) What specific aspect of the EDFA is demonstrated by Fig. 6.10?

7 Explain why the number of modes parameter $l = 2$ is particularly important for high-speed FOCS.

Appendix 6A. Derivation of some Equations in Section 6.4

6A.1 Derivation of Eqs. (6.28c), to (6.28f)

We begin by rewriting the first term on the right-hand side (RHS) of Eq. (6.29a) as

$$a\left[\sum_n [P_{n-1}\{(n-1)^2 + 2(n-1) + 1] - n(n+1)P_n\}\right] \tag{6A.1}$$

where we used $n^2 = [(n-1) + 1]^2 = (n-1)^2 + 2(n-1) + 1$. Thus Eq. (6A.1) becomes

$$a[<n^2> + 2<n> + 1 - <n^2> - <n>] = a[<n> + 1] \tag{6A.2}$$

Next we note that $n(n+1) = [(n+1) - 1](n+1) = (n+1)^2 + (n+1)$. Thus the second term of Eq. (6.29a) becomes

$$b[<n^2> - <n> + <n^2>] = -b<n> \tag{6A.3}$$

Equation (6.28c) then follows directly from Eqs. (6A.1) to (6A.3). In the preceding derivation we made use of the property that $P_k = 0$ for $k < 0$.

Solution of Eq. (6.28c)

$$\frac{dy(z)}{dz} + P(z)y(z) = Q(z) \tag{6A.4}$$

where $P(z)$ and $Q(z)$ are functions of z only (or a constant), but not functions of $y(z)$ in any way. It can be shown that the solution of this equation is given by

$$y(z) = \exp\{-F(z)\} \int Q(z)\exp\{F(z)\}dz + c \exp\{-F(z)\} \tag{6A.5}$$

where c is determined by the boundary condition and where

$$F(z) \doteq \int P(z)dz = \text{functional form} \tag{6A.6}$$

Thus, for example, if $P(z) = az^2$, then
$F(z) = (a/3) z^3$. Now, rewriting Eq. (6.29c) in the following form:

$$\frac{d<n>}{dx} + <n>[b(z) - a(z)] = a(z) \tag{6A.7}$$

thus $P(z) = -[a(z) - b(z)]$ and $Q(z) = a(z)$.

For the traveling-wave amplifier, all integrals run from 0 to z, and we have

$$F(z) = -\int_0^z [a(z') - b(z')]dz' \tag{6A.8}$$

Now let

$$G(z) = \exp[-F(z)] = \exp\left\{\int_0^z [a(z') - b(z')]dz'\right\} \tag{6A.9}$$

and let

$$N(z) = G(z) \int_0^z Q(z') \exp\{F(z')\}dz' = G(z) \int_0^z \frac{a(z')}{G(z')}\,dz' \tag{6A.10}$$

We note that $G(0) = 1$ and that $N(0) = 0$. Therefore Eq. (6A.5) evaluated at $z = 0$ yields $c = \langle n(0)\rangle$ and Eqs. (6.29d) to (6.29f) accrue.

6A.2   Evaluation of $\sigma^2(z)$ for TWOA

We note that C_2 can be found by multiplying each term on both sides of Eq. (6.29a) by $n(n + 1)$ and summing over all n:

$$\frac{dC_2}{dz} = a\left[\sum_n \{n^2(n-1)P_{n-1} - n(n-1)(n+1)\}\right]$$

$$+ b\left[\sum_n \{n(n-1)(n+1)P_{n-1} - n^2(n-1)\}\right] \tag{6A.11}$$

Now, using the identity below Eq. (6A.1), we find that the terms inside the brackets of the upper line of Eq. (6A.11) become

$$[(n-1)^3 + 2(n-1)^2 + (n-1)]P_{n-1} - [n^3 - n]P_n$$

Summing this over n yields

$$\langle n^3\rangle + 2\langle n^2\rangle + \langle n\rangle - \langle n^3\rangle + \langle n\rangle = 2[\langle n^2\rangle + \langle n\rangle]$$

This may be rewritten as

$$2[\langle n^2\rangle - \langle n\rangle] + 4\langle n\rangle = 2C_2 + 4\langle n\rangle$$

Thus the net contribution to the upper line is $aC_2 + 4a\langle n\rangle$.

Next we apply $n = [(+1) - 1]$ and $(n-1) = (n+1) - 2$ to the term in the lower brackets of Eq. (6A.11) to arrive at

$$[(n+1)^3 - 3(n+1)^2 + 2(n+1)]P_{n+1} - [n^3 - n^2]P_n$$

Summing over n yields

$$\langle n^3\rangle - 3\langle n^2\rangle + 2\langle n\rangle - n^3\rangle + \langle n^2\rangle = -2[\langle n^2\rangle - n^2\rangle]$$

and the net contribution from the lower line of Eq. (6A.11) is $-2\,C_2$.

These results lead to the following first-order linear differential equation for C_2:

$$\frac{dC_2}{dz} - 2[a(z) - b(z)] = 4a(z)<n(z)> \tag{6A.12}$$

Then $P(z) = -2[a(z) - b(z)]$ and $Q(z) = 4a(z)[G(z) <n(0)> + N(z)]$ and $F(z) = -2 \int_0^z [a(z') - b(z')]dz'$

which yields

$$\exp[-F(z)] = \exp[2 \int_0^z [a(z') - b(z')]dz' = G^2(z) \tag{6A.13}$$

We next employ these specific formulas in the general solution to a first-order differential equation in the form given in Eq. (6A.5). Specifically, the first term on the RHS of Eq. (6A.5) may be expressed as a sum of two terms indicated as $A(z)$ and $B(z)$ in the following equation:

$$\overbrace{G^2(z) \int_0^z \frac{4a(z')<n(0)>}{G(z')} dz''}^{A(z)} + \overbrace{6G^2(z) \int_0^z \frac{4a(z')N(z')}{G^2(z')} dz''}^{B(z)} \tag{6A.14}$$

where we employed Eq. (6.29d) to replace $<n(z)>$. The solution of Eq. (6A.12) now has the form

$$C_2(z) = A(z) + B(z) + c \exp\{-F(z)\} \tag{6A.15}$$

The first term may be expressed as

$$A(z) = 4a(z)G(z)<n(0)> \left\{ G(z) \int_0^z \frac{a(z'')}{G(z'')} dz \right\} \tag{6A.16}$$

$$= 4<n(0)>G(z)N(z)$$

The simplification of the second term B is somewhat more complicated. We begin by replacing $N(z'')$ in $B(z)$ using Eq. (6A.10) to arrive at

$$B(z) = 4 \, G^2(z) \int_0^z \left\{ \frac{a(z')}{G(z')} \int_0^{z'} \frac{a(z'')}{G(z'')} dz'' \right\} dz' \tag{6A.17}$$

Next we let

$$U \doteq \int_0^{z'} \frac{a(z'')}{G(z'')} dz''$$

then

$$B(z) = 4G^2(z) \int_0^z U \, dU = 4 \, G^2(z) \frac{U^2}{2}$$

Finally, since

$$N(z) = G(z) \int_0^z \frac{a(z')}{G(z')} \, dz'$$

we have

$$B(z) = 2N^2(z) \tag{6A.18}$$

The general solution for $C_2(z)$ is given by Eq. (6A.5) with $y(z)$ replaced by $C_2(z)$. The only portion of this equation that we have not as yet found is the last term on the RHS of that equation. This is obtained by noting that $G(0) = 1$ while $Q(0) = 0$ in Eq. (6A.5) for this case just as they were when we evaluated $<n(z)>$. This leads to

$$c = C_2(0) = <n^2(0)> - <n(0)> \tag{6A.19}$$

Contribution to $C_2(z) = G^2(z)[<n^2(0)> - <n(0)>]$

Finally, employing Eqs. (6A.13), (6A.14), (6A.16), (6A.18), and (6A.19) in Eq. (6A.14), we arrive at Eq. (6.30a).

References

1. C. W. Helstrom, *Probability and Stochastic Processes for Engineers,* 2d ed., Macmillan, New York, 1991.
2. A. Papoulis, *Probability, Random Variables, and Stochastic Processes,* 2d ed., McGraw-Hill, New York, 1984.
3. G. Lachs, "Quantum statistics on multiple-mode, superposed coherent and chaotic radiation," *J. Appl. Phys.* **38** (1967).
4. S. R. Laxpati and G. Lachs, "Closed-form solutions for the photocount statistics of superposed coherent and chaotic radiation," *J. Appl. Phys.* **43** (1972).
5. G. Lachs, "Approximate statistics for coherent and chaotic radiation of arbitrary spectral shape," *J. Appl. Phys.* **42** (1971).
6. R. J. Glauber, "Optical coherence and photon statistics," in C. DeWitt, A Blandin, and C. Cohen Tannoudi, Eds., *Quantum Optics and Electronics Les Houches 1964,* Gordon & Breach, New York, 1965.
7. B. E. A. Saleh and M. C. Teich, *Fundamentals of Photonics,* Wiley, New York, 1991.
8. S. O. Rice, in N. Wax Ed., *Selected Papers on Noise and Stochastic Processes,* Dover, New York, 1954, pp. 153–157.
9. R. J. Glauber, *Photon Fields and Classical Fields, Proceedings of the Symposium on Modern Optics,* Vol. XVII, Polytechnic Press, Brooklyn, N.Y., 1967, pp. 1.
10. G. Lachs, "Effects of photon bunching on shot noise in photoelectric detection," *J. Appl. Phys.* **39** (1968).
11. G. P. Agrawal, *Fiber-Optic Transmission Systems,* Wiley, New York, 1992.
12. M. Schwartz, *Information Transmission, Modulation, and Noise,* McGgraw-Hill, New York, 1959.
13. E. Desurvire, *Erbium-Doped Fiber Amplifiers,* Wiley, New York, 1994.
14. P. Diament and C. Teich, "Evolution of the statistical properties of photons passed through a travelling-wave amplifier," *IEEE J. Quantum Electron.,* **28,** 1325–1334 (1992).
15. J. L. Zyskind, J. W. Sulhoff, and D. J. DiGiovanni, "An investigation of the noise figure and conversion efficiency of 0.98 μm pumped erbium-doped fiber amplifier under saturated conditions," *IEEE Photonics Technol. Lett.* **4,** 1261–1264 (1992) (in particular, Fig. 5).
16. R. I. Laming and D. Payne, "Noise characteristics of erbium-doped amplifier pumped at 980 nm," *IEEE Photonics Technol. Lett.* **2**(6) (1990).
17. A. D. Whalen, *Detection of Signals in Noise,* Academic Press, New York, 1971.

Optical Fiber Systems (General)

7.1 Modulation and Multiplexing

7.1.1. Introduction

In this chapter we consider optical fiber communication systems as a whole. We begin with some modulation considerations followed by discussion of some multiplexing regimes. It turns out that because of the enormous bandwidth available in FOCS, there exists a much broader range of possibilities for FOCS multiplexing than with either wire cable or rf antenna-type (non-confined) communication systems. We shall conclude this chapter with some further discussion of a few simple detection systems that essentially mimic their rf and/or microwave counterparts. Such system performance criteria as error rates and photons per bit will be discussed there. A block diagram for a typical long-range FOCS is shown in Fig. 7.1.

The basic input signals are prepared for transmission in the first block as indicated in Fig. 7.1. Specific aspects will vary from system to system but may include several of the following processes:

1. Filter the various input signals to limit bandwidth presented to modulator.

2. Convert an analog signal to digital format for transmission by sampling, analog-to-digital (A/D) conversion, and encoding for error detection and error correction. Part of this conversion is often performed before the signals arrive at the FOCS transmitter.

3. Time-division-multiplex digital signals, including those generated in part 2 above.

4. In some systems a subcarrier structure is generated as an input to drive the FOCS modulator. The entire subcarrier composite signal then modulates the FOCS optical carrier. We shall describe these types of systems later in this section when we discuss multiplexing in general. We shall

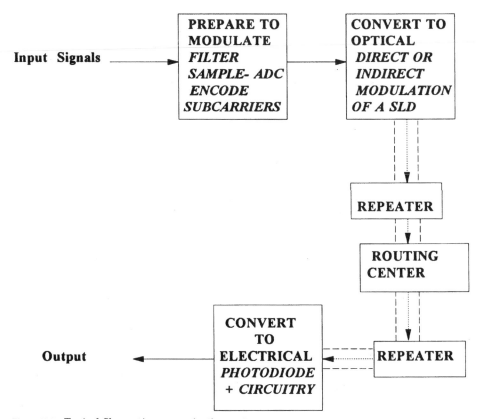

Figure 7.1 Typical fiber-optic communication system.

also encounter a specific example of a proposed system that employs almost all of these characteristics in Chap. 10.

We begin the discussion of this material by considering the repeaters shown in Fig. 7.1. Modulation and multiplexing are considered in separate sections on this chapter. Figure 7.1 is a representative diagram, and an actual FOCS may have more or less repeaters than those shown in Fig. 7.1. There are two basic forms for repeaters: regenerative repeaters (true repeaters) and optical amplifiers. In *regenerative repeaters* the signal is digitally detected, error-corrected, and then retransmitted down the fiber-optic channel to the next node (repeater, receiver, or routing device). The output of each regenerative repeater will be a nearly perfect replica of the transmitted signal (except for SLD noise), provided no digital error has occurred during transmission. Thus, by maintaining sufficient signal strength at the input to each repeater stage, one can satisfy any reasonable error rate criterion. To reduce the expense of constructing a regenerative repeater system, one strives to extend the distance between repeaters as much as possible (while still satisfying the error rate specification for the system).

Optical amplifiers, on the other hand, do not require conversion from optical to electrical and back to optical, as is the case in regenerative repeaters. Traveling-wave optical amplifiers (TWOAs) made of erbium-doped optical fibers are commercially available for operation at wavelengths ~1.5 μm. These amplifiers are typically tens of meters long and are driven by laser injection from the ends. Erbium-doped optical amplifiers are fairly expensive (about $10,000 at the time this book was written), but the state of the art for manufacturing optoelectronic integrated circuits (OEICs) has not as yet presented an economically satisfactory alternative for long-haul ultra-high-speed FOCS. A salient point of traveling-wave amplifiers is that they have very large usable bandwidth because they are not subject to the much more stringent constraint of the constant gain–bandwidth product that is a fact of life in cavity amplifiers. Thus, if the transmission down the FOCS contains signals from a number of separate SLD (each with a distinguishable wavelength within the 1.55-μm dip of the fiber absorption curve), then a single optical amplifier may well serve as repeater for many, if not all, of these channels. Conversely, separate regenerative repeaters would most likely be required for each individual SLD channel. However, regenerative repeaters do a better job of virtually eliminating digital errors than do amplifiers, which amplify the input noise as well as the signal. This has certainly been the case in coax (coaxial cable) communications, and the author of this text believes that this will eventually be the case in FOCS.

Considerable research and development has gone into making as much of the system optical, perhaps even all optical. This includes methods of rapidly switching optical signals from one optical path to another without converting the signal to an electrical format during the process. If the system shown in Fig. 7.1 has this type of switching and optical amplifier repeaters, it said to be "all-optical." However, until optical computing becomes a practical reality, the input information and final output signal from a FOCS will be in the form of electrical signals.

At this point we note a few differences between optical fibers and metallic wire communication channels:

Topic	Wire line	Optical fiber
Losses	High unless	Very small at proper wavelength
EMI	Serious	Nonexistent
Flexible	Yes	Not as much as wire
Weight	Heavy	Much lighter
Linearity of channel	Yes	Some nonlinearity May be important
Dispersion	Yes	Yes; small in single-mode fibers
Bandwidth	1 GHz	100 GHz–1 THz
Simpler	√	—
Higher data	—	√
Cheaper	√ (for now)	—

7.1.2 Direct modulation

Modulation of the output of a laser diode may be accomplished either directly by controlling the SLD driving current, or indirectly by using an external modulator to modify the output of a steady-state laser. Direct modulation in its simplest form may be achieved by simply turning the laser on or off as specified by the binary data stream. Analog intensity modulation, which is very much like amplitude modulation (AM), can be achieved by current control of a SLD with an output power–driving current response that is as nearly linear as possible (see Fig. 4.21). Each detected photon yields an electron, so the response current to this type of modulation will be a replica of the modulation signal. There will be some harmonic distortion caused primarily by the nonlinear response curve of a real SLD, but a number of such systems have been deployed successfully. To couple the signal onto the laser bias current, a *bias-T* network such as the device shown in Fig. 7.2 is often utilized.

The inductor provides a high impedance to the signal input and thereby decouples the signal from other SLD driven by the same power supply. Meanwhile the capacitor blocks the DC bias current from the signal generating circuitry. To utilize the bias-T coupler for an AM source in an intensity-modulated direct-detection system (IMDD), it is necessary to bias the laser to the center of the linear range of the transmitter SLD (see Fig. 4.26). The signal part of the driving current should be directly proportional to the modulating signal. Thus we have the forward-biased diode current i_s

$$i_s = I_b + ax(t)$$

where $x(t)$ = input signal waveform. At the receiving terminal, we simply detect the intensity. In this setup the detected signal will not depend on either the phase or the frequency of the optical carrier.

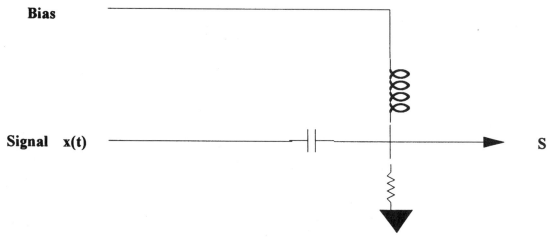

Figure 7.2 Bias-T coupler.

To use intensity modulation for conveying signals that are already in FM or PM (frequency- or phase-modulated)-type analog channels, we need to use subcarrier modulation and place the input waveform $x(t)$ in the appropriate form. Thus, for example, let

$$x(t) = A \cos[\omega_{inst}t] \quad \text{for} \quad \text{FM}$$

where $\omega_{inst} = \omega_c + 2\pi(f_\Delta)s(t)$
ω = "subcarrier" frequency (rf)
$s(t)$ = input waveform (information)

Similarly

$$x(t) = A \cos[\omega_c t + \Phi_\Delta s(t)] \quad \text{for} \quad \text{PM}$$

The modulator input signal $x(t)$ could be any of the usual analog modulation forms such as AM, DSB/SC (double sideband/suppressed carrier), SSB (single sideband), and VSB (video sideband) (television). One could also build $x(t)$ into a frequency-division format and transmit the entire collection of subcarriers over a single fiber-optic channel. Some or all of the subcarriers could even be digital signals in the form of OOK, FSK, PSK, QAM (quadrature amplitude modulation), and so on. However if a large number of channels are to be transmitted using a single SLD source in this fashion, even a slight nonlinearity in the SLD response curve could generate a large amount of crosstalk between the subcarrier channels of $x(t)$. This problem can be dealt with in some of these systems by keeping $x(t)$ small and thereby limiting the magnitude of the current swings. Unfortunately, this will also reduce the signal-to-noise ratio (SNR) at the receiver, and this remedy is probably restricted to channel lengths of a few kilometers. Another possible remedy is to use a memoryless nonlinear response at the receiving terminal which compensates for the nonlinear SLD response. Note that the SLD response is itself a memoryless type of nonlinearity which, in turn, makes this last remedy feasible. It is apparent from Fig. 4.26 that for all of the modulation systems described in this section, *it is very important to maintain a stable operating temperature for the transmitter SLD*. This is particularly true for the nonlinear compensation method we have been considering here.

The bias current may be employed to tune an SLD contiguously over a limited range (see Fig. 4.24) because the index of refraction in the active region depends on the charge carrier density. The diodes used to obtain that figure were observed to have a tuning sensitivity of approximately 3 GHz per milliampere of change in bias current. These SLDs were operating at nominal bias current of ~100 mA and produced a single frequency at a nominal wavelength of ~0.83 μm. This fortuitous property of SLDs can be utilized for direct FM modulation of the laser output. For example, the SLD output can be frequency-modulated by using a bias-T arrangement such as that shown in Fig. 7.2 . We next consider how, using an input signal to the bias-T of the

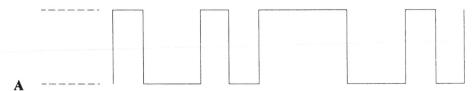

Figure 7.3 Typical binary bit stream.

form shown in Fig. 7.3, we can generate OOK and FSK. Modulation by control of the SLD driving current is known as *direct modulation.*

For OOK modulation, the DC bias current is often set just below lasing threshold, while we set $A = 0$ and B is adjusted to be large enough to produce the desired output power. Because of the inherent FM effect in SLDs, the sudden transitions in input binary signal waveform generate a broadband form of frequency modulation known as "chirping." Chirping can result in considerable distortion because the huge increase in transmission bandwidth could lead to large-scale dispersion distortion. One way to avoid the chirping problem is to use external modulation. We shall discuss this in the next subsection.

Another potentially serious problem that may occur when bias-T couplers are used for direct modulation is the local imbalances that occur in the bit stream, which can cause a slow drift in the DC portion of the SLD drive current. This results in a slow drift in the optical carrier frequency, which can create some problems in FSK systems. On the other hand, this slow drift will in no way affect the performance of an OOK system when simple direct detection is employed. Another problem encountered with some OOK systems is that SLDs used for OOK transmission sometimes display overshoot followed by oscillation. This particular problem often occurs in older diode lasers and seems to be one of the effects of aging in laser diodes.

To generate FSK signals, we use $A = -B$ for the $x(t)$ represented by Fig. 7.3. We note that the magnitude of the input signal swings are limited to much lower values than those encountered in OOK transmission. This reduces chirping considerably, and since a real signal will have nonzero rise times and fall times, the output will be what is called *continuous-phase frequency shift keying* (CPFSK). These signals have better frequency rolloff characteristics than do discontinuous FSK, as would be generated by using two separate SLDs to generate an FSK signal. The detection of such signals is best achieved using optical heterodyne detection (OHD) (to be discussed in the next chapter) followed by discriminator detection. Discriminator detection permits reduction of the spacing between the two FSK frequencies, which can be made small enough to permit the use of "minimum shift-keying" versions of FSK. However, this type of system is particularly susceptible to the carrier frequency drift incurred by statistical fluctuations generated by imbalances in the number of 1s and 0s. The drift could even cause the heterodyned signal to drift out of the passband of the IF (intermediate-frequency) stage. To deal with this problem, it may be necessary to use the same techniques as are used to solve the same problem in coax systems. One technique to avoid the bias drift problem is to employ *alternate mark inversion* (AMI). AMI, however, requires the use of three carrier frequencies to transmit information over a single binary FSK channel. This would require more bandwidth per channel, but as of now there is plenty of bandwidth available in existing FOCS systems.

Both OOK and FSK can be generated by direct modulation of a laser diode. This includes the use of larger signaling alphabets. However, phase modulation, and in particular PSK, cannot be generated by direct-current control of a SLD (except as a subcarrier in intensity modulation). PSK, however, can readily be generated by using an external modulator.

7.1.3 External modulation

In external modulation systems the laser diode is operated as a steady-state source and the modulation is performed externally on the SLD by output usually by an electroptic crystal modulator. A typical arrangement for an external modulator is shown in Fig. 7.4.

It is very important to prevent reflection back into the laser cavity. Even very small reflections back into the laser cavity can destabilize the SLD output and generate large amounts of relative intensity noise. Spurious reflections also may cause large swings in the output frequency, which can devastate optical heterodyne reception. Reflections are eliminated at both entrance exit ports of the modulator by appropriate antireflection (AR) dielectric coatings. The reflectivity of these multilevel dielectric layer structures are wavelength-dependent, and even though the electrooptic modulators can operate over a broad range, the optical wavelength dependence of the AR coatings will limit the modulator operation to a much tighter wavelength region.

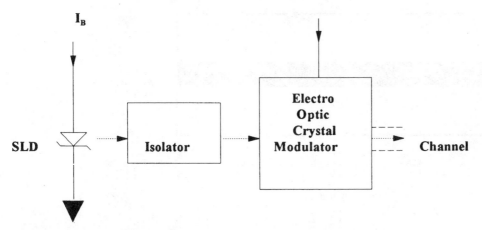

Figure 7.4 External modulator setup.

The function of the isolator is to attenuate all reflections encountered in the channel. The optical signal is linearly polarized at the entrance to the isolator and then rotated 45° in the Faraday effect section. Now, when this beam is reflected back toward the optical source, it passes through the isolator again. However, this time it is moving in the opposite direction from the original pass, and it results in another 45° rotation in the same direction. The polarization of the reflected optical signal is now perpendicular to the linear polarizer at the entrance to the isolator and is therefore severely attenuated. Often the isolator is placed after the modulator; however, a number of commercial diode lasers come with fiber-optic "pigtails" (sections of fiber attached directly to the SLD) that also have 70-dB isolators built in. Finally we note that the isolator itself must have AR coatings on the input side of the device.

Electrooptic devices usually involve materials wherein *the index of refraction (hence the velocity of propagation) can be controlled by an applied electric field.* This effect can be employed to construct a phase modulator out of a single straight-line section of the electrooptic crystal as shown in Fig. 7.5a. Such a modulator can be employed to generate an optical frequency PSK, DPSK QPSK, and other forms of digital phase modulation. These phase-modulated signals can be detected only by using coherent detection (i.e., optical heterodyne or homodyne detection). Another application of the simple phase modulator depicted in Fig. 7.5a is to generate frequency modulation by driving the phase modulator with $\int x(t)dt$, where $x(t)$ represents the information waveform. This arrangement can be employed to externally frequency-modulate a laser beam.

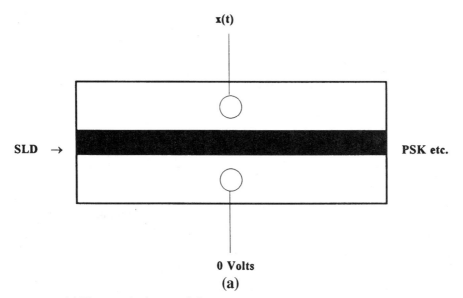

Figure 7.5 (*a*) Electrooptic phase modulator.

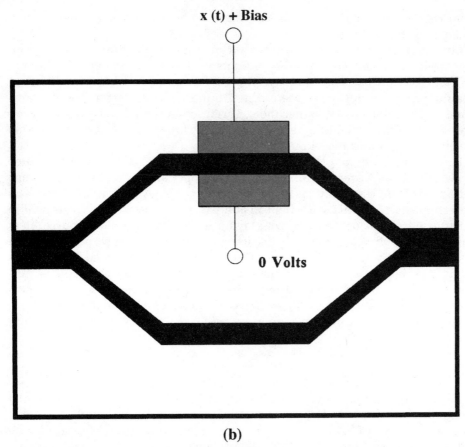

x (t) + Bias

0 Volts

(b)

Figure 7.5 (*Continued*) (*b*) Mach–Zender-type modulator.

A variation of the device shown in Fig. 7.5*a* is the *Mach–Zender Interferometer* [1] shown in Fig. 7.5*b*. In this modulator, the incoming light is split equally in two paths, one of which is phase-modulated by the applied voltage $x(t)$, and this device can be used as a switch by

1. Using a digital $x(t)$ such as that shown in Fig. 7.3
2. Adjusting levels A and B in Fig. 7.3 so that a logical 0 corresponds to 180° phase shift while a logical 1 corresponds to no phase shift at all

Thus, when a logical 0 is to be transmitted, the two legs are out of phase and cancel while the two paths are in phase when a 1 is transmitted. Another way to explain this process (which also happens to conserve energy) is to consider that the in-phase addition results in the fundamental mode propagating to and through the single-mode optical fiber while the out-of-phase addition can be regarded as forming a second-order mode which will not propagate through the single-mode fiber following the modulator. We can see, then, from this viewpoint that the optical fiber is where the energy is consumed when a 0 is transmitted.

Lithium niobate is frequently used for external modulators in FOCS. The two optical channels shown in Fig. 7.5b as well as the single optical channel in Fig. 7.5a can be embedded into the lithium niobate crystal by diffusing titanium into the upper surface of the lithium niobate crystal. These titanium-doped regions have a higher index of refraction than do both the undoped crystal and the surrounding air, so that total internal reflection constrains the light to the indicated channels.

In summary, we note that external modulation is the only way to induce PSK, QPSK, and even QAM directly on an optical carrier. It is vital that reflections from these devices back into the laser be avoided at all costs. The reflections can interact with field inside the diode laser to generate wildly fluctuating signal outputs. By employing a steady-state laser source, it is easier to stabilize the laser diode output. This will ensure production of heat at a steady rate, and the operating temperature will tend to stay constant. Another aspect often overlooked is that the energy per pulse is random in a direct-modulation OOK system. This randomness is a form of noise that can be greatly reduced by external modulation. On the other hand, an external modulation system is more complicated than a direct-modulation system.

7.1.4 Multiplexing systems

Multiplexing means transmitting separate information channels over the same physical channel. The enormous range of useful wavelengths available in modern optical fibers provides an extremely rich tapestry of multiplexing schemes to select from. We shall discuss some of these systems in the remainder of this section, but first let's note some major types of overall systems that employ multiplexing.

1. *Point-to-point transmission.* Transmission over a channel such as is shown in Fig. 7.1 is one type of system that employs multiplexing. In such a system one application of multiplexing is to increase the overall data rate of the fiber-optic channel. For example, multiplexing may be used to adapt a large number of slower data channels for transmission over a single high-speed FOCS channel. In this case the redistribution (demultiplexing) of these separate signals is also a design consideration. The process of restoring the data stream to its original multichannel structure is known as *demultiplexing* and is accomplished at the receiving end after detection and carrier demodulation.

2. *Networking.* In optical communication networks many users are attached to the same fiber-optic network. These networks can be constructed in a variety of interconnection structures. One simple "ring"-type network is shown in Fig. 7.6.

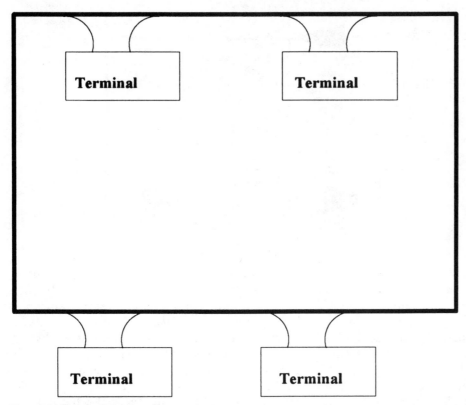

Figure 7.6 Simple ring network.

In the system shown in Fig. 7.6 each user is usually capable of receiving and transmitting information with any other user over this shared network, and "collisions" or channel-usage conflicts are avoided by various control paradigms called *protocols*. Thus information can be shared among many users simultaneously, or it can be transmitted point to point anywhere on the network. An important feature here is flexibility. Repeaters are rarely employed here because of the relatively short distance between terminals.

In token ring networks (see Fig. 7.7) all terminals are themselves repeaters. Each terminal detects and repeats the signal, and collisions are avoided by allowing only one terminal to transmit at a time. This network is controlled by a specific digital sequence called a *token* that indicates which terminal (or node), if any, is transmitting at a particular time. Strictly speaking, networking is not multiplexing since each terminal has at least one transmitter and one receiver. However, the technology of networks often employs many of the same signal-processing concepts as are employed in multiplexing. As an example, consider the fact that the token ring arrangement bears a distinct resemblance to time-division multiplexing.

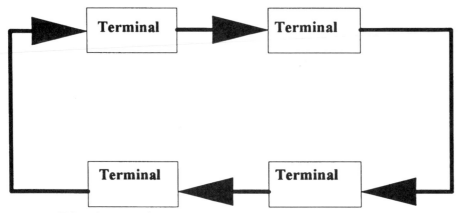

Figure 7.7 Token ring network.

We next discuss a number of multiplexing schemes, beginning with the interleaving of digital signals from a collection of separate input channels that we combine into a single digital data stream. This process, known as *time-division multiplexing* (TDM), is a unique form of multiplexing that is applicable only to digital data communication systems. If multiple analog signals are to be transmitted over digital channels, as is often done today by the telephone companies, it is first necessary to sample each analog input and digitize these samples to via A/D conversion. When TDM is used in networks, the TDM process must take place for each individual terminal separately. By that we mean that the TDM process is performed on the signals at each terminal and then transmitted. It is almost impossible to TDM into the middle of data stream generated at another terminal. On a specific network, a collection of signals may be transmitted from one terminal to all the terminals on the network.

Another multiplexing structure that communication engineers are readily familiar with is *frequency-division multiplexing* (FDM). In line with common, *but not universally accepted,* usage in the FOCS field, we shall reserve the notation FDM for a frequency-division subcarrier structure such as shown in Fig. 7.8 for modulating a single SLD.

In the arrangement shown in Fig. 7.8, a large number of separate channels of all types, digital and/or analog, can be built into an FDM structure of subcarriers. The entire subcarrier structure $x(t)$ is then used to modulate a single diode laser. By using such multiplexing schemes along with SLDs that can be modulated at GHz rates, it is frequently possible to replace microwave relay towers for digital communications with FOCS (of course, one must then lay a fiber-optic cable between terminals). Furthermore, fiber-optic channels will not be affected by weather conditions that occasionally cause outages in microwave relay systems.

We next consider a multiplexing system, of particular importance in FOCS,

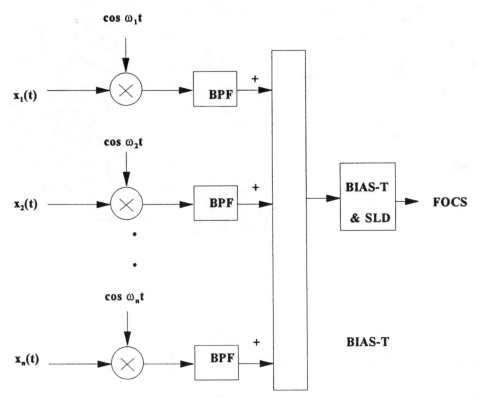

Figure 7.8 FDM setup using subcarriers.

known as *wavelength-division multiplexing* (WDM). In WDM systems *the channels are multiplexed by modulating separate laser diodes with different wavelengths as sources* and then superposing these channels for transmission over a single optical fiber as shown in Fig. 7.9.

Each channel could itself contain a complete group of subcarrier multiplexed channels. Some examples and comments follow:

1. Suppose that one SLD at 1.3 μm is digitally modulated by a number of TDM channels, and let there also exist a 1.55-μm SLD that is analog-modulated by a set of FDM subcarriers. We can form a WDM system by superposing the output of these two lasers and transmit the combination over the same optical fiber. These WDM channels are separated into two separate groups at the receiver by optical filtering, and each group is then separately detected and demultiplexed. The tremendous bandwidth capabilities of the fiber as well as the flexibility of the system make WDM a strong candidate for ultra-high-speed communications.

2. Consider a system with a number of very high-speed digital channels with data rates too large to be interspersed, requiring us to use WDM by modulating a separate laser diode. Now suppose that these SLDs are tuned

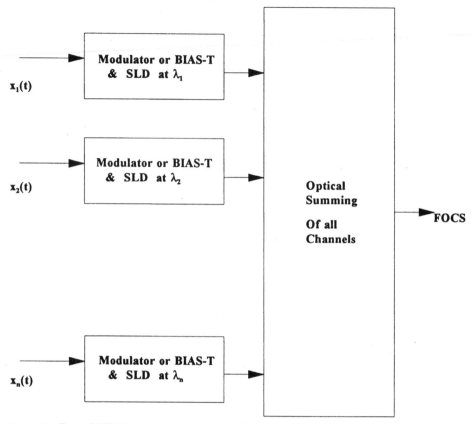

Figure 7.9 Typical WDM arrangement at transmitter.

so that they are spaced just far enough apart in wavelength to ensure that the spectra of the individual WDM channels do not appreciably overlap. Such closely spaced channels can't be separated by optical filtering, but could be separated using optical heterodyne detection (see Chap. 8). The electronic processing of the individual WDM channels is well within existing technology, and this dense packing of high-speed data channels on a single-mode fiber has the potential to achieve Tbit/s data rates over a single fiber.

3. A variation of example 2 above is to use different wavelengths for each node (connection) on a local area network. This would avoid conflicts and still permit each node to interact with any channel by appropriate choice of frequency of the local laser diode.

In designing WDM systems, one must be careful that the superposition of the individual channels yield a net intensity that is small enough to avoid nonlinear effects; otherwise nonlinear distortion could result in massive crosstalk between channels. Technically, WDM is a form of FDM since frequency determines wavelength and vice versa. However, the transmission and detection techniques, particularly the overall bandwidth of FDM and WDM as we have defined it here, are sufficiently different to warrant separate names and notation.

Code-division multiplexing (CDM) has also been suggested for networking on optical fiber systems [2,3]. In CDM, the transmitter SLD is first modulated by a very high-speed unique pseudo-random bit train. This combination acts as a carrier which is then further modulated by the slower (but not slow) digital data stream for transmission. The pseudo-random code for each channel is selected to be orthogonal to the pseudo-random codes for all the other channels. Thus, each channel has a unique and separable carrier structure based on these pseudo-random structures. At the receiving terminal the incoming signal consisting of all the channels is multiplied by a synchronized version of the pseudo-random code assigned to each channel to be demultiplexed followed by digital detection.

In network applications this CDM system is sometimes called *code-division multiple access* (CDMA). Typically in network applications a number of distinct pseudo-random channels share the same spectral region, and CDMA channels can be added to the network or received at each node. By suitable receiver design it is possible to receive any individual channel for which the pseudo-random code sequence is available. One problem with CDM systems is that it takes time to synchronize the pseudo-random carrier at the receiver, and there might be a significant delay in initiating communications on a specific channel.

7.2   Detection of Digital Signals

7.2.1   Introduction

In this section we shall employ a simplified approach to obtain approximate numerical values for the probability of error as well as to formulate an estimate of the photons per bit required to achieve a prespecified error rate. To present a picture of the detection process that is compatible with existing FOCS, we shall utilize the integrated photodiode current as the primary detection variable instead of the number of photons counted in T seconds. Specifically, we define[1] the integrated current x as

$$x = \frac{1}{T} \int_0^T i(t)dt \tag{7.1}$$

We shall assume also that perfect bit and/or baud synchronization has been achieved. In this chapter we consider only thermal noise and shot noise,

---

[1]The $1/T$ factor in Eq. (7.1) does not affect the signal-to-noise ratio (SNR); thus it will not affect the probability of error. It does, however, simplify the notation beginning with Eq. (7.5b) in the next subsection.

which are both gaussian random processes due to central-limit-theorem (CLT) arguments. Since integration is a linear process, the decision variable x will be a gaussian random variable. Therefore, we may express the probability density of x as

$$f_x(x) = \frac{1}{\sqrt{2\pi\sigma_k^2}} \exp\left[-\frac{(x - \bar{x}_k)^2}{2\sigma_k^2}\right] \tag{7.2}$$

where $\bar{x}_k$ and σ_k^2 are the mean and variance of x, respectively, when the digital symbol k is transmitted. This notation allows for larger signaling alphabets, although in this section we shall restrict our attention to OOK-type binary communication systems. Furthermore, we shall ignore any crosstalk generated by the impulse response of the photodiode circuitry initiated by photoelectrons detected near the edges of the baud. Dispersion from the previous baud will also be assumed to be negligibly small. Now, the total current, including the noise contributions from shot noise and thermal noise, is given by

$$i(t) = i_{sig}(t) + i_{sh}(t) + i_{th}(t) + i_{dark}(t) \tag{7.3a}$$

where the subscripts are the same as those we used in Chap. 6. As in Chap. 6, we assume that *each noise current on the RHS of Eq. (7.3a) is a statistically independent, zero-mean, ergodic process. We then obtain*

$$\bar{x}_k = \mathbf{E}\left\{\frac{1}{T}\int_0^T i(t)dt\right\} = \frac{1}{T}\int_0^T i_{sig}(t)dt \tag{7.3b}$$

$$\sigma_k^2 = \mathbf{E}\left\{\frac{1}{T}\int_0^T \left\{i_{sh}(t) + i_{th}(t) + i_{dark}(t)\right\}^2 dt\right\} \tag{7.3c}$$

where all averages of the integrated cross-product terms vanish. Thus, for example

$$\mathbf{E}\left\{\frac{1}{T}\int_0^T i_{sh}(t) \times i_{th}(t)dt\right\} = 0 \tag{7.3d}$$

Then the conditional variance when symbol k is transmitted as

$$\sigma_k^2 = \overline{i^2_{sh,k}} + \overline{i^2_{th}} + \overline{i^2_{dark}} = \overline{i^2_{N,k}} \tag{7.3d}$$

7.2.2 Probability of error

The notation in the preceding section will now be employed to determine the error rate for some simple binary communication systems. We have already introduced the basic binary error rate formula given in Chap. 6.

$$P_e = P(0)P(e\,|\,0) + P(1)P(e\,|\,1) \tag{7.4}$$

In the remainder of this chapter we shall assume that, unless stated otherwise, $P(0) = P(1) = \frac{1}{2}$. Many binary detection systems employ "integrate and dump" circuits to obtain the decision variable; therefore, our primary task will be to find both $P(e \mid 0)$ and $P(e \mid 1)$ from the conditional probability densities of the integrated current x. The output of the integrator is sampled at the end of the baud interval, and all the initial conditions in the integrate-and-dump circuit are rapidly reset to zero. We note, however, that many receivers, such as those systems using "raised cosine" pulses for data transmission, employ specialized linear filters in which the integration is performed by the convolution of the input with the impulse response of these specially designed filters. One important property of these filters is that, although they may have steep frequency rolloff skirts, the impulse response is such that they can be sampled so as to virtually eliminate intersymbol interference. This results in a detection system that obviates the need to initialize the detection circuits at the end of each baud. To simplify the notation here, we shall denote the detection variable by i instead of x for the remainder of this chapter. If the specialized pulse shapes and matching filters of the type that we have just described are not detection variables, use of this variable i would represent a change in notation. Regardless of whether integrate and dump or such specialized filters are used, the output samples are gaussian random variables with

$$f(i \mid k) = \frac{1}{\sqrt{2\pi\sigma_k^2}} \exp\left[-\frac{(i - i_{s,s_k})^2}{2\sigma_k^2} \right] di \qquad (7.5a)$$

where $i_{s,k}$ is the *integrated* output of a noiseless system when the symbol k is sent.

$$i_{s,k} = \mathbf{E}\left\{ \frac{1}{T} \int_0^T i_{s,k}(t) h_F(T - t) dt \right\} \qquad (7.5b)$$

where $h_F(\cdot)$ is the impulse response of the detection filter. In the case of an integrate-and-dump detector, $h_F(\cdot) = 1$. The variance is given by

$$\sigma_k^2 = \overline{i_{N,k}^2} = \overline{i_{sh,k}^2} + \overline{i_{th}^2} + \overline{i_{dark}^2} \qquad (7.5c)$$

where the subscript on the noise current variance $\overline{i_{N,k}^2}$ signifies that this particular variance may depend on which signal is actually transmitted. We note that while the dark noise and thermal noise levels are not dependent on the identity of the transmitted signal, the shot noise portion $\overline{i_{sh,k}^2}$ does depend on the intensity of the received light field. In an OOK FOCS the shot noise level during an ON-state baud will be larger than in an OFF-state transmission. On the other hand, thermal noise is not data-dependent, and in systems where thermal noise dominates, we have

$$\overline{i_{N,k}^2} = \overline{i_{th}^2} = \overline{i_{N,0}^2} = \overline{i_{N,1}^2} \qquad (7.5d)$$

Let us now consider a detection system for an optical binary communication system in which Eq. (7.5d) is valid. We note that the thermal noise (as well as shot noise) causes the detection variable to be a gaussian random variable. The only difference between the two conditional densities will be due to the difference between the conditional mean values denoted by $i_{s,1}$ and $i_{s,0}$. Some typical conditional densities are plotted in Fig. 7.10.

A suitable detection procedure for the conditional densities depicted in Fig. 7.10 is threshold detection, which is defined by

$$\text{Assume} \begin{cases} 0 & \text{if} \quad i \le i_0 \\ 1 & \text{if} \quad i > i_0 \end{cases} \tag{7.6}$$

where i_0 denotes the detection threshold.

Now let $P(e\,|\,1)$ denote the probability of error when a 1 is transmitted. This type of error will occur whenever the detector output is less than i_0 even though a pulse (i.e., a 1) was transmitted. Then we may write

$$P(e\,|\,1) = \int_{-\infty}^{i_0} p_1(i)di \tag{7.7a}$$

which, in terms of the notation here, becomes

$$P(e\,|\,1) = \int_{-\infty}^{i_0} \frac{1}{\sqrt{2\pi\sigma_1^2}} \exp\left[-\frac{(i - i_{s,1})^2}{2\sigma_1^2}\right] di \tag{7.7b}$$

To put Eq. (7.7b) into a standard format, we perform the following change of variable. Let

$$Y = \frac{(i - i_{s,1})}{\sigma_1} \qquad \text{then} \qquad di = \sigma_1 dY$$

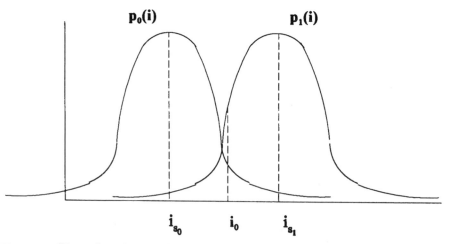

Figure 7.10   Binary detection (equal variance).

This leads to

$$P(e \mid 1) = \int_{-\infty}^{a} \frac{\exp\{-(Y^2/2)\}}{\sqrt{2\pi}} \, dY \tag{7.8a}$$

where $a = (i_0 - i_{s,1})/\sigma_1 < 0$ since the threshold shown in Fig. 7.10 is such that $i_0 < i_{s,1}$.

Now Eq. (7.8a) gives the probability density of a zero-mean unit-variance gaussian random variable. We can exploit the symmetry of this density as follows:

$$P(e \mid 1) = \int_{-\infty}^{a} \frac{\exp\{-(Y^2/2)\}}{\sqrt{2\pi}} \, dY = \int_{-a}^{\infty} \frac{\exp\{-(Y^2/2)\}}{\sqrt{2\pi}} \, dY \tag{7.8b}$$

Then we have

$$P(e \mid 1) = Q(a) = Q\left\{ \frac{i_{s,1} - i_0}{\sigma_1} \right\} = Q\left\{ \frac{(i_{s,1} - i_0)}{\sqrt{i_{N,1}^2}} \right\} \tag{7.9}$$

where

$$Q(u) \doteq \int_{u}^{\infty} \frac{\exp\{-(Y^2/2)\}}{\sqrt{2\pi}} \, dY \tag{7.10a}$$

Similarly, an error occurs when a 0 is transmitted and $i > i_0$. Then

$$P(e \mid 0) = \int_{i_0}^{\infty} p_0(i)di = Q\left\{ \frac{(i_0 - i_{s,0})}{\sqrt{i_{N,0}^2}} \right\} \tag{7.10b}$$

Then, from Eqs. (7.4), (7.9), and (7.10b), we obtain

$$P_e = P(1) \, Q\left\{ \frac{(i_{s,1} - i_0)}{\sqrt{i_{N,1}^2}} \right\} + P(0) \, Q\left\{ \frac{(i_0 - i_{s,0})}{\sqrt{i_{N,0}^2}} \right\} \tag{7.11}$$

An important part of the receiver design problem is to determine the decision threshold level that minimizes the error rate. The detection of OOK signals propagating through the atmosphere is complicated by random fluctuations in signal strength. This phenomenon, known as "fading," is caused by multipath interference at the receiver. OOK transmission in a fading environment usually necessitates the use of an dynamic threshold tracking system. However, fading is not a problem in single-mode fiber-optic communication systems,[2] and there is no need to track the detection threshold. Now, when the a priori probabilities are equal [i.e., $P(0) = P(1) = \frac{1}{2}$, then the opti-

---

[2]We note that random pressure fluctuations on the fiber-optic cable itself can cause random fluctuations in the type and orientation of the polarization of the received optical signal.

mum detection threshold turns out to be the intersection of the two conditional probability densities $p_0(i)$ and $p_1(i)$. On the other hand, when the a priori probabilities for the transmitted signals are not equal, the optimum detection threshold will shift toward the direction, thus increasing the likelihood of selecting the signal with the larger a priori probability.

An example of the intersection threshold is illustrated in Fig. 7.11, in which the ideal threshold is denoted by i_0. The area of the filled region represents $P(e \mid 1)$, and since we have assumed equal variances for both of the conditional densities, the symmetries of Fig. 7.11 are such that $P(e \mid 0) = P(e \mid 1)$. Thus we have a *binary symmetric channel,* which means that the probability of error is the same regardless of which symbol is transmitted.

We next demonstrate that the error rate will be increased when we employ a threshold detection level other than the previously described optimum threshold. Once again we shall assume equal variance for the conditional densities and equal a priori probabilities for our discussion of Fig. 7.12. This figure demonstrates how the error rate will increase if an alternate threshold is employed instead of the crossing point of the conditional densities. The area of the lightly filled section represents the decrease in error rate for the 1 decision, while the sum of the two filled areas represents the increase in the error rate of the 0 decision. Since $P(0) = P(1)$, the darkened area (divided by 2 to account for the a priori probability) represents the increase in error rate caused by this shift of threshold.

If, on the other hand, we had lowered the threshold from the ideal setting, there would have been a small decrease in 0 decision errors along with a larger increase in the 1 decision error rate. Thus here, too, P_e would increase. In the specific case considered here, we have equal variances for both the transmission states, and the ideal threshold is the midpoint between the means $i_{s,1}$

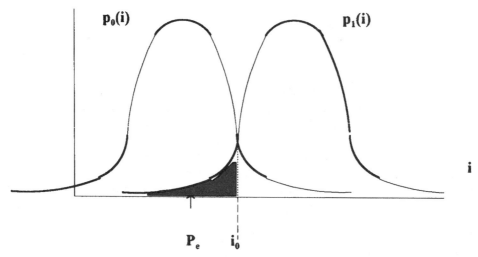

Figure 7.11   Example of an ideal threshold.

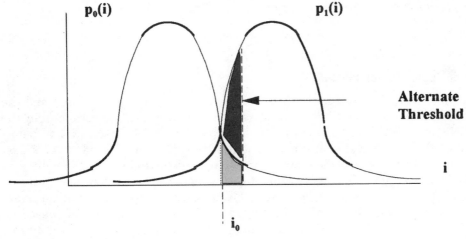

Figure 7.12   Alternate threshold.

and $i_{s,0}$ specifically, the optimum threshold is $i_0 = (i_{s,0} + i_{s,1})/2$. When the variances determined from the two conditional densities are not equal and $P(0) = P(1) = \frac{1}{2}$, then optimum i_0 will still occur at the crossover point, val that is, where $p_0(i) = p_1(i)$; however, the optimum threshold will not occur at the midpoint between the two conditional means.

In our next example we shall find the error rate in terms of the signal-to-noise ratio (SNR). We begin by considering the special case where $P(0) = P(1) = \frac{1}{2}$, $i_{s,0} = 0$, and $\overline{i^2_{N,0}} = \overline{i^2_{N,1}} = \overline{i^2_N}$. This corresponds to an OOK transmission system in which the receiver is dominated by thermal noise. The threshold for this case will be

$$i_0 = \frac{i_{s,1}}{2} \tag{7.12a}$$

Now let

$$\text{SNR} = \frac{S_{\text{peak}}}{N_{\text{mean}}} = \frac{i^2_{s,1}}{\overline{i^2_N}} = 2 \times E(\text{SNR})^\circ(\text{SNR})_1$$

where the subscript in $(\text{SNR})_1$ denotes that this symbol refers to the SNR when a 1 is transmitted. This particular notation will be used only for the ON state of an OOK system, specifically, the transmission of a pulse (1). It does not apply to FSK and PSK FOCS. However, the relation between the ON-state SNR and the average SNR will be used in Sec. 7.2 when we convert the SNR parameter into an average photons-per-bit parameter. The preceding assumptions, along with Eq. (7.11), yield

$$P_e = Q\left\{ \frac{i_{s,1}}{2\sqrt{\overline{i^2_N}}} \right\} \tag{7.12b}$$

In terms of the preceding notation, this becomes

$$P_e = Q\{\tfrac{1}{2} \sqrt{(SNR)_1}\} \qquad (7.12c)$$

7.2.3 Calculation approximation for Q(u)

Equation (7.12c) provides a direct connection between the error rate and the signal-to-noise ratio (SNR). Therefore, to obtain a numerical estimate for the SNR required to achieve a specific error rate, it will be necessary to determine the specific value of the argument of $Q(u)$ that corresponds to that particular error rate specification.[3] Somewhat crude estimates of this value can be obtained by employing a graph of $Q(u)$ versus u such as that shown in Fig. 7.13. In App. 7A we describe a relatively simple algorithm that provides more accurate values for the aforementioned value of u. This method is based on the asymptotic expansion in Eq. (7.13a), which turns out to be a good approximation for the very low error rate specification encountered in modern FOCs.

$$Q(u) \approx \frac{\exp[-(u^2/2)]}{u\sqrt{2\pi}} \left[1 - \frac{1}{u^2} + \frac{(1\cdot3)}{u^4} - \frac{(1\cdot3\cdot5)}{u^6} + \cdots \right] \qquad (7.13a)$$

---

[3] In effect, this process amounts to setting $u = Q^{-1}(P_e)$.

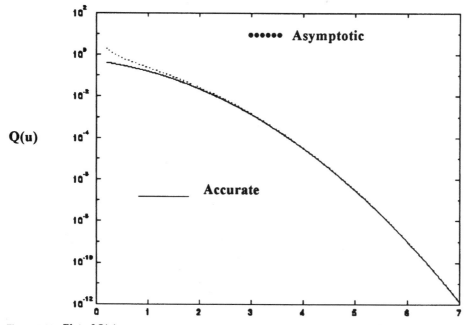

Figure 7.13 Plot of $Q(u)$ versus u.

Although the complete asymptotic form for $Q(u)$ is presented in this equation, the data labeled as *asymptotic* in Fig. 7.13 were obtained using the following lowest-order asymptotic approximation:

$$Q(u) \approx \frac{\exp[-(u^2/2)]}{u\sqrt{2\pi}} \qquad (7.13b)$$

The data for the accurate version of $Q(u)$ were obtained from the following equation:

$$Q(u) = \frac{1}{2}\,\mathrm{erfc}\left(\frac{u}{\sqrt{2}}\right) \qquad (7.14)$$

where[4]

$$\mathrm{erfc}(z) = \sqrt{2\pi}\int_z^\infty e^{-t^2}\,dt$$

Data plotted in Fig. 7.13 show that the short version of the approximation is well within graphical accuracy for arguments ≥ 3.

To obtain an estimate of the photons/bit parameter in Sec. 7.2.5, we need to determine the receiver SNR required to achieve a specific error rate. For the remainder of this subsection we describe a method to accomplish this task and provide several numerical examples of the technique. As a first example, we note that $Q(6) \approx 9.86 \times 10^{-10} \sim 10^{-9}$. Now

$$\frac{\exp[-(6^2/2)]}{6\sqrt{2\pi}} \approx 1.01 \times 10^{-9} \qquad \text{and} \qquad 1.01 \times 10^{-9}[1 - \frac{1}{6^2}] \approx 9.85 \times 10^{-10}$$

It can be seen from the preceding approximations that using only the first two terms of the asymptotic expansion provides a good fit to $Q(u)$. In modern FOCS with typical error criteria on the order of $P_e \sim 10^{-9}$ or better, it is necessary to obtain P_e to only one or two significant figures. At these low error rates a tiny change in the input SNR can easily alter P_e by an order of magnitude. In practice, it's the exponent that is of most importance. We shall use only the first-order approximation to solve example problems in this text.

Next we shall reverse this process to determine the signal-to-noise ratio that produces a prespecified error rate P_e. This will be accomplished by combining Eq. (7.12c) with the algorithm described in App. 7A. This process will be demonstrated in Example 7.1.

Example 7.1 We have already demonstrated that $Q(6) \approx 10^{-9}$. Now, combining this with Eq. (7.12c), we obtain $\frac{1}{2}(\mathrm{SENR})_1 = 6$, which leads to $(\mathrm{SNR})_1 = 12^2 = 144 =$

---

[4]The function erfc is based on erf as defined in most reference tables [7,8] as opposed to the definition of erf used in a number of graduate electrical engineering texts [5,9].

21.58 dB. Now, at the end of App. 7A we demonstrate that $Q(5.880) \approx 2 \times 10^{-9}$, which leads to $(SNR)_1 = 138.3 = 21.41$ dB. Thus, by increasing the signal by 0.17 dB, you can cut your error rate in half.

These results help explain why rough estimates of the error rate are usually sufficient. The difference in SNR required to achieve $P_e = 10^{-9}$ and 2×10^{-9} is only 0.17 dB.

7.2.4 Photons/bit criterion

So far our only measure of receiver sensitivity, specifically, the minimum signal strength required to achieve a desired error rate (P_e), has been the evaluation of the SNR in the preceding examples. The specific signal-level requirement for that SNR depends on determining the noise power, which itself may depend on the input signal level. A parameter often used in specifying minimum acceptable power levels at the receiving terminal of a FOCS is the minimum average number of photons per bit (ppb) required to achieve the desired P_e. To derive a formula for this system parameter, we now define the following notation:

p_b = Average number of photons per bit required to satisfy P_e

R = Data rate or number of bits per second

E_p = Energy/photon = $\dfrac{hc}{\lambda_0}$

S_{min} = Minimum input power required to achieve this value of p_b

Thus we have

$$S_{min} = Rp_b \frac{hc}{\lambda_0} \tag{7.15}$$

For example, suppose we are transmitting 1 Gbit/s at $\lambda_0 = 1.55$ μm. Now, if this particular system requires 100 photons per bit (which, by the way, is an excellent system in practice), we obtain the following from Eq. (7.15):

$$S_{min} = 10^{-9}(10^2) \frac{1.9865 \times 10^{-25}}{1.55 \times 10^{-6}} = 12.8 \text{ nW}$$

To place this in perspective, suppose we are using a 10-mW source at the transmitter and we wish to determine how far we can transmit this signal without repeaters while meeting the error rate specification. We can tolerate a net loss $(\mathcal{L})$ of

$$\mathcal{L} = 10 \log_{10}\left\{\frac{10 \times 10^{-3}}{12.8 \times 10^{-9}}\right\} = 58.9 \text{ dB}$$

Now suppose we use a fiber with a loss coefficient $\alpha = 0.2$ dB/km. Then we find the maximum spacing between repeaters L_{max} for this system to be

$$L_{max} = \frac{58.9}{0.2} = 294 \text{ km}$$

The reader should bear in mind that this result was completely predicated on a single criterion based entirely on obtaining sufficient signal power at the receiver to satisfy the error rate specification. Other important system considerations such as distortion caused by dispersion (which can be significant over such a long distance), and the effects of obtaining digital pulse (baud) synchronization from the noisy receiver signal, have not been included in this analysis.

In the preceding example a value for the number of photons per bit was assumed at the beginning in order to provide a numerical example. So far no attempt has been made to connect this parameter with the SNR required to attain the desired error rate. We now demonstrate a method for finding that SNR with examples based on OOK systems employing avalanche photodiodes (APDs) as detectors. The SNR for an APD is given by Eq. (6.28), which has been slightly modified as follows:

$$\text{SNR} = \frac{\overline{G}^2 I_p^2}{2eB(I_p + I_d)\overline{G}^{2+x} + \dfrac{4kTB\Gamma}{R_L}} = \frac{i_{s,k}^2}{i_{N,k}^2} \tag{7.16}$$

Now, for sufficiently large $\overline{G}$, we can ignore the thermal noise, and we shall also omit the dark-noise contribution to obtain

$$\frac{\overline{i_{s,k}^2}}{i_{N,k}^2} \approx \frac{\overline{G}^2 I_p^2}{2eB(I_p)\overline{G}^{2+x}} = \frac{I_p}{2eBF} \tag{7.17}$$

where we have used the excess noise factor $F = \overline{G}^x$ generated by the avalanche gain process. Note also that we have discarded the relatively small thermal noise which is not amplified by the avalanche process, and we have canceled out a common factor of $\overline{G}^2$ to obtain the last ratio on the right-hand side of Eq. (7.17).

Now we impose OOK modulation on the binary communication system and assume that $i_{s,0} = 0$ for the OFF state. When a 1 (the ON-state pulse) is sent, we obtain the following SNR:

$$(\text{SNR})_1 \approx \frac{I_p}{2eBF} = \frac{\eta_d e(<m_C>/T)}{2eBTF} = \frac{\eta_d e<m_C>}{2e(BT)F} \tag{7.18}$$

where $<m_C>$ is the average number of detected photons in T seconds under the condition that a 1 was sent, η_d is quantum efficiency of the detector, and B is the bandwidth of the channel. The subscript on $(\text{SNR})_1$ denotes that this SNR applies specifically to an ON-state transmission. We also note that

$$I_p = \frac{\eta_d e<m_C>}{T} = \text{charge per second}$$

was used in Eq. (7.18). This equation forms the basis for the following connection between $<m_C>$ and the ON-state SNR:

$$<m_c> = \frac{2(BT)F}{\eta_d} \, (\text{SNR})_1 \qquad (7.19)$$

The time–bandwidth product BT in Eq. (7.17) is a direct function of the digital pulse shape. The minimum value for this quantity as obtained from Fourier analysis is $BT = 0.5$, while a raised cosine pulse shape has $BT = 0.6$.

Example 7.2   Consider a system with the following parameters: $\eta_d = 0.8$, $F(\overline{G}) = 4$, $BT = 0.6$, and $(\text{SNR})_1 = 144$. The value of the SNR was selected to match the $P_e = 10^{-9}$ condition that we obtained in Sec. 7.2.3 for the equal-variance model that we looked at earlier. Using these values in Eq. (7.19), we obtain

$<m_C> = 864$ photons/logical 1 $\Rightarrow$ p$_b$ = 432 photons/bit   for OOK

This result expands to $\sim 10^3$ photons/bit (ppb) for an actual APD receiver.

In the next example, determine the number of photons per bit required to achieve $P_e \leq 10^{-9}$ for an ideal APD-based receiver.

Example 7.3   Let

1. $F = 1$ (this implies *no excess noise* although the usual shot noise is still present)
2. $\eta_d = 1$ (*100% percent quantum efficiency*)
3. $BT = 0.5$ (*minimum value possible from Fourier transform theory*)

Then, using Eq. (7.19), we obtain $<m_C> = 2(0.5)(1)(144) = 144$ photons/logical 1 $\Rightarrow$ 72 ppb

Although this result represents a significant improvement over the setup used in Example 7.2, it still raises the question as to *why this result differs so much from the 20 photons per logical 1 that we obtained for ideal photon-counting detection*. The answer to this question lies in recognizing that some of the assumptions that were made for ideal photon-counting analysis have not as yet been adapted to integrated-current-based detection. In particular, Example 7.3 was based on the assumptions that

$$\sigma_0^2 = \sigma_1^2 \qquad \text{and} \qquad i_0 = \frac{i_{s,1}}{2} \qquad \text{since} \qquad i_{s,0} = 0$$

when we derived the SNR = 144 requirement to satisfy $P_e = 10^{-9}$. One consequence of the equal-variance assumption is that it leads directly to a binary symmetric channel. On the other hand, it may be recalled that the ideal photon-counting detector never made an error when a 0 was sent. We also note that when a 0 is sent in an OOK system, the primary noise at the receiver is thermal noise while avalanche-effect-amplified shot noise determines the variance when a 1 is sent. Now, for an APD detector with large predetection amplification, it is reasonable to assume that not only does

$$\sigma_0^2 \neq \sigma_1^2 \qquad \text{but also} \qquad \sigma_0^2 \ll \sigma_1^2$$

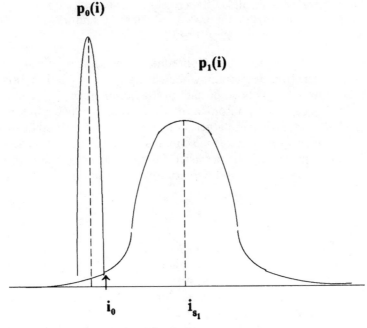

Figure 7.14 Threshold when $\sigma_0^2 \ll \sigma_1^2$.

This also implies that the decision threshold $i_0 \neq i_{s,1}/2$. Furthermore, since $\sigma_0^2 \ll \sigma_1^2$, the conditional density curves look more nearly like those shown in Fig. 7.14 than the conditional densities depicted in Fig. 7.10.

Figure 7.14 shows that the threshold i_0 is only slightly larger than zero. A somewhat more careful analysis of the intersection between these two conditional gaussian densities leads to the following selection of threshold for $P(0) = P(1)$:

$$i_0 \approx i_{s,1} \frac{\sigma_0}{\sigma_1} \left\{ 1 + \frac{2\,\sigma_1^2}{i_{s,1}^2} \ell n\!\left(\frac{\sigma_0}{\sigma_1}\right) \right\}^{1/2} \tag{7.20}$$

where we assumed $i_{s,0} = 0$ in addition to $\sigma_0^2 \ll \sigma_1^2$.

Now we found earlier, in Eq. (7.9), that

$$P(e\,|\,1) = Q\!\left\{ \frac{i_{s,1} - i_0}{\sigma_1} \right\}$$

or (assuming $i_0 = 0$)

$$P(e\,|\,1) \approx Q\!\left\{ \frac{i_{s,1}}{\sigma_1} \right\}$$

Then, using Eq. (7.12) and the definition for $(\mathrm{SNR})_1$, we obtain

$$p(e\,|\,1) = Q\!\left\{ \sqrt{(\mathrm{SNR})_1} \right\} \tag{7.21}$$

Example 7.4 To demonstrate the results of the preceding analysis, let $P_e = P(e \mid 1)$ $= 10^{-9}$. Then, as in previous examples, we have $\sqrt{(\text{SNR})_1} \approx 6.0$ and $(\text{SNR})_1 \approx 36$. Thus $<m_C> = 2(0.5)(1)36 = 36$ photons/logical $1 \Rightarrow 18$ ppb $= p_b$.

This result is clearly much closer to the 10 photons per bit that we obtained for OOK with a photon-counting detection system. Finally we note that this result applies only to OOK systems with perfect predetection gain ($F = 1$ and $\overline{G} \gg 1$), as was also the case for photon-count detection. In both photon-counting and integrated-current-based detection systems, these values for p_b could be improved by using other modulation schemes such as PSK. In particular, considerable improvements can be achieved (at the price of complexity) by employing modern error correction and detection coding techniques, as we shall see in Chap. 10.

7.2.5 Sensitivity for optical preamplifier detection

We now consider an OOK system employing a combination of an optical preamplifier and p–i–n photodiode detection system. Equation (6.34) presents the optimum SNR for this detection system under conditions in which the optical preamplifier (OP) has a large gain and all the Er^{3+} atoms are in the upper state. To arrive at a numerical value for the minimum photons/bit parameter, we need to recast the right-hand side of Eq. (6.34) in the following form:

$$\frac{P_s}{4n_{eq}h\nu B_e} = \left\{ \frac{P_s T}{h\nu} \right\} \frac{1}{4n_{eq}(B_e T)} \tag{7.22}$$

where T is the baud or pulse duration, P_s is the average signal power entering the photodiode (assumed here to have a 100 percent quantum efficiency ($\eta = 1$), n_{eq} is given by Eq. (6.29i) (note that some typical values for n_{eq} are displayed in Fig. 6.8), and B_e is the 3-dB bandwidth of the photodiode electrical response. The only change that has been made here is to multiply both numerator and denominator by the pulse duration T. With this notation, the bracketed term clearly represents the average energy per pulse as expressed in terms of the average number of photons per pulse.

Example 7.5 We now consider OOK transmission with an optical preamplifier at the receiving terminal. The term in brackets in Eq. (7.22) represents the average number of photons per bit. Now we once again consider the unequal variance case represented by Fig. 7.14 and employ the ideal value for the SNR as we employed in Example 7.4, $(\text{SNR})_1 \approx 36$. We also assume that $n_{eq} = 1$ for ($G \gg 1$) and $(B_e T) = 0.5$. This leads to 72 photons for a 1 and an average $p_b = 36$ ppb.

The 36-photon/bit sensitivity obtained in Example 7.5 is in good agreement with analogous results published elsewhere for Optical Preamplifier/p–i–n photodiode receivers. These results are tabulated by Desurvire [5, p.186] and show receiver sensitivities in the range of $pb \sim 38$ to 43 photons/bit, depending on receiver details.

7.2.6 Predetection gain and receiver sensitivity

We conclude this chapter with a discussion of how the results of Examples 7.2 through 7.5 demonstrate the effects of predetection gain on FOCS receiver sensitivity. One common thread for all these examples is that only OOK modulation was considered. The calculations for Examples 7.2 and 7.3 are predicated on equal variance for both signal states, while $\sigma_0{}^2 \ll \sigma_1{}^2$ (see Fig. 7.14) was assumed for Examples 7.4 and 7.5.

Equal variance will arise whenever thermal noise dominates over the shot noise in the receiver. In that case the noise is not dependent on the incoming signal. However, when the primary detection system is a based on a large-gain APD, the variance is definitely dependent on the input signal. This is especially the case for OOK modulation where the APD-amplified shot noise is much larger for the ON state than the OFF state. We included APD in these two examples to demonstrate both the effects of excess noise in PD detection and the fallacy of using equal variance to model OOK detection in large-predetection-gain configurations. Note that the primary difference between Examples 7.2 and 7.3 is the absence of excess noise in the latter case.

In Example 7.4 we correct the equal-variance fallacy and employ $\sigma_0^2 \ll \sigma_1^2$ to derive a receiver sensitivity of 18 photons/bit. It should be noted here that this result was obtained using an unrealistic excess-noise factor of $F = 1$ or 0 dB for the APD detector. Although we did not specify an actual APD gain, we assumed that it was large enough to negate the effects of thermal noise. A more realistic value for an APD with $\overline{G} = 100$ (40 dB power gain) would be $F = (100)^{0.7} \sim 25$ or 14 dB, for a GaAs-type APD detector.[5] This would result in ~450 photons/bit, which clearly results in a much poorer receiver sensitivity than can be obtained using optical preamplifier detection as can be seen by comparing Examples 7.4 and 7.5. We note that for optimum conditions the excess noise for the preamplifier is 3 dB. In Fig. 6.9 experimental noise figure data are presented, and while some of the noise figures are larger than 3 dB, all are much less than the number quoted above for a 40-dB-gain APD receiver. Clearly the sensitivity of the optical preamplifier receiver is superior to that of an APD-based receiver. However, the APD receiver is considerably less expensive to construct and maintain.

The primary purpose of predetection gain is to negate the effects of electronic noise generated at the receiver, particularly the effects of thermal noise. With this in mind, it is clear that there is nothing to be gained by

---

[5]GaAs is the source of the 0.7 exponent on the excess-noise factor, and we assume here that the transmission wavelength is either $\lambda_0 = 1.3$ μm or $\lambda_0 = 1.55$ μm. The bandgap of silicon is too large for these wavelengths.

employing both an OP and an APD as a combination at the receiver. The OP is enough to do the job, and it is only necessary to follow the OP with a simple p–i–n photodiode. In fact, Desurvire [5] shows that not only does the combination of an APD in conjunction with an OP fail to improve receiver performance over OP/p–i–n detection, it may actually deteriorate the performance because of the excess noise factor of the APD.

In the next chapter we shall describe optical heterodyne detection (OHD), which also provides a predetection gain mechanism. This system can detect frequency modulation and phase modulation in addition to amplitude modulations. As we shall see in Chap. 8, FSK and PSK result in equal-variance detection statistics. Under ideal conditions a 3-dB penalty is incurred in OHD which is identical to the optimum preamplifier. However, we shall see in Chap. 10 that there are aspects of OHD that, if properly utilized, lead to better receiver sensitivity than is attainable with an optical preamplifier.

Problems

1 Let $P_e = 10^{-12}$ for a supercomputer interconnect system, and find the following:
   (**a**) The u required to achieve this error rate by using the asymptotic form and the relaxation-method-based technique we described in class.
   (**b**) The SNR for an OOK with $i_{s,0} = 0$ and equal variance for both the 1 and 0 states.
   (**c**) The number of photons per bit, given $F = 1$, $\eta = 0.7$, $BT = 0.6$, and equal variance still in effect.
   (**d**) Repeat part **c** with $\eta = 1$ and BT = 0.5.
   (**e**) Now try the ultimate OOK where $\sigma_0^2 = 0$, $\eta = 1$, $BT = 1$, and $F = 1$, and find SNR and photons per bit.

2 Consider the FOCS system displayed in Fig. P7.2. It has the following specifications. The system transmits at 10 Gbits/s at $\lambda_0 = 1.55$ μm and requires 1000 photons/bit (ppb) to satisfy the error rate criterion. The amplifier input must have at least 2000 ppb. Let the total length of the system be 400 km, the loss coefficient $\alpha = 0.2$ dB/km, and $P_{in} = 10$ mW.
   (**a**) Find L_1.
   (**b**) Find L_2.
   (**c**) Determine the gain of the amplifier that just barely satisfies the specifications.

3 Now consider Fig. P7.2 with all specifications the same as in problem 2, with the following two exceptions. Let the gain of the optical amplifier preset 50 dB, and in this

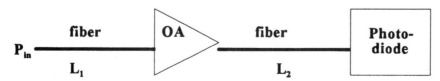

Figure P7.2 Flowchart of a hypothetical FOCS system.

problem there is no prespecified total length for the system. The task here is to find the maximum length of the total system that will satisfy the specifications.

4 Let λ_0 1.3 μm and the loss coefficient $\alpha = 0.4$ dB/km.

   (a) Repeat problem 3 using these specifications.

   (b) Has the achievable distance been halved by this process? Explain.

5 Suppose that an optimum optical preamplifier with 40 dB gain is inserted directly before the photodiode receiver in Figure P7.2. Now repeat problem **3** using this setup. *Before you begin, consider how the presence of this OP alters the problem.*

6 Let the noise in a receiver for in an OOK system be dominated by thermal noise.

   (a) How much more transmitter power (expressed in decibels) is required reduce the error rate from 10^{-9} to 10^{12}?

   (b) Now assume that an optical preamplifer with large gain is inserted into the receiver. Repeat part *a* using $p(e \mid 1)$ for this system.

Appendix 7A. Numerical Evaluation of $Q(u)$ and $Q^{-1}(u)$

7A.1 Numerical approximations for $Q(u)$

The key to the numerical techniques described here is the asymptotic form for $Q(u)$ given by Eq. (7.13a), which we repeat here for convenience:

$$Q(u) \approx \frac{\exp[-(u^2/2)]}{u\sqrt{2\pi}}\left[1 - \frac{1}{u^2} + \frac{(1\cdot3)}{u^4} - \frac{(1\cdot3\cdot5)}{u^6} + \cdots\right] \qquad (7A.1)$$

As an example, we note that [6] $Q(6) \approx 9.86 \times 10^{-10} \sim 10^{-9}$. Now

$$\frac{\exp[-16^2/2]}{6\sqrt{2\pi}} \approx 1.01 \times 10^{-9} \quad \text{and} \quad 1.01 \times 10^{-9}\left[1 - \frac{1}{6^2}\right] \approx 9.85 \times 10^{-10}$$

It can be seen from these numbers that in the vicinity of $P_e = 10^{-9}$ that an approximation based on the first two terms in Eq. (7A.1) provides a very good fit to $Q(u)$. In modern FOCS with typical error criteria on the order of $P_e \sim 10^{-9}$, and as shown in Example 7.1, it is necessary to obtain P_e to only one or two significant figures. At these low error rates, a tiny change in SNR can easily alter P_e by an order of magnitude. In practice, it's the exponent in P_e that is of primary importance. We also note that even the first term in Eq. (7A.1) alone is reasonably accurate, and we shall use only the first-order approximation in examples and to solve problems in this text:

$$Q(u) \approx \frac{\exp[-(u^2/2)]}{u\sqrt{2\pi}} \qquad (7A.2)$$

As a rule, the accuracy of an asymptotic approximation improves as the argument of the function increases for any fixed number of terms in the asymptotic expansion. However, as can be seen from the factorial-type terms that appear in the numerators in Eq. (7A.1), the asymptotic series will eventually diverge as the number of terms increases while the argument remains constant. For this reason asymptotic series are sometimes called *divergent series*. However, it is important to note here that an asymptotic expression using only a few terms of the expansion often provides relatively simple and accurate approximations to many functions for large values of the argument.

7A.2 Numerical evaluation of $Q^{-1}(u)$

The specific equation that we shall use is obtained by taking the natural logarithm of both sides of Eq. (7.14). Then we rearrange the resulting equation so that the term with the strongest dependence on the variable u stands alone on one side of the equation and arrive at

$$u^2 = 2\{-\ell n(u) - \tfrac{1}{2}\ln(2\pi) - \ln(p_e)\} \tag{7A.3}$$

To demonstrate the algorithm, we now work three examples, the results of which will be used in Example 7.1 to demonstrate that accuracy to two significant figures is sufficient for $P_e \sim 10^{-9}$ (or smaller).

This is an iterative technique and the first approximation to u (denoted by u_1) is obtained by ignoring the small $\ln(u)$ term on the right-hand side of Eq. (7A.3).

1. Let $P_e \sim 10^{-9}$. Then we have

$$u_1^2 = -\ell n(2\pi) - 2\ln(10^{-9}) = -1.83879 + 41.4465 = 39.6087$$

or

$$u_1 = 6.2935$$

Now, for our second estimate u_2, we substitute the $\ln(u_1)$ term into Eq. (7A.3). This yields

$$u_2^2 = -2\ln(6.2935) + 39.6087 = 35.9297; \qquad \Rightarrow u_2 = 5.9941$$

$$u_3^2 = -2\ln(5.9941) + 39.6087 = 36.083 \qquad \text{and} \qquad u_3 = 6.002$$

Thus, in only two iterations, we hone in quickly on an answer. In the remaining two examples we shall terminate the process with the evaluation of u_2 in order to demonstrate how small a change in SNR can change P_e by a factor of 2.

2. Let $P_e = 2 \times 10^{-9}$. Then

$$u_1^2 = -\ell n(2\pi) - \ln(2\pi - 2\ln(2 \times 10^{-9}) = -1.8379 + 40.0602 = 38.222$$

$$u_1 = 6.182422$$

$$u_2^2 = -2\ln(6.182422) + 389.2222 = 34.579; \qquad \Rightarrow u_2 = 5.880$$

3. Let $P_e \sim 10^{-10}$. Then

$$u_1^2 = -\ln(2\pi) - 2\ln(10^{-10}) = -1.8379 + 46.0517 = 44.2138$$

$$u_1 = 6.49346$$

$$u_2^2 = -2\ln(6.49346) + 44.2138 = 40.4248; \quad \Rightarrow u_2 = 6.359.$$

These results are used in Example 7.1 to demonstrate how a small variation in the input power can have a large effect on the error rate at very low error rates.

References

1. B. E. A. Saleh and M. C. Teich, *Fundamentals of Photonics,* Wiley, New York, 1991.
2. G. J. Foschini and G. Vannucci, "Using spread-spectrum in high-capacity fiber-optic local network," *J. Lightwave Technol.* **6,** 3709 (1988).
3. P. R. Prucnal, M. A. Santoro, and T. R. Franl, "Fan spread spectrum fiber-optic local area network using optical processing," *J. Lightwave Technol.* **4,** 547–554 (1986).
4. W. B. Davenport, Jr. and W. L. Root, *An Introduction to the Theory of Random Signals and Noise,* McGraw-Hill, New York, 1958, p. 145.
5. E. Desurvire, *Erbium-Doped Fiber Amplifiers,* Wiley, New York, 1994.
6. C. W. Helsrrom, *Probability and Stochastic Processes,* Macmillan, New York, 1991, p. 593.
7. M. Abramowitz and I. A. Stegun, *Handbook of Mathematical Functions with Formulas, Graphs, and Mathematical Tables,* National Bureau of Standards, 1964, p. 297.
8. W. H. Beyer, *Handbook of Tables for Probability and Statistics,* 2d ed., Chemical Rubber Co., Cleveland, Ohio, 1968, p. 126.
9. A. Papoulis, *Probability, Random Variables, and Stochastic Processes,* 2d ed., McGraw-Hill New York, 1984.

Coherent Lightwave Communications

8.1 Introduction

8.1.1 Heterodyning at nonoptical frequencies

In this chapter we shall concentrate on *coherent detection* (CD), which consists of the optical equivalent to heterodyne detection (HD) as well as homodyne detection. We begin this chapter with a brief review of common radio-frequency-type heterodyne detection. Originally heterodyne detection was developed to simplify the tuning of radio receivers. A number of other benefits that accrue with HD were discovered, and today almost all radio, television, satellite, microwave relay, cellular telephone, and radar receivers employ HD receivers. A block diagram for a typical "super"heterodyne receiver is shown in Fig. 8.1.

Let $r(t) = A \cos(\omega_c t + \phi_c)$ represent the received signal and $x(t) = B \cos(\omega_L t)$ represent the local oscillator output. Then the product of these two sinusoids $z(t)$ is

$$z(t) = \frac{AB}{2}\left\{\cos\{(\omega_{if})t + \phi_c\} + \cos\{(\omega_c + \omega_L)t + \phi_c\}\right\}$$

where $\omega_{if} = |\omega_L - \omega_c|$.

In most heterodyne detection systems the IF stage is a sharply tuned bandpass filter centered at f_{if}; both the bandwidth and the center frequency of the IF stage remain constant as the receiver is tuned from one channel to another. The IF stage is the site in a heterodyne receiver where all unwanted adjacent stations [as well as the last term in the $z(t)$ equation] are eliminated. For the particular signals specified here, the output of the IF stage $y(t)$ will be

$$y(t) = \frac{AB}{2}\cos\{\omega_{if}t + \phi_c\}$$

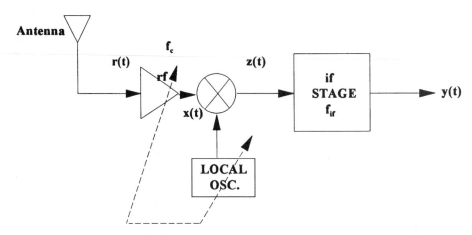

Figure 8.1 Standard "superhet" receiver.

which is then detected and processed in a manner consistent with the modulation scheme of the channel. A key characteristic of heterodyne receivers is the simple manner in which they are tuned by adjusting the local oscillator frequency so that the difference frequency between the local oscillator and the desired channel coincides with the passband of the IF stage. Of course, the dial is usually calibrated to display the center frequency of the channel.

The system structure as described so far would actually permit two stations to pass through the IF stage; one of which is at $\omega_{c1} = \omega_L + \omega_{if}$ and the other is at $\omega_{c2} = \omega_L - \omega_{if}$. Whichever of these stations is an unwanted channel is called the "image" channel. The image channel can be eliminated only if it is blocked before it reaches the mixer stage. Otherwise the output of the IF stage will contain superposition of the two channels, which is almost impossible to separate. However, the image channel can be eliminated quite easily by designing the rf amplifier stage to act as a crude tunable bandpass filter. The rf stage is tuned in tandem with the local oscillator frequency, as indicated by the dashed line in Fig. 8.1, in such a way that the passband of the rf stage is centered on the carrier frequency of the desired station. It turns out that the bandwidth constraints on the rf stage filter are much looser than those of the (nontunable) IF-stage amplifier-filter combination. The rf stage can be a fairly crude filter because it is only necessary to select the desired signal from an unwanted image channel signal which is spaced $2f_{if}$ from the desired channel. This spacing is usually much larger than the channel bandwidth. We note that no attempt is made at the rf stage to separate out closely packed adjacent channels for, as stated earlier, that task is left to the IF stage. We shall see in Chap. 9 that image channel rejection is a much more complicated task at optical frequencies than at rf frequencies. It turns out that image channel rejection by predetection optical filtering is more difficult because of the need to separate two optical signals only a few gigahertz apart centered at an optical frequency at approximately 2×10^{14} Hz. Fortunately there are signal processing methods (albeit complicated) for removal of the image channel after heterodyne detection.

Another significant difference between optical and rf heterodyne detection systems concerns the nature of amplification. All the amplifiers displayed in Fig. 8.1 are what are known as *postdetection amplifiers*. Both the incoming signal and local electronic noise are equally amplified by these amplifiers. As we shall soon see, however, optical heterodyne detection does provide conversion gain, which is a form of predetection gain. One important advantage common to all heterodyne detection systems is that the intermediate frequency is usually better suited to further signal processing than is the spectral range in the neighborhood of the carrier frequency. This is particularly important for optical heterodyne detection where the demodulation and detection processes are accomplished at microwave frequencies using well-established microwave circuit technology, including monolithic microwave integrated chips. As we shall see in this chapter, optical heterodyne detection employs this property to detect optical PSK and QPSK channels.

8.1.2 Coherent detection of optical signals

The heterodyning of optical frequency signals is not possible using diode switching techniques. At optical frequencies heterodyne detection is obtained by superposing the signal with a local laser output and utilizing the nonlinear aspects of the photon absorption process to perform the mixing operation. Optical heterodyning was actually achieved even before the advent of the laser by superposing two very narrow natural spectral lines that were closely spaced in frequency. After the advent of the laser, but before modern fiber optics, the primary interest in optical heterodyning was for detecting highly directional weak signals such as those encountered in lidar (optical frequency radar). Superposition of local lasers with the received signal in these early examples of optical heterodyne detection was achieved with the aid of a beam splitter and required extremely accurate optical alignment of the two fields. In FOCS the optical alignment problem is avoided by using devices known as *optical couplers* to superpose the optical fields. Optical couplers are readily available commercially and consist more or less of closely pressed optical fibers. One nice feature of commercial optical couplers is that they come supplied with "pigtails" (attached optical fibers) for both inputs and outputs. A block diagram of a typical optical heterodyne setup is shown in Fig. 8.2.

On the basis of Fig. 8.2 we have

$$s_0(t) = A \cos(\omega_0 t + \phi_S) + B \cos(\omega_L t + \phi_L)$$

where the parameters A, ω_0, and ϕ_S are the amplitude, frequency, and phase of the received optical signal, respectively, while B, ω_L, and ϕ_L are the corresponding parameters for the local laser. We now define the intermediate fre-

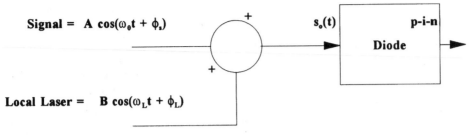

Figure 8.2 Optical heterodyne setup.

quency ω_i by $\omega_i = |\omega_L - \omega_0|$, and we note that typically in OHD we have $\omega_0 \approx \omega_L$ and $\omega_i \ll \omega_L$, and therefore we also have typically that $\omega_i \ll \omega_0$. Then, using $\omega_0 = \omega_L - \omega_i$ as well as a well-known trigonometric identity for $\cos\{[\omega_L t + \phi_L] - [\omega_i t + \phi_L - \phi_S]\}$, we obtain

$$\cos(\omega_0 t + \phi_S) = \cos(\omega_L t + \phi_L)\cos[\omega_i t + \phi_L - \phi_S] + \sin(\omega_L t + \phi_L)\sin[\omega_i t + \phi_L - \phi_S]$$

Thus we can express the superposition of signal and local laser as represented by $s_0(t)$ as

$$s_0(t) = [B + A\cos(\omega_i t + \phi_s + \phi_L)]\cos(\omega_L t + \phi_L)$$

$$+ [A\sin(\omega_i t + \phi_s + \phi_L)]\sin(\omega_L t + \phi_L) \qquad (8.1a)$$

Now, since $\omega_i \ll \omega_L$ and $\omega_i \ll \omega_0$, we can use the narrowband approximation

$$s_0(t) = E(t)\cos[\omega_L t + \beta(t)] \qquad (8.1b)$$

where

$$E(t) = \sqrt{[B + A\cos(\omega_i t + \phi_L - \phi_S)]^2 + [A\sin(\omega_i t + \phi_L - \phi_S)]^2} \qquad (8.1c)$$

$$\beta(t) = \tan^{-1}\left\{\frac{A\sin(\omega t + \phi_L - \phi_S)}{B + A\cos(\omega_i t + \phi_L - \phi_S)}\right\} \qquad (8.1d)$$

As can be seen from Eq. (6.1), the instantaneous output current for the diode is then proportional to $I(t) \propto E^2(t)$. We shall soon utilize this nonlinear connection between the optical frequency electric field and the response current observed by the photodiode to describe statistical properties of coherent detection.

8.1.3 Direct detection systems

We shall employ the preceding notation in the next four subsections, where we examine some simplified detection models for several types of coherent detection[1] systems. In particular, we shall employ photon-counting detection to simplify our analysis, and furthermore, we shall disregard entirely the effects of phase noise. Our goal in the remainder of this section is to describe a few general trends in the behavior of several systems rather than obtain specific numerical values for the system performance of actual systems.

Consider an intensity-modulated direct-detection (IMDD) OOK communication system in which the amplitude of the transmitted A_0 signal is constant for the entire T-second baud duration. In terms of the notation used to arrive at Eqs. (8.1) and (8.2), let $B = 0$, $\omega_i = 0$, and

$$A_0 = \begin{cases} 0 & \text{if } 0 \\ A & \text{if } 1 \end{cases}$$

where A is the signal amplitude at the receiver when a 1 is transmitted. Recall that the photodiode current is proportional to the magnitude squared of the electric field. In this case the integrated current response of the detector is given by

$$\int_0^T I(t)dt = A_0{}^2 \int_0^T dt = A_0{}^2T = \begin{cases} 0 & \text{if } 0 \\ A^2T & \text{if } 1 \end{cases} \tag{8.2}$$

This situation is the same OOK system that we described earlier, where the decision process is

$$\begin{array}{lll} \text{If} & m = 0 & \text{select} \quad 0 \\ \text{If} & m > 0 & \text{select} \quad 1 \end{array}$$

Now the average number of photons per bit p_b is given by

$$p_b = \frac{1}{2}(A^2T) + \frac{1}{2}(0) = \frac{A^2T}{2}$$

where the $\frac{1}{2}$ factor in this expression comes from the a priori probabilities. It does not arise from $<\cos^2>$, which doesn't apply here [note how $i(t)$ is proportional to the magnitude squared of the electric field in Eq. (6.1)].

Using the homogeneous Poisson counting model along with the assumption that errors can occur in this ideal photon-counting detection system when a 1 is sent, we can write

$$P_e = \frac{1}{2} e^{-<m|1>} = \frac{1}{2} e^{-A^2T} = \frac{1}{2} e^{-2p_b} \tag{8.3a}$$

---

[1]In this text the term *coherent detection* incudes both heterodyne detection and homodyne detection.

Recall that to achieve $P_e = 10^{-9}$, we require that $2p_b = 20$. Thus, on the average, $p_b = 10$ photons/bit. This is, of course, a repeat of what we found in Chap. 6. We reintroduce it here in order to compare this result with several variations of coherent detection systems, and to apply it to several new situations where this detection scheme (although impractical) yields some system performance parameters in a simple manner.

8.1.4 Homodyne detection

Let us now return to the general heterodyne-type picture and consider the following special case: $A_0 = \pm A$(BPSK), $B = A$, and $\omega_i = 0$.

We shall also assume that both ϕ_L and $\phi_S = 0$. Physically, this implies that we phase-lock the local oscillator to the received signal. Then we also assume that $\omega_L = \omega_0$. Note that phase locking could in principle be achieved with a harmonic of the incoming signal, but this process would be very difficult in FOCS. This particular version of coherent detection is known as *homodyne detection*. In the ideal homodyne case, the envelope of the superposition of the local laser and received signal is given by

$$|E(t)| = A \pm A = \begin{cases} 0 & \text{if} \quad 0 \\ 2A & \text{if} \quad 1 \end{cases}$$

and

$$b(t) \propto (A \pm A)^2 = \begin{cases} 0 & \text{if} \quad 0 \\ 4A^2 & \text{if} \quad 1 \end{cases}$$

For photon counting

$$<m> = \begin{cases} 0 & \text{if} \quad 0 \\ 4A^2T & \text{if} \quad 1 \end{cases}$$

Therefore this somewhat unusual and unrealistic detection process is equivalent to OOK-type detection even though the transmission is a PSK format. *Note, however, that for this detection process to function at all, we require a phase-locked local oscillator with exactly the same amplitude and phase as the incoming signal.* This detection system is not feasible at optical frequencies, particularly in the presence of phase noise and relative intensity noise. Nevertheless, we shall analyze this system using results obtained for the previous OOK calculation. Thus we have

$$p_b = \tfrac{1}{2}(4A^2T) + \tfrac{1}{2}(0) = 2A^2T$$

Now, for PSK, there is always a pulse with amplitude A transmitted for each data pulse; hence $p_b = A^2T$. Thus

$$<m\,|\,1> = 4A^2T = 4p_b \qquad \text{and} \qquad P_e = \tfrac{1}{2}e^{-<m\,|\,1>} = \tfrac{1}{2}e^{-4p_b} \qquad (8.3b)$$

This ideal homodyne detection system is 3 dB better than a direct-detection OOK system and results in $p_b = 5$ photons/bit to achieve $P_e = 10^{-9}$.

8.1.5 Realistic homodyne system

Once again we assume optical frequency phase lock and that $\omega_i = 0$, but now we make the more realistic practical assumption that $B >> A$ (we shall see in this chapter that $B >> A$ is assumed for almost all cases of CD). We now consider a BPSK FOCS, which means that we can write the transmitted amplitude in the usual PSK form as $A_0 = \pm A$. Then we have

$$E^2(t) = (B \pm A)^2 = B^2 + A^2 \pm 2BA$$

The average number of photons for the two transmission states may be summarized by

$$<m> = (B^2 + A^2 \pm 2BA)T = (B^2 + A^2)T \pm 2BAT$$
$$\qquad\qquad\qquad\qquad \text{bias} \qquad\quad \text{antipodal BPSK}$$

It is important to note here that the bias term does not contribute to the signal part of the output, but it does contribute to the shot noise level. Since $B >> A$, it follows that

$$B^2 + A^2 \approx B^2 \qquad \text{and} \qquad \sigma^2 = <m_{\text{total}}> \approx B^2 T$$

Some representative densities (with the means downshifted by B^2T) are shown in Fig. 8.3.

Figure 8.3 clearly represents a binary symmetric channel with a probability of error given by

$$P_e = \frac{1}{\sqrt{2\pi\sigma^2}} \int_0^\infty \exp\left[\frac{-(i + \mu)^2}{2\sigma^2}\right] di$$

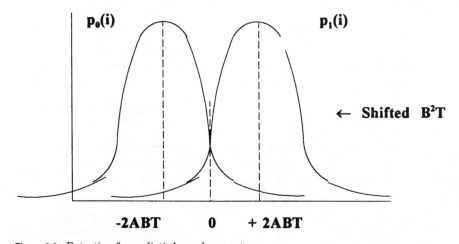

Figure 8.3 Detection for realistic homodyne system.

where

$$\mu = -i_{s,0} = 2ABT \qquad \text{and} \qquad \sigma^2 = B^2T$$

Now, as in Chap. 7, let

$$Y = \frac{i + \mu}{\sigma} \qquad \text{and} \qquad \sigma \, dY = di$$

$$P_e = \frac{1}{\sqrt{2\pi}} \int_{\frac{\mu}{\sigma}}^{\infty} \exp\left[-\frac{Y_2}{2}\right] Q\left\{\frac{\mu}{\sigma}\right\} = Q\,\frac{2ABT}{B\sqrt{T}} = Q\{2A\sqrt{T}\}$$

where the local laser amplitude B cancels out to yield the final form of the preceding equation. We now employ Eq. (7.15) to obtain

$$P_e \approx \frac{e^{-2A^2T}}{2\sqrt{2\pi A^2 T}} \tag{8.4a}$$

As in the case of ideal homodyne detection, the average number of photons per bit $p_b = A^2T$ for BPSK. Therefore, we have

$$P_e \approx \frac{e^{-2p_b}}{2\sqrt{2\pi p b}} \tag{8.4b}$$

Now the exponential term in Eq. (8.4b) provides dominant dependence on p_b, as can be seen from

$$\ln(P_e) = -2p_b - \tfrac{1}{2}\ln(2\pi p_b) - \ln(2) \approx -2p_b \qquad \text{(for } p_b \gg 1\text{)}$$

Often the denominator in Eq. (8.4b) is simply ignored, as indicated in the preceding expression, to arrive at the following approximation for the direct relation between the error rate and p_b:

$$P_e \approx e^{-2p_b} \tag{8.4c}$$

Some of the following important trends are apparent in the material considered so far:

1. The same energy is transmitted for each baud in a BPSK, while OOK systems transmit energy only half of the time.

2. The performance of the more realistic version of homodyne detection is 3 dB poorer than the ideal version, while the performance of realistic homodyne is $\approx$ performance of OOK. The comparison between homodyne BPSK and OOK is affected by item 1 above.

3. P_e is independent of B as long as $B \gg A$. This will be the case for all versions of coherent detection, including optical heterodyne detection, as we shall see later on in this chapter.

4. The local laser amplitude B does, however, contribute to the shot noise in CD, and in fact, it forms the principal ingredient of the noise process at the receiver, provided B is large enough so that the receiver-generated thermal and dark noise can be neglected.

5. Homodyne detection amplifies the signal part by a factor equal to $2B$. These tendencies will be apparent in the next subsection.

8.1.6 Heterodyne detection

In optical heterodyne detection the frequency of the local laser is offset from the frequency of the input signal by ω_i. Thus, unlike homodyne detection, $\omega_L \neq 0$, and $\omega_i \neq 0$. Furthermore, since the received signal is typically much weaker than the local oscillator, we have $B >> A$. It is the large amplitude of the local laser as well as the mixing term in heterodyne detection that provides conversion gain to the signal without amplifying the receiver-generated noise. Thus conversion gain in heterodyne detection is a form of predetection gain. In heterodyne detection systems no attempt is made to phase-lock the local laser to the incoming optical signal in the optical frequency domain. For PSK systems, however, phase lock is often achieved at microwave frequencies following the IF stage. Note for future reference that the phases of both the local oscillator and the transmitter SLD are perturbed by phase noise.

To compare heterodyne detection with homodyne detection, we shall investigate the heterodyne detection version of BPSK. One of the major advantages of heterodyne detection is the fact that virtually all signal processing takes place at the microwave frequency range, including synchronization of both the carrier phase and the baud interval, both of which are required for digital detection. We also note here that the rolloff skirts on channel filters can be made much steeper in the microwave domain than is possible in the optical domain.

As was the case in the previous subsection, the BPSK signal is characterized by $A_0 = \pm A$. We now substitute this into Eq. (8.1a) and square both sides of the equation to obtain

$$E^2(t) = B^2 + A^2 \cos^2[\omega_i t + \phi_L - \phi_S] \pm 2AB \cos[\omega_i t + \phi_L - \phi_S]$$

$$+ A^2 \sin^2[\omega_i t + \phi_L - \phi_S]$$

which reduces to

$$E^2(t) = B^2 + A^2 \pm 2AB \cos(\omega_i t + \phi_L - \phi_S) \qquad (8.5)$$

We now present a more complete block diagram in Fig. 8.4 for the optical heterodyne reception of a BPSK channel. The IF stage in this system serves as both an amplifier and a bandpass filter centered at ω_i. The bandwidth of this block is somewhat larger than the transmission bandwidth because of the effects of phase noise (to be discussed in the next section). For ordinary

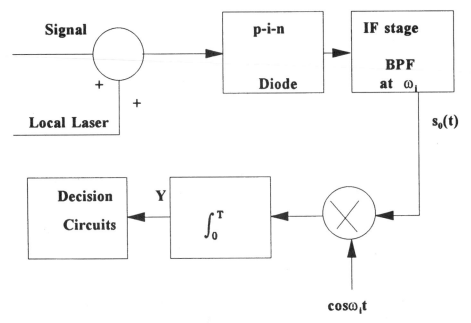

Figure 8.4 Optical heterodyne detection for BPSK.

BPSK it is necessary to extract the phase reference from the IF-stage output. To simplify the notation, we shall assume that phase-locked-loop techniques can extract a perfect reference and cos $\omega_i t$, thereby eliminating both ϕ_L and ϕ_S from further consideration in this idealized system model. We note that some of the post-IF-stage circuitry may be different for other modulation schemes.

An alternative phase-shift modulation and detection scheme to the preceding straightforward PSK system is to apply differential phase-shift-keying (DPSK) modulation and detection techniques. The primary detection setup for DPSK is shown in Fig. 8.5.

In the detection system shown in Fig. 8.5 each PSK pulse acts as a phase reference for the succeeding PSK pulse. Consequently DPSK systems do not need to extract an explicit phase reference to synchronize a local oscillator. It is necessary, however, to employ differential coding and concomitantly adjust the receiver detection process, so that a phase reversal represents a 1 and no change represents a 0 (or vice versa). As is true for all digital communication systems, both the transmitter and the receiver must play the same game.

In rf and microwave versions of these systems, there is only ~1 dB penalty, at low error rates, for using a time-delayed noisy phase reference for digital detection. Optical heterodyne detection is absolutely required for DPSK systems. In particular, the time delay utilized in DPSK detection must be accurate to within about 1 percent of the period of center frequency of the IF stage.

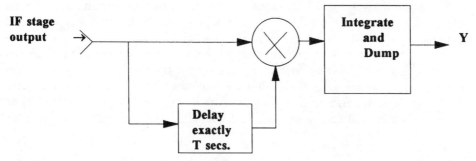

Figure 8.5 Block diagram DPSK detection system.

One conceivable way to perform DPSK detection directly on an optical DPSK signal is to add one pulse to the delayed pulse in the style of the homodyne detection systems described in the preceding two subsections. The great difficulty in using this seemingly simple detection scheme at optical frequencies is the accuracy required for the one-baud time delay as well as the need to maintain coherence from one baud to the next.[2] This time delay must be accurate to within about 1 percent of the period of the optical frequency carrier.

To determine the error rate performance for the ordinary BPSK heterodyne detection signal (in the absence of phase noise), we examine the output of the IF stage

$$s_0(t) = [\,\pm 2AB \cos \omega_i t + n(t)]K_d$$

where $n(t)$ is filtered white gaussian noise. Its source is shot noise, and it is generated primarily by the local laser in the heterodyne detection process. Detection of BPSK signals is achieved by multiplying the output of the IF stage by the $\cos \omega_i t$ reference signal and passing the product through an integrate-and-dump circuit. The output of the integrate-and-dump circuit, denoted by Y in Fig. 8.4, is the decision variable. We first examine the statistics of the primary detection parameter Y more closely before we define a decision algorithm:

$$Y = \int_0^T s_0(t) \cos \omega_i t \, dt = \pm 2AB \int_0^T \cos^2 \omega_i t \, dt + \int_0^T n(t) \cos \omega_i t \, dt = \pm ABT + Z$$

where

$$Z = \int_0^T n(t) \cos \omega_i t \, dt \qquad \text{and} \qquad \int_0^T \cos^2 \omega_i t \, dt = \int_2^T$$

---

[2] The 2-baud coherence requirement also applies to heterodyne detection.

The random variable Z amounts to the in-phase component of the shot noise, which, in turn, is a zero-mean gaussian variable with a normalized variance of $\sigma_Z^2 = T/2$. Now, as we have seen, the heterodyne process amplifies the signal by a factor of B, and we have also seen that the primary source of receiver noise is shot noise generated by the local oscillator. The other important point to remember here is that the signal-to-noise ratio (SNR) does not depend on B as long as it is large enough to mask the other sources of noise. All this suggests that we select the following variance for Z:

$$\sigma_Z^2 = \frac{B^2 T}{2}$$

As further justification of this, we note that $\sigma_Z^2 \propto I_{dc} \propto (B^2 + A^2) \approx B^2$. Therefore, the conditional probability densities for the detection variable Y are gaussian with the following variance and conditional means:

$$\sigma_Z^2 = \frac{B^2 T}{2} \quad \text{for both PSK states;} \quad E[Y|1] =$$

$$+ ABT \quad \text{and} \quad E[Y|0] = -ABT$$

Except for a factor of 2 for the conditional mean, these conditional densities are depicted in Fig. 8.3. This figure also clearly shows that for $P(0) = P(1) = 1/2$, the optimum decision threshold is 0, which suggests the following decision rule for the OHD BPSK receiver:

$$Select \begin{cases} 0 & if \ \ Y \le 0 \\ 1 & if \ \ Y > 0 \end{cases}$$

We can determine the probability of error for this binary symmetric communication system by computing the probability that $Y > 0$ when a 0 is sent, specifically, $E[Y|0] = -ABT = -\mu$ for short. Then P_e is given by

$$P_e = \frac{1}{\sqrt{2\pi\sigma^2}} \int_0^\infty \exp\left[-\frac{(i+\mu)^2}{2\sigma^2} \right] di$$

Now, in the same way that we obtained Eq. (8.4a), we find

$$P_e = Q\left\{\frac{\mu}{\sigma}\right\} = Q\left\{\frac{\sqrt{2}ABT}{B\sqrt{T}}\right\} = Q\left\{\sqrt{2A^2 T}\right\} \tag{8.6a}$$

Since a pulse of amplitude A is always transmitted in BPSK, we have $p_b = A_2 T$. Thus, with the aid of the asymptotic approximation to $Q(\bullet)$, we use Eq. (7.15) in much the same manner as it was used for realistic homodyne detection to obtain

$$P_e \approx e^{-A^2 T} = e^{-p_b} \tag{8.6b}$$

Comparing this result with Eq. (8.4c), we see that ideal heterodyne detection is 3 dB poorer in performance than the realistic homodyne detection system. Furthermore, heterodyne detection is also approximately 3 dB poorer than the performance of an ideal photon-counting detection system for OOK transmission. Another way to see where the 3-dB penalty for using OHD comes from is to note that the heterodyne detection process generates the following changes in the mean and variance:

$$\int_0^T \cos^2 \omega_i t \, dt = \frac{T}{2} \quad \text{causes} \quad <Y_{het}> = \frac{1}{2}<Y_{hom}>$$

$$\int_0^T n(t)\cos \omega t \, dt \quad \text{causes} \quad \sigma_{het}^2 = \frac{1}{2}\sigma_{hom}^2$$

Thus the ratio

$$\left\{\frac{<Y_{het}>}{\sigma_{het}}\right\}^2 = \frac{1}{2}\left\{\frac{<Y_{hom}>}{\sigma_{hom}}\right\}^2 \Rightarrow 3 \text{ dB}$$

Although heterodyne detection is about 3 dB poorer than homodyne detection, we note that heterodyne detection does not require that the local laser be phase-locked at optical frequency to the received FOCS signal.

Heterodyne and homodyne detection systems are collectively known as *coherent detection* (CD). We conclude this subsection by listing (and briefly discussing) several important aspects of CD systems.

1. The signal that emerges from the IF stage is a bandpass-type signal that must be processed further for final detection. To receive a particular incoming signal, the local laser is tuned so that its output frequency differs from the incoming signal (or channel in the case of WDM) frequency by ω_{if}. To ensure that heterodyne detection operates properly, the IF-stage output must be bandpass signal. This requires that the ratio ω_{if}/B_T, where B_T is the channel bandwidth, should be sufficiently large to permit the IF-stage output to act like a bandpass signal. When values of this ratio become too small, serious distortion can result. At the end of this chapter we shall examine an actual FSK system in which $\omega_{if} = 4.6$ GHz and $\omega_{if}/B_T \approx 2$. This low ratio strained the microwave frequency capabilities of the receiver. In Chap. 10 we shall describe a system in which $\omega_{if} = 40$ GHz and $\omega_{if}/B_T \approx 10$. In the analysis associated with that system, it was found that serious distortion arose when we lowered the ω_{if}/B_T ratio significantly below 10.

2. It is important to note here that the bandwidth of the transmitted signal (at high data rates) is determined solely by the modulation scheme. It is in no way affected by the tuning of the local laser at the receiver. On the other hand, ω_{if} is, for the most part, determined by receiver characteristics. The local SLD must be tuned so that the difference frequency matches ω_{if} of the IF stage; however, as discussed in paragraph 1 above, the receiver system must be designed so that ω_{if}/B_T is sufficiently large to permit further process-

ing. The transmitter portion is in no way affected by ω_{if}. In fact, it is the other way around; it is the bandwidth of the transmitted signal that plays a large role in selecting at least lower bounds on ω_{if}.[3]

3. Heterodyne detection is virtually a necessity for reception of densely packed WDM channels. By *densely packed,* we mean that the channels are too closely spaced to be resolved by optical filtering. One such detection system that is capable of avoiding image channel problems is described at the end of Chap. 9.

4. One "brute force" version of FSK utilizes two separate diode lasers, widely spaced in wavelength, for the two transmission states. The transmitter would then send the output of only one of these lasers at each baud interval selected according to the binary data stream. Such a signal could be detected by employing two separate OOK detectors, each preceded by an individual optical filter tuned to an appropriate wavelength, in order to distinguish between 0 and 1. It seems that these dual sources—filters and detectors—could be more profitably utilized in two separate OOK channels in a WDM configuration over the same optical fiber. It is likely that a single optical preamplifier could be used for both the brute-force single channel version as well as the two-OOK-channel system.

5. Consider a CD system along with FM discriminators that permit the use of FSK modulation with the two signal frequencies much closer together than optical filtering would allow. In such a system it is possible to employ the binary FSK signal frequencies sufficiently close to each other so as to accrue some of the advantages of *minimum shift keying.* Another aspect of closely packed frequencies is that CPFSK (continuous phase FSK) can be readily generated by direct modulation of laser diodes. This type of modulation avoids most of the frequency chirping which affects direct-modulation OOK systems.

6. It is interesting to note that the local laser amplitude B does not appear in any of the error rate formulas. This will be true as long as B is sufficiently large so that the shot noise it generates completely masks (overwhelms) the receiver electronic noise. Clearly the local laser amplitude B multiplies the signal part of the output, which is a form of amplification known as *conversion gain.* It is an example of predetection gain here that is peculiar to optical heterodyne detection for it does not occur in rf heterodyne detection. Also note that OHD does not amplify receiver-generated electronic thermal and dark noises.

7. Ideal coherent detection yields predetection gain without the excess-noise factor that is generated in APD detection systems. If you were to make the APD excess-noise factor F unity ($x = 0$) and the APD gain large enough, you would at least achieve the same SNR as with CD. However, APD detec-

---

[3]In Chap. 10 we shall study a system in which all these considerations play a very important role.

tion would not by itself yield all the advantages of CD that accrue from signal processing at a much lower IF frequency ω_i (which usually is in the microwave frequency range).

8. Any optical noise coming in with the received signal such as transmitter laser output fluctuations will be amplified by the heterodyne process just as the signal is amplified. In particular, the heterodyne process does nothing to inhibit phase noise; in fact, it exacerbates the problem, as we shall see in the next section.

8.2 Phase Noise

8.2.1 Introduction

All the results and formulas derived in the previous section were obtained for additive gaussian shot and/or thermal noise in the absence of phase noise. As we shall see, phase noise requires special attention because it isn't additive noise. Although it turns out that the output phase of laser is itself a gaussian random process, the phase term occurs inside the argument of a sinusoidal function, which, of course, is not a linear function. Thus, not only is its effect on a system described by nongaussian statistics; its effects on a FOCS are very complicated. Most of the difficulties generated by phase noise cannot be alleviated by simply increasing the transmitter power level.

The primary sources of phase noise in a FOCS are the diode lasers themselves. In coherent detection, phase noise is generated independently by both the local and the transmitter lasers. We shall discuss the physical reasons for phase noise in Sec. 8.2.2, and describe some aspects of the statistical properties of phase noise in Sec. 8.2.3. The effects of phase noise on system performance will be described in Sec. 8.3 along with some techniques for reducing its effects on coherent detection systems.

8.2.2 Source of phase noise

The source of phase noise within the laser is randomly occurring spontaneous emission events. *Each random photon adds a quantum of energy with a completely random phase* to the existing laser cavity field. Some points of interest here are

1. The spontaneous emission rate is not directly affected by the electromagnetic field inside the laser cavity.

2. In a steady-state lasing situation the population of the upper laser transition state is a constant. Thus spontaneous emission events and the associated photon-number distribution form a homogeneous Poisson random process. The rate of events is usually so high that the superposition is accurately approximated by a gaussian density. We shall soon see, however, that this is not the most important statistical property of phase noise.

Now, one manifestation of phase noise is the bandwidth of the output of an unmodulated single-frequency laser. It should not come as a surprise that phase fluctuations also generate frequency fluctuations, which, in turn, affect the laser bandwidth. Experimental measurements on diode laser outputs, however, displayed much larger bandwidths than predicted by the theory prevalent at that time these measurements were made (see Fleming and Mooradian [1]). Henry [2] investigated this phenomenon, and his paper on this subject turned out to be a major contribution to the understanding the nature of phase noise in semiconductor lasers. The remainder of this subsection is based on that paper.

We begin by taking a closer look at the phase noise process. Consider Fig. 8.6. In this figure (as in Henry's paper) I represents optical intensity expressed in terms of the number of photons in the cavity. The length of the phasor is therefore proportional to the square root of the intensity. Thus the field amplitude is $\sqrt{I}$, while the intensity and hence the amplitude for a single spontaneous photon is = 1. The phase angle between the spontaneous photon and the field in the cavity is totally random and uniformly distributed over 2π radians. This angle is denoted by θ_i for the particular photon depicted in Fig. 8.6. The resultant phasor has parameters $\sqrt{I+\Delta I}$ for the amplitude and $\Phi + \Delta\Phi$ for the phase, also depicted in Fig. 8.6. In a typical laser cavity $I \gg 1$, and, as we shall see in the subsequent text, the bigger I is, the less the effects of phase noise.

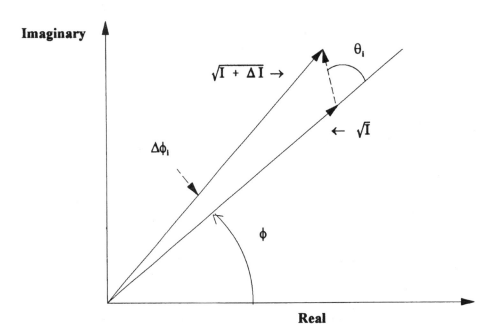

Figure 8.6 Spontaneous photon added to laser cavity. (*After C. H. Henry [2].*)

There are two separate contributions by spontaneous photons to the phase fluctuations in SLD outputs. One is the direct contribution to the resultant by the orthogonal component which is proportional to $\sin \theta_i$. The other is proportional to $\cos \theta_i$ and is parallel to the cavity field. Surprisingly, this in-phase component of the spontaneous photon turns out to have the greatest effect on phase noise. Initially the in-phase component produces a change in the amplitude of the resultant, which, in turn, causes a slight change in the index of refraction of the active region of the laser diode. The index of refraction affects the phase propagation constant of the lasing field traveling along the entire length of the cavity, and this affects both the phase and frequency of the laser output. The total phase change is given by [2]

$$\Delta\phi = \frac{\alpha}{2I} + \frac{1}{\sqrt{I}} \{\sin \theta_i - \alpha \cos \theta_i\} \qquad (8.7a)$$

where $\alpha \doteq \Delta n''/\Delta n'$, and where $\Delta n''$ and $\Delta n'$ are the real and imaginary parts of the index of refraction, respectively (which, in turn, makes α a real quantity). Estimates of the parameter α by Henry based on experimental data show that the in-phase contribution to the net phase greatly increases the laser diode bandwidth over that predicted by a theory based on fixed index of refraction. The data referred to by Henry indicate that α is on the order of 5 or 6 [2].

A formal derivation of this value for α requires details of the mathematical model for laser dynamics. The equations for the laser model provide the coupling between the time derivative of phase $[dt]\phi$ and the time derivative of the optical intensity $[dt]I$ inside the laser cavity. This connection is required to properly evaluate the in-phase portion of the spontaneous photon contribution. Now recall that θ_i is uniformly distributed over 2π radians. Thus we obtain

$$<\Delta\phi> \text{ per photon} = \frac{\alpha}{2I} \qquad \text{and} \qquad <\Delta\phi> = \frac{\alpha Rt}{2I}$$

where Rt = average number of spontaneous photons in t seconds. It is shown in App. 8A that the variance of the phase fluctuations is given by

$$<\Delta\phi^2> = \frac{Rt(1 + \alpha^2)}{2I} \qquad (8.7b)$$

Henry's theoretical model based on the preceding result is in good agreement with experimental data. It yields an approximately 50-fold increase in bandwidth over that previously obtained using the old theory. Equation (8.7b) could have been obtained more simply by noting the independence of individual spontaneous emission events and multiplying our single-photon result by Rt. This simpler derivation, however, overlooks the effects of the Poisson fluctuations in the actual number of spontaneous events. It works well in this case because I is so large in laser diodes, but the simplified model may run into difficulty when applied to some much smaller quantum-well optical sources.

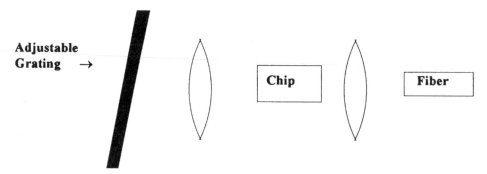

Figure 8.7   External cavity laser diode setup.

It is important to note here that the phase fluctuations are diminished as the number of photons in the lasing mode represented by I increases. To increase I, it is necessary to increase the energy stored in the lasing mode inside the SLD by increasing the size (most likely the length) of the cavity. This process will simultaneously increase the Q of the cavity. Some numerical values for diode laser bandwidth follow. The bandwidths of the lasers used to generate the data presented in Fig. 4.28 were on the order of 15 MHz. Multiple-section lasers of the type shown in Fig. 4.29 have been reported with output bandwidth less than 1 MHz. Some laboratory sources such as depicted in Fig. 8.7 use external optics to create a larger cavity.

Diode lasers with the open-structure cavity shown in Fig. 8.7 have been operated with bandwidths under 100 kHz. While such lasers have been used in laboratory environments for some time now, primarily because they are capable of being tuned over a very broad range of wavelengths without mode hopping, they are obviously too clumsy (and sensitive to vibrations) for incorporation into commercial FOCS.

8.2.3 Coherence time

Since each spontaneous photon has a statistically independent random phase as well as the same energy (hence equal length vector contributions), the resultant pictured in Fig. 8.6 undergoes a two-dimensional random walk. The angle of the resultant with the real axis $\theta(t)$ is the phase of the output radiation. The spontaneous emission rate in terms of photons per second is often very large (on the order of 10^{12} in a typical diode laser [3]), and the contribution of each spontaneous event to the net cavity phase is very small. In addition, the elapsed time between spontaneous photon emission events is random, with a short average interevent time. These characteristics, along with the inherent property of spontaneous emission that each event is statistically independent of all the other spontaneous events, result in $\theta(t)$ being accurately remodeled as a Wiener process [4]. In the context of the phase noise version of the Wiener process that we are considering here, the total "phase walk" that occurs in a τ-second interval can be expressed in the following form:

$$\theta(\tau) = 2\pi \int_0^\tau \mu(t)dt \tag{8.8a}$$

where $\mu(t)$ is a white gaussian noise process with spectral density N_0 and autocorrelation function

$$R_\mu(t - t') = N_0\delta(t - t')$$

Thus, from Eq. (8.8a)

$$E\{[\theta^2(\tau)]\} = (2\pi)^2 \, \mathbf{E}\left\{\int_0^\tau \int_0^\tau \mu(t)\mu(t')dtdt'\right\}$$

$$E[\theta^2(\tau)] = (2\pi)^2 N_0 \int_0^\tau \int_0^\tau \delta(t - t')dtdt' = (2\pi)^2 N_0 \tau \tag{8.8b}$$

where N_0 = average noise power per hertz $\rightarrow$ watts per hertz.

We next show that phase noise results in the output of a single-frequency diode laser having a lorentzian power spectrum. Let the SLD output be given by

$$s(t) = A \cos[2\pi f_0 t + \theta(t) + \phi] \tag{8.9}$$

where $\theta(t)$ is the phase noise contribution and ϕ is a random variable that is uniformly distributed over 2π radians. This uniform density for ϕ represents our ignorance, or, as is often the case, our total lack of interest in the over absolute phase of the SLD output. The absolute phase represents a specific time reference, and it plays no role in direct detection. Optical heterodyne detection of FSK and DPSK also is independent of the absolute-phase reference.

Of greater importance to us, however, is the fact that this reasonable assumption about the uniform phase distribution serves to make the random process represented by Eq. (8.9) stationary. The argument of the cosine function in Eq. (8.9) might look something like Fig. 8.8.

To obtain the power spectrum, we first find the autocorrelation function for the SLD output:

$$R_s(\tau) = E[s(t)\,s(t + \tau)]$$

The $s(t)$ of Eq. (8.9) is now substituted into the preceding autocorrelation function to arrive at

$$R(\tau) = \frac{A^2}{2} E\left\{\cos[2\pi f_0\tau + \theta(t + \tau) - \theta(t)]\right\} \tag{8.10a}$$

where we have averaged $R(\tau)$ over the random parameter ϕ to eliminate the sum frequency term, which drops out because

$$E[\cos(\bullet\bullet\bullet\bullet\bullet\bullet + 2\phi] = 0$$

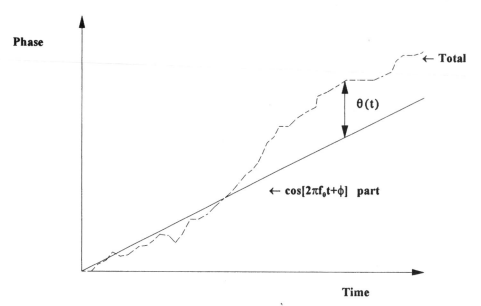

Figure 8.8 Argument of cosine function with phase noise.

We note here that the statistical independence of spontaneous emission events cause the random-phase walk to be statistically independent for non-overlapping time intervals. Thus we find that the random-phase walk that occurs in the time interval $(t, t + \tau]$ is statistically equivalent to the random walk that takes place in the interval $(0, \tau]$. This leads to the realization that $\theta(t + \tau) - \theta(t) = \theta(\tau)$ is statistically equivalent to the random-phase walk that takes place in the time interval $(t, t + \tau]$. Thus we may express Eq. (8.10a) in the following form:

$$R(\tau) = \frac{A^2}{2} E \left\{ \cos \left[2\pi f_0 \tau + \theta(\tau) \right] \right\} \tag{8.10b}$$

or equivalently

$$R(\tau) = \frac{A^2}{2} \text{Re} \left\{ e^{j\omega_0 \tau} E[e^{j\theta(\tau)}] \right\} \tag{8.10c}$$

Now, since Eq. (8.8a) represents a linear operation on the white gaussian noise random process $\mu(t)$, the phase noise term $\theta(\tau)$ is a zero-mean gaussian random variable. Then the characteristic function for $\theta[\psi_\theta(\xi)]$ is given by

$$\Psi_\theta(\xi) = E[e^{j\theta\xi}] = \exp \left[-\frac{\xi^2 \sigma_\theta^2}{2} \right] \tag{8.11a}$$

We next evaluate Eq. (8.11a) at $\xi = 1$ and employ Eq. (8.8b) to arrive at

$$\Psi_\theta(1) = E[e^{j\theta}] = \exp\left[-\frac{\sigma_\theta^2}{2}\right] = \exp\left[-\frac{(2\pi)^2 N_0 |\tau|}{2}\right] \qquad (8.11b)$$

Finally the autocorrelation function may be written as

$$R(\tau) = \frac{A^2}{2} \cos[2\pi f_0 \tau] \exp\left[-\frac{(2\pi)^2 N_0 |\tau|}{2}\right] \qquad (8.12a)$$

The Fourier transform of Eq. (8.12a) provides the following power spectral density $G(f)$ for the SLD output:

$$G(f) = \frac{A^2}{4\pi^2 N_0} \left\{ \frac{1}{1 + [(f - f_0)/\pi N_0]^2} + \frac{1}{1 + [(f + f_0)/\pi N_0]^2} \right\} \qquad (8.12b)$$

Thus the power spectrum of the SLD output turns out to have a lorentzian shape. The first term inside the brackets represents a peak at $f = f_0$ with a 3-dB bandwidth of πN_0 Hz. The other term in the brackets represents the usual even symmetry of Fourier transforms of autocorrelation functions. Thus we have the following relation between the SLD bandwidth and the internal phase noise generation parameter N_0:

$$B_L = 2\pi N_0 \qquad \text{and} \qquad N_0 = \frac{B_L}{2\pi} \qquad (8.13)$$

This equation clearly shows that it is useful to minimize B_L. One important aspect of the effect of phase noise on FOCS is the loss of phase coherence during a digital pulse interval. We begin with Eq. (8.12a) and define the normalized autocorrelation function:

$$\gamma(\tau) \doteq \frac{R(\tau)}{R(0)} \qquad (8.14a)$$

then

$$|\gamma(\tau)| = \exp\left\{-\frac{|\tau|}{\tau_{coh}}\right\} \qquad (8.14b)$$

Where the coherence time τ_{coh} is given by

$$\tau_{coh} = \left[\frac{(2\pi)^2 N_0}{2}\right]^{-1} = \left[\frac{(2\pi) B_L}{2}\right]^{-1} = \frac{1}{\pi B_L} \qquad (8.14c)$$

We see then that the coherence time is inversely proportional to the laser source bandwidth. For coherent detection, the bandwidth of the SLD sets an upper bound on the baud duration, which, in turn, sets a lower bound on the data rate. This is caused by the need to maintain sufficient phase coherence for the duration of the baud. In the case of DPSK, coherence is required over at least 2 baud. We next look into some of the effects of phase noise on FOCS performance.

8.2.4   General effects of phase noise on FOCS

The random phase of the SLD output is a gaussian random process, but because it occurs in the arguments of nonlinear functions such as sin($\bullet$) and cos($\bullet$]) its effect on a FOCS is decidedly nongaussian. First, phase noise is not additive, and in many circumstances increasing the signal strength will not eliminate the effects of phase noise. Consider the coherence time parameter τ_{coh} discussed in the preceding subsection. As an example of how coherence time can affect the performance of a system, consider a DPSK CD FOCS (pardon the alphabet soup) where the coherence time and the baud time are of comparable size. The phase reference for DPSK detection is the previous baud. Under the assumed coherence time condition, the baud to be detected is almost completely uncorrelated to the phase of the preceding baud and the resulting error rate would be very large, indeed. Any attempt to improve the situation by increasing the received signal strength, say, by reducing repeater spacing, will not remedy the situation. Actually it takes very little loss of coherence to prevent the attainment of a low error rate in the range of $P_e \sim 10^{-9}$ regardless of signal strength.

A rather surprising consequence of the finite coherence time is the existence of a lower bound on the data rate. The straightforward method of reducing the loss of coherence during a data pulse is to reduce the duration of the data pulse (baud). This can be accomplished by increasing the baud rate. At first this might seem to be an advantage, but this constraint could result in either a minimum data rate that is not feasible or possibly the need for a very expensive receiver. For example, suppose the phase noise of the SLD sources in a particular local area network (LAN) is such that the minimum data rate is, say, 5 Gbits/s. If the individual LAN users requires service at only 50 Mbits/s or less, it would be an unnecessary expense to purchase receivers capable of operation at the much higher data rate. We shall describe the effects of phase noise on several specific transmission systems in the next section.

We next note that when an input signal containing phase noise passes through a linear filter, the random-phase fluctuations can generate random fluctuations in the amplitude response of the filter. This phenomenon turns out to be very difficult to analyze because the random phase occurs within the argument of a sinusoid. To obtain a qualitative feel for this phase-to-amplitude conversion phenomenon, we introduce the following simplified mode for the detection process. Now consider an optical heterodyne receiver with an IF stage whose impulse response $h_F(t)$ and is assumed to have the following form:

$$h_F(t) = \exp\left\{\frac{t}{\tau_0}\right\} \cos\omega_{if}t \qquad (8.15a)$$

where τ_0 is the characteristic decay time and ω_{IF} is the center frequency of the filter. We note here that τ_0 will be inversely proportional to the bandwidth of the IF-stage filter-amplifier combination. Next we let the phase-noise-perturbed input to the IF stage be of the form

$$s(t) = A \sin[\omega_c t + \theta(t)]$$

where $\theta(t)$ is the random phase of the SLD output. Then the output $y(t)$ of the IF stage may be written as

$$y(t) = A \int_0^t \sin[\omega_c x + \theta(x)] h_F(t - x) dx \qquad (8.15b)$$

In the absence of phase noise is modeled by $\theta(t) = 0$ and the output of the IF stage would be a sinusoid at the same frequency with altered amplitude and phase, as expressed by

$$y(t) = A |H_F(f_c)| \sin [\omega_c t + \sphericalangle H(f_c)]$$

where $H(f)$ is the transfer function of the IF stage, which, of course, is equal to the Fourier transform of the impulse response $h_F(t)$. As an example, consider an OHD system described by the following assumptions:

1. Let the bandwidth of the IF stage be very broad, say, ~2 GHz. This implies that the characteristic decay time τ_0 will be <1 ns.

2. We next assume reasonable values for the SLD output bandwidth, say, ~15 MHz for both the transmitter and local laser SLDs. Thus, as we shall see in the next section, the bandwidth applicable to the phase noise at the input to the IF stage is 30 MHz. These numbers imply that $\tau_{\text{coh}} \sim 1/(3 \times 10^7)$.

The very short duration of the impulse response in this case permits us to approximate the output of the IF-stage filter $y(t)$ by

$$y(t) \approx A \int_{t-\tau_0}^t \sin[\omega_c x + \theta(x)] h_F(t - x) dx \qquad (8.15c)$$

Since τ_0 is so short while the net 30 MHz bandwidth of the SLD-generated phase noise is so small, we can treat the phase noise as a constant $\theta(t)$ during a τ_0 time interval. The output of the filter at $y(t)$ under these conditions is approximately

$$y(t) \approx A |H_F(f_c)| \sin [\omega_c t + \sphericalangle H(f_c) + \theta(t)] + n(t) \qquad (8.15d)$$

where $n(t)$ is the usual additive gaussian noise. We note that even after downconversion from optical to microwave frequencies via the optical heterodyne detection (OHD) process, the filter output $y(t)$ still contains randomphase $\theta(t)$. However, we also note that the effect of phase noise appears only as a random-phase term, and there are no amplitude fluctuations for this set of parameters. Thus, if OOK transmission is employed along with OHD, one could perform incoherent digital detection (say, integrate-and-dump detection on the envelope) of the IF-stage output without any direct effect from phase noise. Nevertheless, phase noise has made its presence felt (albeit indirectly) by requiring a larger IF-stage bandwidth than would be otherwise needed, and this, in turn, results in the presence of more noise at the detection circuits. We also note here that the random phase in Eq. (8.15d) is still present and the system will still experience the effects of reduced coherence time. Thus, for example, if DPSK modulation-detection is employed, phase noise yields a lower bound on data rate.

We next examine qualitatively how phase noise can be converted to random amplitude fluctuations. Figure 8.8 displays a typical phase walk driven by spontaneous emission. Now the preceding example used an extremely wideband IF-stage filter with a very brief impulse response duration τ_0, and the numerical values of these parameters are equivalent to approximating the random phase of the SLD output as a constant over the τ_0 time interval. Now we shall go to the next order of approximation and assume a larger but still very short impulse response duration τ_1 such that a τ_1 long portion of the random-phase curve can be approximated by a straight-line section with characterized by two parameters, $a(t)$ and $b(t)$, which are constant for the τ_1 second interval but are both random on a longer time scale. Then we have

$$\theta(t) = a(t) + b(t)x \qquad (8.15e)$$

This can be achieved by reducing the IF-stage bandwidth sufficiently until the desired impulse response duration τ_1 is achieved. Note that x is the dummy variable of integration in Eq. (8.15b), specifically

$$y(t) \approx A\,|H_F\{f_c + b(t)\}|\sin\{\omega_c t + \sphericalangle H\{f_c + b(t)\} + a(t)\} \qquad (8.15f)$$

As we shall soon see, $a(t)$ represents the phase-effect portion of the phase fluctuations, while $b(t)$ represents instantaneous frequency fluctuations. Under these intermediate conditions we can approximate Eq. (8.15c) as follows:

$$y(t) \approx A\int_{t-\tau_0}^{t}\sin\{\{\omega_c + b(t)\}x + a(t)\}h_F(t-x)dx \qquad (8.15g)$$

Then the output $y(t)$ will still be sinusoidal; however, its frequency will have shifted by $b(t)$. This yields the output

$$y(t) \approx A\,|H_F\{f_c + b(t)\}|\sin[\{\omega_c + b(t)\}t + \sphericalangle H\{f_c + b(t)\} + a(t)] \qquad (8.15h)$$

Now, since the amplitude response $|H_F\{f_c + b(t)\}|$ is a function of frequency which includes $b(t)$ inside the argument, the random fluctuations in frequency will manifest amplitude fluctuations. These amplitude fluctuations have been induced in the filter response by the phase noise fluctuations. There are also phase fluctuations in the filter output as represented directly by the presence of the $a(t)$ term inside the sine argument of the sine function in Eq. (8.15h), as well as indirectly by the frequency-dependent phase-shift portion of the filter transfer function.

The presentation here on the effects of phase noise as a function of IF-stage bandwidth is fairly crude and certainly oversimplified. Nevertheless, the conclusion that having a wideband IF stage avoids the phase-to-amplitude conversion for phase noise is valid at least for the initial conversion from optical to microwave frequency ranges. Care must be taken in later signal processing to retain this advantage. A closed form for the exact solution of the phase noise detection problem has been very elusive, and probably the most frequently quoted results on this subject are based on computer simulations by Foscini et al. [7] and others [5,6]. As we shall see in the next section, where we consider the effects of phase noise on some specific FOCS, there exist several techniques for reducing, if not eliminating, phase noise effects entirely.

8.3   Analysis of FOCS with Phase Noise

In the preceding section we examined some of the general properties of phase noise. We saw there that phase noise is not additive noise and that it can be changed into amplitude noise when passed through a linear filter. In this section we shall consider how phase noise affects the performance of some specific communication systems. One of the most important effects of phase noise is the upper limit it places on the usable duration of digital pulses. This phenomenon is responsible for lower limits on data rates (often impractical). We shall see in this section that phase noise also can generate a decibel penalty in system performance and for some conditions prevent attainment of the desired error rate. The primary reference for the material contained in this section is the classic paper by Salz [8].

8.3.1   FSK without phase noise

It is helpful to examine the calculation of P_e for FSK without phase noise before we consider the general case. The two basic FSK transmission signals are

$$s_0(t) = A_T \sin\left\{\left(\omega_c - \frac{\Delta\omega}{2}\right)t + \theta(t)\right\} \tag{8.16a}$$

$$s_1(t) = A_T \sin\left\{\left(\omega_c + \frac{\Delta\omega}{2}\right)t + \theta(t)\right\} \tag{8.16b}$$

where A_T is the amplitude at the transmitter and $\Delta\omega$ is the frequency spacing between the two FSK signals. Let R represent the baud rate and B_{TL} be the phase noise bandwidth of the transmitter laser. In the analysis that follows

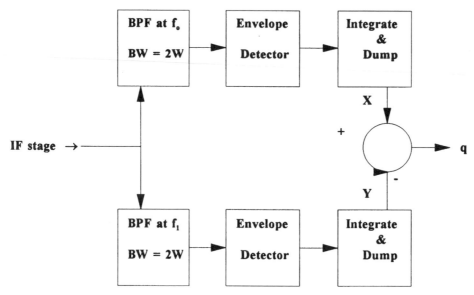

Figure 8.9 Dual-filter FSK detection system.

we assume that $R/B_{TL} \gg 1$ so that the phase drifts slowly enough to be considered constant for the baud duration. A block diagram of one type of FSK detection circuit is shown in Fig. 8.9. The input to this detection system is the output of the IF stage, so we begin our analysis by examining the form of the signals after heterodyne conversion.

The output of the receiver photodiode can be obtained from

$$E(t) = A \cos[\omega_{cl}t + \theta(t)] + B \cos[\omega_{Lo}t + \Phi(t)] \qquad (1 = 0,1)$$

and

$$E^2(t) = A^2 + B^2 + 2AB \cos[\omega_{il}t + \Delta(t)]$$

where we have eliminated the sum frequency term and where

$$\omega_{il} = |\omega_{cl} - \omega_{Lo}| \qquad \text{and} \qquad \Delta(t) = \theta(t) - \Phi(t)$$

We noted earlier in this chapter that as long as the local laser amplitude is sufficiently large, the magnitude of B does not affect the signal-to-noise ratio as seen by the detection circuits, and we shall simply delete this symbol from our further calculations (or set $B = 1$ for convenience). The task at the receiver is to decide which of the following signals is present at the IF-stage output during each baud:

$$2A \cos[\omega_0 t + \Delta(t)] \qquad or \qquad 2A \cos[\omega_1 t + \Delta(t)]$$

where we have shortened the frequency notation from ω_{il} to ω_l. To employ the detection scheme suggested by Fig. (8.9), the FSK signal frequencies must be sufficiently spaced (i.e., $\Delta\omega$ must be sufficiently large at the transmitter) so that there is no appreciable signal output at the ω_1 side when a 0 is transmitted (i.e., ω_0 is sent), and vice versa when a 1 is transmitted. We also assume that for the analysis in this section to be valid, $\Delta\omega$ must be large enough so that the noise outputs from the two filters shown in Fig. (8.9) are, to a good approximation, statistically independent.

The detection process essentially involves determining which filter rings up the most energy during the baud. This energy is directly proportional to the sum of the squares of the in-phase and quadrature parts. Since the input noise to the detector is shot-noise-generated white noise and both legs in the detection scheme have the same spectral shape (except, of course, for the center frequency), both filter outputs will have the same variance. Thus this system is a binary symmetric channel, and therefore we need compute the probability error only when a 0 or ω_0 is the transmitted baud. We now define the following parameters. Let x_i, x_q be the in-phase and quadrature parts of the ω_0 leg and y_i, y_q be the in-phase and quadrature parts of the ω_1 leg; then

$$X \doteq x_i^2 + x_q^2 \qquad \text{and} \qquad Y = y_i^2 + y_q^2$$

An error occurs in this case if $q = X - Y < 0$. Thus

$$P_e = P[X - Y < 0] = P[Y > X] = P\left\{ \left[\left[x_i^2 + x_q^2 \right] - \left[y_i^2 + y_q^2 \right] \right] < 0 \right\}$$

Now each of the x_i, x_q, y_i, y_q are statistically independent gaussian random variables (GRVs). Since ω_0 is transmitted, the y_i and y_q are zero-mean GRV while the x_i and x_q have nonzero means. Hence we have

$$x_i = AT \cos \Delta + \int_0^T n(t) \cos \omega_0 t \, dt \qquad x_q = AT \sin \Delta + \int_0^T n(t) \sin \omega_0 t \, dt$$

$$y_i = \int_0^T n(t) \cos \omega_1 t \, dt \qquad y_q = \int_0^T n(t) \sin \omega_1 t \, dt$$

The density of Y (noise alone) is a Rayleigh density which leads to

$$P[e \mid X] = \int_x^\infty p_1(Y) dY = \int_x^\infty \frac{Y}{2\sigma^2} \exp\left\{ \frac{-Y^2}{2\sigma^2} \right\} dY = \exp\left\{ \frac{-X^2}{2\sigma^2} \right\} \qquad (8.16c)$$

Now
$$P_e = \int_0^\infty P[e \mid X] \, p_1(X) dX$$

Furthermore, the probability density of X [$p_1(X)$] is a Rician density [note that Eq. (6.10c) is a Rician density in $\sqrt{I}$. Then, after some manipulation, we obtain, with the aid of an integral table (see Gradshteyn and Rhyzik [13], Eq. 6.63-4, p. 718)

$$P_e = \exp\left[-\frac{(AT/2)^2}{2\sigma^2}\right] \int_0^\infty \frac{q_1}{\sigma^2} \exp\left[-\frac{q_1^2}{\sigma^2}\right] I_0\left[\frac{q_1 AT}{2\sigma^2}\right] dq_1 \qquad (8.16d)$$

$$P_e = \frac{1}{2} \exp\left[-\frac{E}{2\sigma^2}\right] \qquad (8.16e)$$

where the average energy per pulse $E = A^2T/2$ and where the ½ factor is due to the heterodyning process. Then, in accord with our setting $B = 1$ causes $\sigma^2 = 1$, we obtain, as our final form for FSK without phase noise

$$P_e = \frac{1}{2} \exp\left\{-\frac{A^2T}{2}\right\} \qquad (8.16f)$$

This is 6 dB poorer than the so-called quantum limit obtained for OOK. Of this, 3 dB comes from heterodyne detection, and the other 3 dB comes from the use of orthogonal signals instead of antipodal signals. Some advantages of FSK are that neither phase-lock nor threshold-level tracking is required, and we note here that FSK channels have proved to be reliable for communicating in poor signal environments. The variation of FSK known as CPFSK can be generated by direct modulation of a SLD. This version is often used in high-speed coherent detection system experiments using discriminator-based digital detection. We describe one such system in depth at the end of this chapter.

8.3.2 FSK performance with phase noise

The analysis of FSK with phase noise is far more complicated than our previous analysis. In fact, the only reason for including the previous subsection here is to aid in the understanding of FSK performance in a phase noise environment. The principal difference is that *phase noise requires bandpass filters with considerably larger bandwidths than would be required to pass the primary information bearing signals.* Hence the IF-stage filter lets more noise reach the digital detection portion of the receiver than in the absence of phase noise. Some of the more complicated portions of the development are relegated to App. 8B at the end of the chapter. The following development is based on Salz [8].

Now, as was the case in the previous subsection, energy is detected for each leg shown in Fig. 8.9 and the results are compared at the end of a baud period and reset. Assume that ω_l is the IF-stage output frequency that would be received in the absence of phase noise. The output of the ω_l filter with phase noise expressed in quadrature form is

$$S(t) = x(t) \cos[\omega_l t] + y(t) \sin[\omega_l t]$$

where

$$x(t) = 2A \cos\{\Delta(t)\} + n(t)$$

and
$$y(t) = 2A \sin\{\Delta(t)\} + n(t)$$

$$\Delta(t) = \theta(t) - \Phi(t)$$

where $\theta(t)$ is the phase noise of the transmitter laser and $\Phi(t)$ is the phase noise of the local laser. Thus $\Delta(t)$ has the same meaning as it did in the previous subsection, but here it consists primarily of the composite phase noise from both transmitter and local laser sources. The bandwidth and N_0 for the $\Delta(t)$ phase noise is determined by the sum of both the transmitter and local laser bandwidths.

To simplify the tasks ahead, we assume that the magnitude of the two filter transfer functions $|H(\omega)|$ shown in Fig. 8.9 are ideal filters where

$$|H_\ell(f)| = \begin{cases} 1 & \text{for} \quad f_\ell - W \le |f| < f_\ell + W \\ 0 & \end{cases} \qquad \text{otherwise}$$

This assumption simplifies the task of obtaining specific numerical results for system performance. These ideal filters are crude representations for the actual filters, and therefore the results obtained using these filters are at best only approximations. However, our main goal in this development is to obtain a qualitative feel for the effects of phase noise on FSK FOCS using coherent detection. Some consequences of assuming rectangular spectral filters with $2W$ bandwidths are

$$E[n^2] = E[n_0^2] = E[n_1^2] = 4W$$

where we have conveniently assumed the noise power spectral density for shot noise to be $\eta/2$ (recall $\eta \propto B^2$ and that B cancels out in the error rate formulas). Next we average over each of the preceding equations for $x(t)$ and $y(t)$ and obtain

$$E[x(t)] = 2A \cos \Delta(t) \qquad \text{and} \qquad E[y(t)] = 2A \sin \Delta(t)$$

The rectangular filter response yields the following autocorrelation functions which apply to the quadrature components $x(t)$ and $y(t)$ for both of the transmission states:

$$R_{n,1}(\tau) = R_{n,0}(\tau) = 4W \frac{\sin 2\pi W\tau}{2\pi W\tau} \qquad (8.17a)$$

We shall soon employ this particular autocorrelation function to obtain a Karhunen–Loeve expansion.

Now, as in the previous subsection, we assume that frequency ω_0 is transmitted so that a 0 is transmitted and therefore the upper leg of Fig. 8.9 contains signal plus noise while the lower leg contains only noise. Then we may express the detection parameters X and Y as

$$X = \int_0^T \{x^2(t) + y^2(t)dt\} \tag{8.17b}$$

$$Y = \int_0^T \{v_1^2(t) + v_2^2(t)dt\} \tag{8.17c}$$

where X represents the output of the 0 channel which has nonzero means for the in-phase and quadrature components (when averaged over the additive gaussian noise). Similarly, the output Y yields zero means for the comparable components denoted for that path by $v_1(t)$ and $v_2(t)$, respectively. Furthermore, since there is no phase lock involved in the detection process, the reference for in-phase and quadrature components is arbitrary. Of course, these two parts must be 90° out of phase with each other. As before, an error occurs when $Y > X$. The probability of error for this binary symmetric channel is then

$$P_e = P[X < Y] \tag{8.17d}$$

Now, in contradistinction to Sec. 8.3.1, we include phase noise as expressed by $\Delta(t)$ within the model. Note, however, that while $\Delta(t)$ appears in the means of the quadrature components (averaged over the gaussian noise only), we have not as yet taken any averages over random phase.

The quantities X and Y required for FSK detection are quadratic in nature, and variables of this type have long been of interest in analyzing radar systems. One tool that has been shown to be very useful in these analyses is Karhunen–Loeve equation (KLE) expansions. To evaluate Eq. (8.17d), we now employ some well-known solutions for the Karhunen–Loeve integral equation [Eq. (6.14c)] using the autocorrelation function given in Eq. (8.17a) as the kernel [9]. This particular autocorrelation function arises from the rectangular power spectrum generated by the shot noise as it passes through the two ideal filters in Fig. 8.9. Now it is well known [9,10] that the eigenfunctions of this particular KLE are prolate spheroidal functions, which for large values of $2WT$ can be approximated by Eq. (8.18a). The eigenvalues of the KLE can be approximated also for large values of $2TW$ by a constant inside the passband and zero outside the passband. In this particular example the eigenvalues are given approximately by Eq. (8.18b).

$$\Psi_k(t) \approx \frac{\sin[\pi(2Wt - k)]}{\pi(2Wt - k)} \tag{8.18a}$$

$$\lambda_k \approx \begin{cases} 2 & \text{for} \quad k \leq n = 2Wt \\ 0 & \text{for} \quad 1 > n = 2Wt \end{cases} \tag{8.18b}$$

All of these approximations improve as $2WT$ increases.

Let $\Psi_k(t)$ be the normalized eigenfunctions of the Karhunen–Loeve equation (KLE) so that we may express X and Y in the following form:

$$x(t) = \sum_k x_k \psi_k(t) \quad \text{and} \quad y(t) = \sum_k y_k \psi_k(t)$$

where
$$x_k = \sum_k x_k \psi_k(t) \quad \text{and} \quad y_k = \int_0^T y(t)\psi_k(t)dt \qquad (8.18c)$$

The solutions of the KLE have the property that the coefficients x_k and y_k are uncorrelated. Then, because of the gaussian nature of $x(t)$ and $y(t)$, the individual coefficients are all mutually statistically independent. We also note that these are *not* zero-mean random variables. By taking advantage of the orthogonality of the $\Psi_k(t)$, we can show that

$$X = \sum_k \left[x_k^2 + y_k^2 \right] \qquad (8.19a)$$

In the forthcoming analysis we shall utilize characteristic functions as a major tool for obtaining P_e. To avoid confusion with our notation for the eigenfunctions $\Psi_k(t)$, we shall now denote the characteristic function by $C_X(\xi)$.

$$C_X(\xi) = E[e^{j\xi X}] = E\left[\exp\{j\xi \sum_k [x_k^2 + y_k^2]\} \right]$$

$$C_X(\xi) = \prod_k E[\exp\{j\xi[x_k^2 + y_k^2]\}] = \prod_k \left\{ \frac{1}{1 - 2j\xi\lambda_k} \exp\left[\frac{j\xi(\bar{x}_k^2 + \bar{y}_k^2)}{1 - 2j\xi\lambda_k} \right] \right\} \qquad (8.19b)$$

where
$$\bar{x}_k = \int_0^T 2A \cos[\Delta(t)]\Psi_k(t)dt \qquad (8.19c)$$

and
$$\bar{y}_k = \int_0^T 2A \sin[\Delta(t)]\Psi_k(t)dt \qquad (8.19d)$$

The form for $C_X(\xi)$ in Eq. (8.19b) is almost identical to our generating function for the photon-count distributions for superposed coherent and chaotic radiation, as can be seen from Eq. (6.17c) repeated here:

$$Q(s) = \prod_k \left\{ \frac{1}{1 + s\chi_k <m_T>^{m+1}} \exp\left\{ \frac{-s\phi_k^2 <m_c>}{1 + s\chi_k <m_T>} \right\} \right\}$$

Note that we can obtain $C_X(\xi)$ from $Q(s)$ if we make the following substitutions:

1. $s \rightarrow -j\xi$

2. $\chi_k <m_T> \rightarrow \lambda_k$ [recall that our χ_k values were based on the normalized autocorrelation function $\gamma(t)$]

3. $\Phi_k^2 <m_c> \rightarrow \bar{x}_k^2 + \bar{y}_k^2$

The characteristic function for the output from the ω_1 path (lower path) in Fig. 8.9, when ω_0 is transmitted (Y), has the following form because of the zero-mean situation:

$$C_Y(\xi) = \prod_k \left\{ \frac{1}{1 - 2j\xi\lambda_k} \right\} \tag{8.19e}$$

The detection variable is $q = X - Y$. Since the two channels are assumed to be far enough apart in frequency, their outputs are statistically independent. Thus

$$C_q(\xi) = C_X(\xi)C_Y(-\xi) \tag{8.19f}$$

where the minus sign in $C_Y(-\xi)$ can be traced to the minus sign before Y in $q = X - Y$. The net characteristic function is then

$$C_q(\xi) = E_{\Delta(t)} \left\{ \frac{\exp\{j\xi \sum_k [(\bar{x}_k^2 + \bar{y}_k^2)/(1 - 2j\xi\lambda_k)]\}}{\prod_k (1 - 2j\xi\lambda_k)\prod_k(1 + 2j\xi\lambda_k)} \right\} \tag{8.19g}$$

We are now ready to compute an approximate error rate for an optical-heterodyne-detected FSK system. It is shown in App. 8B that the probability of error in terms of the decision variable q is given by

$$P_e = -\frac{1}{2\pi j} \int_{-j\infty}^{j\infty} \frac{\exp[Pz/(1 - z)]}{z(1 - z)^n(1 + z)^n} \, dz \tag{8.19h}$$

where $P = A^2T$. This last substitution in effect eliminates the phase noise terms contained in the numerator of the exponent in the numerator of the characteristic function presented in Eq. (8.19g):

$$\sum_k [\bar{x}_k^2 + \bar{y}_k^2] = \text{power per pulse} = A^2T \tag{8.19i}$$

An examination of the defining equations for $\bar{x}_k$ and $\bar{y}_k$ [i.e., Eqs. (8.19c) and (8.19d)] shows that these terms are definitely affected by the phase noise $\Delta(t)$. Now, when we substitute the approximate $\Psi_k(t)$ of Eq. (8.18a) into Eqs. (8.19c) and (8.19d), these approximate eigenfunctions have the effect of shortening the effective integration time in much the same fashion as the impulse response did in Eq. (8.15c). To demonstrate this trend, consider Fig. 8.10.

For our purposes here we consider only the $k = 0$ case since the $\Psi_k(t)$ for other values of k only represent time-shifted versions of $\Psi_0(t)$. Only those $\Psi_k(t)$ that correspond to $k \leq 2TW$ will have an effect on the detection statistics because they will have peaks within the T-second baud interval while all other $\Psi_k(t)$ will have peaks lying outside the detection interval. There is a complete orthonormal set of $\Psi_k(t)$ for each numerical value of TW (actually an infinite number, if one allows for coordinate rotation). Our main interest here is in the manner that TW—or, more to the point—bandwidth of the FSK detection filters affect the performance of OHD FSK FOCS. Now we note that the approximate KLE solutions represented by Eqs. (8.19a) and (8.19b) are accurate only for large values of the time–bandwidth product TW. It turns out, however, that they are surprisingly good for moderate values. We note

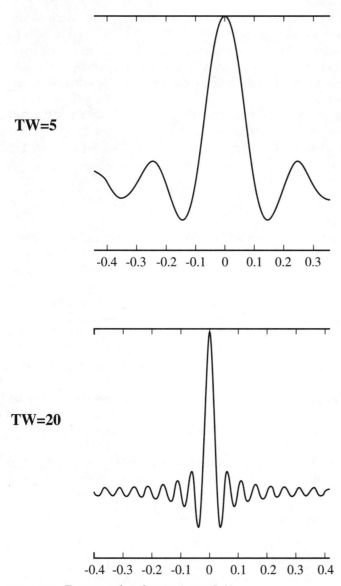

Figure 8.10   Two examples of approximate $\Psi_0(t)$.

that the accuracy of the approximation improves as TW increases, and that they are poor approximations for situations where $2TW < 4$.

It can be seen from Fig. 8.10 that the effective duration of $\Psi_k(t)$ decreases as the time–bandwidth product TW increases, as does the accuracy of the approximate eigenfunctions represented by Eq. (8.18a). In some of the numerical examples given at the end of this subsection in connection with

Salz's paper [8], these approximations are stretched a bit, and some of the results obtained there are crude approximations. However, our goal here is to develop an intuitive feel for the various phenomena that occur in coherent detection, and this is better accomplished with specific functions than with numerical computer solutions. If more precise numbers are needed, this latter technique may be required.

We now return to Eq. (8.19h) and note that the integrand has an nth-order pole at $z = -1$ and an essential singularity at $z = +1$. The probability of error is given by the residue of the nth-order pole at $z = -1$. Thus for integer $n = 2TW$, we have

$$P_e = -\frac{1}{(n-1)!}\left[\frac{d^{n-1}}{dz^{n-1}}\left\{\frac{\exp\left[Pz/(1-z)\right]}{z(1-z)}\right\}\right] \tag{8.19j}$$

Clearly the error rate is a function of both the energy per pulse as represented by the parameter P and the time–bandwidth product, which is a major constituent of the parameter n. Now, when TW is sufficiently large, the phase noise parts of $\bar{i}_k$ and $\bar{y}_k$ as given by Eqs. (8.19c) and (8.19d), respectively—namely, $\Delta(t)$—will be approximately constant over the shortened effective-time extent of the $\Psi_k(t)$. Then $\sin[\Delta(t)]$ and $\cos[\Delta(t)]$ may be factored out of their respective equations. Thus, by employing Eq. (8.19i) and the fact that $\sin^2[\Delta(t)] + \cos^2[\Delta(t)] = 1$, we find that the energy parameter P, hence the error rate P_e, is independent of phase noise (as will be all the detection statistics in this subsection).

We shall not go through the details of evaluating the preceding residue except to note that our old acquaintance, the Laguerre polynomial, appears as part of the answer. We shall, however, use the numerical results obtained by Salz [8, Fig. 11, p. 2186] and plotted in Fig. 8.11.

To utilize Fig. 8.11, we first need to define some of the terms employed in it. To begin with, we define a net phase noise bandwidth parameter that reflects the contributions of both transmitter and local lasers to the overall phase noise. Thus we let

$$B_L = B_{L,1} + B_{L,2}$$

Then
$$n = 2TW = \frac{1}{R}(R + k_{fe}B_L) = 1 + \frac{B_L k_{fe}}{R} = 1 + \frac{2k_{fe}}{\gamma}$$

where $R = 1/T$, $\gamma = 2R/B_L$, and k_{fe} generates the excess bandwidth needed to accommodate the effects of phase-noise-generated SLD drift on the IF frequency.

The numerical values displayed in Fig. 8.11 are based on the dual-filter detection system shown in Fig. 8.9, where, for the specific data plotted in the figure, Salz sets $k_{fe} = 10$ and thus the bandwidth of each filter to

$$2W = R + 10(B_{L,1} + B_{L,2})$$

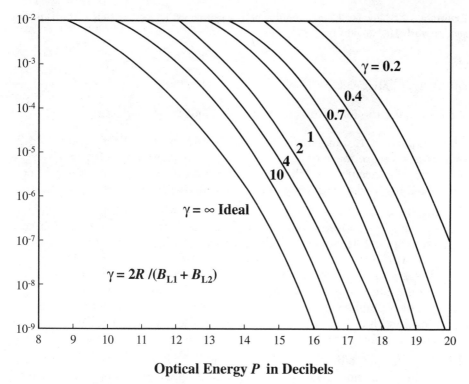

Optical Energy P in Decibels

Figure 8.11   Performance of dual-filter FSK. (*From Ref. 8, p. 2186. Copyright © 1985, AT&T. All rights reserved. Reprinted with permission.*)

We now present two numerical examples that utilize Fig. 8.11 to demonstrate how phase noise places a lower limit on data rate. For both examples, we select

$$B_{L,1} = B_{L,2} = 15 \text{ MHz}$$

Note here that this is a realistic value for SLD bandwidth since these were the actual bandwidths of the diode lasers that were used to generate the data displayed in Fig. 4.28.

Example 8.1   Suppose we desire to obtain only a 1-dB penalty from the ideal situation where there is no phase noise. Note that the ideal curve ($\gamma = \infty$) in Fig. 8.11 intersects the $P_e = 10^{-9}$ at 16 dB. Visual interpolation estimates that $\gamma \sim 7$ coincides with the intersection between 17 dB and $P_e = 10^{-9}$. This yields

$$\gamma = 7 = \frac{2R}{30 \times 10^6} \Rightarrow R \geq 105 \text{ Mbits/s}$$

and $2W = 105 + 10(30)\text{MHz} = 405 \text{ MHz}$ and $n = 2TW = 1 + 2k/\gamma = 3.86 \approx 4$.

Example 8.2   Now let a 2-dB penalty be the criterion. Then we use the 18-dB inter-section with 10^{-9} to obtain

$$\gamma = 2 \quad \text{and} \quad \rightarrow R > 30 \text{ Mbits/s}$$

and $2W = 330$ MHz and $n = 11$.

We now make the following observations about these results:

1. It is important to note here that the results obtained for Example 8.2, where $n = 11$ are more dependable than those of Example 8.1, where $n \approx 4$ prevails. The accuracy of the eigenvalue and eigenvector approximations employed in the derivation increases as n increases.

2. During the derivation of the primary formulas we noted that the effects of phase noise were eliminated in the detection process. This development, however, ignores the effects of the conversion of phase noise to amplitude noise by the linear filters. The approximations used in arriving at Eq. (8.19i) seem to work here (at least in Example 8.2) partially because of the flat magnitude response of the ideal filters. However, that does not fully explain away higher-order phase noise effects. We now invoke the earlier explanation that larger n and larger bandwidth essentially eliminate phase noise effects because of the short impulse response of such filters.

3. Keeping the preceding comments in mind, FSK is not particularly disturbed by phase noise. The principal effect of phase noise is an increase in bandwidth necessitated by the IF frequency variations caused by phase noise in both local and transmitter SLDs. One reason for this is that the primary detection for this form of FSK receiver is envelope detection followed by integrate-and-dump circuits.

4. We shall see, however, that phase noise has a much greater effect on FOCS that employ phase modulation. In particular, the effects of phase noise is a major problem in various DPSK, M-PSK, and QAM systems. These systems absolutely require OHD and are particularly sensitive to the effects of diminished coherence during a baud.

8.3.3   PSK and DPSK heterodyne detection

In this subsection we shall examine the effects of phase noise on the performance of OHD on PSK and DPSK FOCS. The principal tools used here are based on analyses and graphs of system performance obtained by Salz [8]. The derivation of these results requires many assumptions, and we shall make no attempt to derive the formulas which generate the computed data plotted in these figures. As was the case with previous topics covered in this chapter, we shall concentrate on the general trends implied by these graphs. Although some numerical examples will be given, the student should bear in mind that the numerical results are approximations.

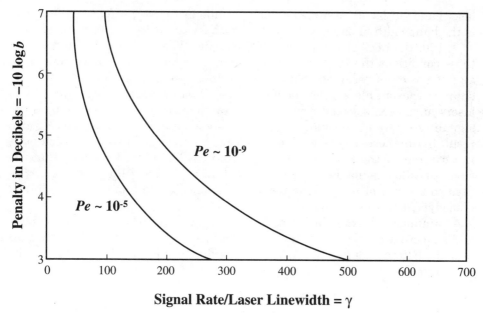

Figure 8.12 PSK heterodyne detection with identical laser bandwidths. (*From Ref. 8, p. 2177.* *Copyright © 1985, AT&T. All rights reserved. Reprinted with permission.*)

The principal results of Salz's analysis of PSK performance are summarized in Fig. 8.12.

Salz employs an optical frequency phase-locked loop (PLL) to synchronize the receiver, and his results are predicated on the performance of that PLL. We note that in his model the PLL is driven by a reference signal that is perturbed by phase noise. To obtain the results displayed in the figure, it is first necessary to find the probability density of a parameter ξ used to analyze the PLL performance. It is defined as

$$\xi = \frac{1}{T} \int_0^T \cos\{\Phi(t)\}dt$$

where $\Phi(t) = \theta(t) - \hat{\theta}(t)$, $\theta(t)$ is the phase noise driving the circuit, and $\hat{\theta}(t)$ is the PLL estimate of the instantaneous phase. Once again we encounter a parameter that depends on the integration of a term containing the phase noise as the argument of a sinusoidal function. This is further complicated because in addition to a direct dependence on phase noise, $\Phi(t)$ depends on the PLL estimate $\hat{\theta}(t)$. Furthermore, $\hat{\theta}(t)$ itself is a random process since it is obtained from a noisy driving signal. The data displayed in Fig. 8.12 is based on results obtained using many approximations.

In Fig. 8.12 we have the decibel penalty generated by phase noise plotted as a function of the same parameter γ that we encountered earlier in connection with Fig. 8.11. However, for the data plotted in this figure, identical laser bandwidths are assumed, which permits us to use $\gamma = R/B_{L,1}$ where $B_{L,1}$ is the bandwidth of one laser. We note that for the $P_e = 10^{-9}$ curve that for $\gamma < \sim 100$, the decibel penalty is off the scale. What this implies is that for those conditions the 10^{-9} error rate cannot be achieved no matter how much signal power is received. However, by increasing γ sufficiently, we can achieve reasonable system performance. Once the transmitter and local lasers have been selected, the only way to increase γ is to increase both the signaling rate and consequently the receiver bandwidth. This process can result in unnecessarily large data rates and receiver bandwidths, which may, in turn, render the system economically unfeasible. Finally we note that the smallest value of the decibel penalty is 3 dB. This lower limit can be attributed to the use of heterodyne detection as opposed to (nonpractical) optical homodyne detection.

As a numerical example, let each laser have a 15-MHz output bandwidth, and suppose we agree accept a 4-dB overall penalty, which is really only 1 dB over the heterodyne minimum. Then, according to Fig. 8.12, $\gamma \sim 300$, and the minimum data rate required to achieve $P_e = 10^{-9}$ is

$$R > 300(15.0 \times 10^6) = 4.5 \text{ Gbits/s}$$

This is, of course, a very high rate and a distinct disadvantage in most systems. If, on the other hand, we were to use lasers with only 1-MHz bandwidth (well within the state of the art), we would obtain $R > 300$ Mbits/s, which is eminently reasonable. One aspect of the graphs plotted in Fig. 8.12 is that these curves are pessimistic evaluations of system performance because of all the approximations employed in the derivation. Also the results use upper-bound estimates for P_e (which are safer than lower-bound ones) and not P_e itself.

We now consider DPSK with optical heterodyne detection. In the discussion following Fig. 8.5, we pointed out that direct optical detection of DPSK was not a practical alternative because of the extreme precision requirements on the full baud delay in the detection circuit. However, at microwave frequencies such as those found at the IF-stage output in optical heterodyne detection, it is possible to attain sufficient baud delay accuracy for satisfactory DPSK detection. Nevertheless, as we shall soon see, the DPSK requirement for coherence over two baud intervals results in much more stringent conditions on the signaling rate than ordinary PSK detection.

The IF-stage output signal $V(t)$ and the corresponding T-second delayed version previous baud $V_d(t)$ may be written in the following form:

$$V(t) = a_0 2A \cos[\omega_{if} t + \Delta(t)] + n(t)$$

$$V_d(t) = a_{-1} 2A \cos[\omega_{if}(t-T) + \Delta_d(t - T)] + n_d(t - T)$$

where $a_0 = \pm 1$ depending on the present baud and a_{-1} refers in a similar sense to the preceding baud. It is taken for granted that the data are differentially encoded to suit the DPSK structure. The detection parameter (q in this subsection) is the integral of the product of these two signals. These signals can be decomposed separately into in-phase and quadrature components as follows:

$$V(t) = V_i(t) \cos \omega_{if}t + V_q(t) \sin \omega_{if}t$$

$$V_d(t) = V_{di}(t) \cos \omega_{if}t + V_{dq}(t) \sin \omega_{if}t$$

Using the same notation here as Salz did, we now define the following detection variables:

$$q_0 = \int_0^T V(t)V_d(t)dt = a_0 a_{-1}q$$

where

$$q = \int_0^T [V_i(t)V_{di}(t) + V_q(t)V_{dq}(t)]dt$$

Now since all the quadrature components—$V_i(t)$, $V_{di}(t)$, $V_q(t)$, and $V_{dq}(t)$—are slowly varying with respect to $\omega_{if}t$, we have approximately

$$\int_0^T [V_i(t)V_{dq}(t)] \cos \omega_{if}t \sin \omega_{if}t \, dt \approx 0$$

The decision scheme is then
$$\begin{cases} \text{no phase change} & \text{if} > q \; 0 \\ 180° \text{ phase change} & \text{if } q < 0 \end{cases}$$

Clearly this is a binary symmetric channel. Let us now assume that no phase change is transmitted. Then we have $P_e = \Pr[q \leq 0]$.

The characteristic function technique that was employed to obtain the FSK graphs displayed in Fig. 8.11 does not work for DPSK. The DPSK equivalent of Eq. (8.19h) turns out to have essential singularities at both $z = +1$ and $z = -1$. Thus we cannot close the contour integral to either side, and the usual residue techniques won't work here. Salz uses a number of approximations, including Chernoff bounds, to obtain an ideal upper bound on P_e. (*This implies that the actual system can perform better than this bound.*) The results of Salz's endeavors are presented in Fig. 8.13.

Once again we observe the existence of minimum baud rates below which we cannot attain $P_e = 10^{-9}$. Now let us again assume 15-MHz bandwidth for each laser and a 4-dB net penalty to arrive at

$$\frac{R}{B_L} = 400 \Rightarrow 4 \text{ dB loss} \qquad \text{and then} \qquad R > 6 \text{ Gbits/s}$$

If 1-MHz bandwidth lasers are used, we obtain $R > 400$ Mbits/s.

These results indicate that DPSK requires a higher minimum signaling rate than does ordinary BPSK to overcome the effects of phase noise in optical heterodyne reception. However, the detection processes are very different for the two systems. In BPSK a PLL is used to extract a reference signal,

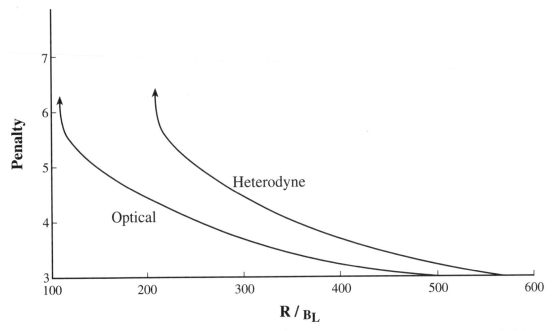

Figure 8.13 Phase noise penalty for DPSK at $P_e = 10^{-9}$ for lasers with identical bandwidths. (*From Ref. 8, p. 2200. Copyright © 1985, AT&T. All rights reserved. Reprinted with permission.*)

while in the DPSK detection system the previous baud is used as a phase reference. Both detection systems use noisy phase references, but the PLL reference is processed by complicated circuits while the DPSK reference is merely a delayed version of the previous baud. As a consequence, the approximations used in the analyses of these systems are quite different and direct comparison between these systems is very difficult. Additionally, the final results in both cases are only bounds on the error rate. Considering the performance of these systems at lower frequencies, one would expect the BPSK to perform somewhat better while the DPSK system should be easier to implement.

8.3.4 ON/OFF keying with phase noise

A typical IF-stage output signal for OOK heterodyne detection systems is

$$V(t) = aA \cos[\omega_{if}t + \Delta(t)] + n(t) \tag{8.20a}$$

where $a = 0$ or 1 as specified by the data, and where $\Delta(t)$ is the phase noise combination of signal source and local SLDs. This signal is then processed by a system such as that shown in Fig. 8.14.

The primary detection parameter q is then

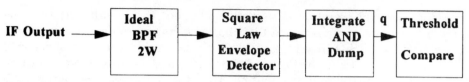

Figure 8.14 OOK heterodyne detection system.

$$q = \int_0^T [x^2(t) + y^2(t)]dt$$

where $x(t) = V_i(t) = 2A\,\cos[\Delta(t)] + n_1(t)$ and $y(t) = V_q(t)\,]2A\,\sin[\Delta(t] + n_2(t)$. We next substitute these $x(t)$ and $y(t)$ into Eq. (8.18c) to obtain the following quadratic forms, which can be used to analyze the integrated envelope of OOK digital signals with rectangular pulse shapes for the 1 state.

$$q_1 = \sum_k x_k^2 + y_k^2 \tag{8.20b}$$

$$q_0 = \sum_k n_{1k}^2 + n_{2k}^2 \tag{8.20c}$$

The Ψ_k are the Karhunen–Loeve eigenfunctions; q_1 is the energy of the signal plus noise processed for when the 1 state is transmitted, and q_0 is the corresponding energy of noise alone when a 0 is transmitted. These structures lead to the characteristic functions for variable q_1 [see Eq. (8.20b)] and q_0 [see Eq. (8.20c)].

$$C_{q1}(\omega) = \prod_k \frac{1}{1 - 2j\omega\lambda_k} \exp\left[\frac{j\omega(\bar{x}_k^2 + \bar{y}_k^2)}{1 - 2j\omega\lambda_k} \right] \tag{8.20d}$$

$$C_{q,0}(\omega) = \prod_k \frac{1}{1 - 2j\omega\lambda_k} \tag{8.20e}$$

where $\bar{x}_k$ and $\bar{y}_k$ are given by Eqs. (8.19c) and (8.19d) are used in our previous analysis of FSK with optical heterodyne detection. Note that each leg of Fig. 8.9 is very similar to Fig. 8.14. Also note that for the OFF state we have $\bar{n}_{1k} = \bar{n}_{2k} = 0$. Once again we assume that the eigenvalues are approximately

$$\lambda_k = \begin{cases} 2 & \text{for} \quad k \le n = 2WT \\ 0 & \text{for} \quad k > n = 2WT \end{cases}$$

In this case we cannot find P_e in the same manner as we did for FSK because there is only one singularity in Eq. (8.20d) and furthermore it is an essential singularity [when converted into a form similar to that of Eq. (8.19j)]. Also note that in this case we are not using the difference between q_1 and q_0 with a zero threshold for detection as was the case in FSK. Therefore, the technique for obtaining the probability of error as expressed by Eq. (8B.10) (in App. 8B) is not applicable here. However, when we substitute the preceding eigenvalues into Eqs. (8.20d) and (8.20e), they result in the characteristic functions of well-known distributions that occur in the analysis of multipulse radar. Specifically, Eq. (8.20d) is associated with the noncentral chi-squared (χ^2) distribution [11] and Eq. (8.20e) is the characteristic function for the central chi-squared distribution. Then the process for evaluating P_e is as follows:

1. Find the probability densities $f(q_0)$ and $f(q_1)$ from the characteristic functions $C_{q,0}(\xi)$ and $C_{q,1}(\xi)$ given above.

2. Then set the detection threshold at the crossover of these densities. We denote the detection threshold as "th" and find P_e from

$$P_e = \tfrac{1}{2}\left\{\int_{\text{th}}^{\infty} f_0(q_0)dq_0 + \int_0^{\text{th}} f_1(q_1)dq_1\right\} \tag{8.20f}$$

Salz employs Chernoff bounds to obtain upper bounds for P_e, and then he uses these bounds to compute the penalty for $\gamma \neq \infty$. The results indicate that the performance of OOK using OHD is approximately the same as FSK after one takes into account that an OOK system transmits no energy when a 0 is sent.

8.4 Delay-and-Multiply Discriminator

The material covered in Sec. 8.3 concentrated on examining the effects of phase noise on the performance of FOCS employing coherent detection. In our coverage we did not consider any techniques for reducing these effects other than the sometimes awkward alternative of increasing data rate. In fact, we pointed out that in many situations where phase noise is present, increasing the received power does not improve system performance at all. However, we did note in Sec. 8.2.4 that the effects of phase noise can be reduced by increasing IF-stage bandwidth to the point where the characteristic decay time of the IF-stage impulse response is sufficiently short—actually much shorter than the coherence time of the optical signals. On the other hand, we also noted that this increased bandwidth might cause a significant increase in the error rate because of the extra noise that reaches the digital decision circuits.[4]

In this section we consider one specific method for reducing the effects of phase noise in frequency-modulated FOCS that use OHD. In particular, we shall see that by employing *delay-and-multiply discriminators* (DAMDs) for the primary FM detection, the effects of phase noise can be limited to random-phase walks over extremely short time durations. Therefore, although the phase is still random, it is very nearly constant over these short intervals

---

[4]This is especially true when the detection process includes nonlinear operations such as envelope detection and limiters.

and thus behaves like a gaussian random variable with a very small variance. Recall that in ideal FM detection only the time variation of the phase is manifested in the discriminator response, and not the absolute phase itself. Specifically, in Sec. 8.4.1 we shall describe the operation of this specific discriminator; in Sec. 8.4.2 we consider the phase noise reduction aspect of DMD as well as some aspects of DMD applied to FSK systems, and in Sec. 8.4.3 we examine an actual 2-Gbit/s FSK fiber-optic communication system that was built and tested by Bellcore. There the DMD will be employed specifically for the demodulation and detection of binary FSK digital modulation. In Chap. 10 we shall examine a novel system that uses analog frequency modulation of an optical carrier by digitally modulated subcarriers to achieve a receiver sensitivity of less than 6 photons per bit.

8.4.1 Basic discriminator analysis

A typical delay-and-multiply discriminator is displayed in Fig. 8.15. The time delay t_d here is much smaller than the baud time delay that we encountered in DPSK. As we shall soon see, the delay t_d will be on the order of one-fourth of the period of the carrier frequency, which for heterodyne detection is equal to the center frequency of the IF stage, denoted in this section as ω_i (or f_i). In particular, t_d will be extremely small for typical microwave ranges f_i that arise in OHD. Thus, for example, when $f_i = 5$ GHz, then t_d will be ~50 ps. This extremely short delay time will be the principal limitation on the effects of phase noise on the output from DAMD-based receivers.

Let the input signal waveform $x(t)$ frequency-modulate an SLD (most likely by direct modulation) to generate the transmitted signal Then, after heterodyne detection, the IF-stage output $s(t)$ is split and multiplied by the delayed version $s_d(t)$ as shown in Fig. 8.15. These two signals may each be written in typical FM format as

$$s(t) = A \cos\left[\omega_i t + 2\pi f_\Delta \int^t x(v)dv + \theta(t)\right] \tag{8.21a}$$

$$s_d(t) = A \cos\left[\omega_i(t - t_d) + 2\pi f_\Delta \int_{t-t_d}^t x(v)dv + \theta((t - t_d))\right] \tag{8.21b}$$

where f_Δ is the maximum instantaneous frequency deviation from the optical carrier frequency (or f_i after OHD), and where, for convenience, we have normalized the magnitude of $x(t)$ to unity. The discriminator output $z(t)$ is a low-pass filtered version of the product of these two waveforms. Then, since the product of cosines yields the sum and difference of the cosine arguments in $s(t)$ and $s_d(t)$, respectively, we obtain after the low-pass filter:

$$z(t) = \frac{A^2}{2} \cos\left[\omega_i t_d + 2\pi f_\Delta \int_{t-t_d}^t x(v)dv + \theta(t_d)\right] \tag{8.21c}$$

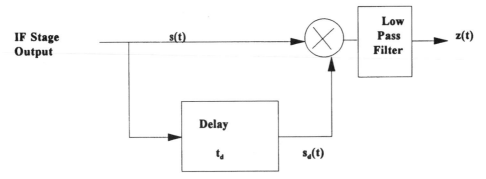

Figure 8.15 Delay-and-multiply discriminator.

where we have used $\theta(t_d) = \theta(t) - \theta(t - t_d)$. Now we select the delay time t_d from the following set of values where

$$\omega_i t_d = (2k - 1)\frac{\pi}{2} \longrightarrow t_d = 2K - \frac{1}{4f_i} \tag{8.21d}$$

where k is a positive but small integer. We note that $k = 1$ is almost always used for the detection of an analog frequency-modulated optical carrier. In the next subsection we shall examine how the choice of specific values for k affects system performance of an FSK system perturbed by phase noise. For now we note that every one of these t_d values results in an odd multiple of a quarter-wavelength delay at ω_i, which, in turn, converts the cosine function in Eq. (8.21c) into a sine function as follows:

$$z(t) = (-1)^k \frac{A^2}{2} \sin\left[2\pi f_\Delta \int_{t-t_d}^{t} x(v)\,dv + \theta(t_d)\right] \tag{8.21e}$$

The next step in this analysis is to examine the argument of the sine function in Eq. (8.21e). When t_d is sufficiently small compared to the inverse bandwidth of the modulation signal $x(t)$, we can approximate the integral inside the argument as

$$\int_{t-t_d}^{t} x(v)\,dv \approx t_d x(t) = \frac{2k - 1}{4f_i} x(t) \tag{8.21f}$$

Then the discriminator output may be expressed as

$$z(t) = (-1)^k \frac{A^2}{2} \sin\left[\frac{(2k - 1)\pi}{2} \frac{f_\Delta}{f_i} x(t) + \theta(t_d)\right] \tag{8.21g}$$

Typical phase noise bandwidths for SLDs used in optical heterodyne detection are <20 MHz. Thus, for typical $t_d \sim 50$ ps or less, we have phase noise such that $\theta(t_d) \approx$ a constant << 1 (tiny t_d leads to very short random walks) and the $\theta(t_d)$ term will have little effect on the discriminator response. We look into this topic further in the next subsection. We note that since in most circumstances we shall encounter

$$\frac{(2\pi - 1)\pi}{2} \frac{f_\Delta}{f_i} \, |x(t)| \ll 1$$

the argument of the sine function in Eq. (8.21g) is so small that we can apply the $\sin X \approx X$ approximation to the discriminator output to arrive at

$$z(t) \approx (-1)^k \frac{A^2}{2} \left\{ \frac{(2k - 1)\pi}{2} \, \frac{f_\Delta}{f_i} \, x(t) + \theta(t_d) \right\} \tag{8.21h}$$

Now the value of f_Δ can be controlled by adjusting the maximum of $|x(t)|$ prior to modulation. If Eq. (8.21h) is to be valid, then the assumed normalization of $|x(t)|$ imposes the following condition on the parameter f_Δ:

$$f_\Delta \ll \frac{2}{\pi} f_i = 0.637 f_i$$

In spite of the dependence of DAMD operation on the $\sin X \approx X$ approximation for analog detection, the response of the DAMD has proved to be more linear than the balanced discriminator. For example, even when $f_\Delta = 0.2 f_i$, the error of the approximation is ~1.7 percent.

Next we consider the application of DAMD to the OHD of FSK FOCS. The frequencies of the two transmission states in FSK can be expressed as $f_c \pm (f_d/2)$, where f_c is the laser diode operating frequency (before modulation) and f_d is the frequency difference between the two FSK tones. Direct modulation may be achieved by letting $x(t)$ be the usual rectangular pulse train (see Fig. 7.3) with the signal levels set symmetrically about zero. This binary waveform is then AC-coupled onto the SLD bias current via a bias-T coupler (see Fig. 7.2). Even for equiprobable a priori inputs this system may experience some center frequency drift when the input data stream contains local imbalances in the number of 1s and 0s. There does exist, however, a number of well-known techniques for dealing with this problem that have been developed for similar problems that arise in ordinary cable systems. One method for alleviating this DC drift difficulty is to use *alternate mark inversion* (AMI), which can be readily adapted to FOCS.[5]

Now for FSK transmission the IF-stage output frequency will be $f_i + f_d/2$ when a 1 is transmitted and $f_i - f_d/2$ when a 0 is sent. Let a decision parameter Y be defined as

$$Y \doteq \int_0^T z(t)dt \tag{8.21i}$$

This sets up a detection scheme with a decision threshold at zero, and we have following decision scheme:

$$k = \text{odd select} \begin{cases} 1 & \text{if } Y < 0 \\ 0 & \text{if } Y > 0 \end{cases}$$

---

[5] AMI employs three tones for binary transmission, and since there is an abundance of usable bandwidth in modern fibers, the primary difficulty of the three tones is increased noise levels generated by the larger bandwidths.

or

$$k = \text{even select}\begin{cases} 0 & \text{if } Y < 0 \\ 1 & \text{if } Y > 0 \end{cases} \quad\quad (8.21j)$$

where we note that k is a fixed preselected parameter for a particular DAMD. Actually the polarity of the decision process will depend on whether the local laser is at a higher or lower frequency than the received signal. Equation (8.21j) applies to the case where the local laser is at the higher frequency and the reverse polarity holds when the signal is at the higher frequency. In either case the detection threshold will be zero, which simplifies the decision process considerably. The maximum instantaneous deviation frequency for this FSK system is $f_\Delta = f_d/2$, and the discriminator output in terms of the difference frequency is

$$z(t) \approx (-1)^k \frac{A^2}{2}\left\{\pm \frac{(2k-1)\pi}{4} \frac{f_d}{f_i} + \theta(t_d)\right\} \quad\quad (8.21k)$$

where the $\pm$ represents the information in the binary data stream and where we have assumed that the frequency remains constant over the baud interval.

8.4.2 Phase noise reduction

In this subsection we examine how the use of a DAMD to detect an FM-modulated optical carrier reduces the effects of phase noise. We first consider the very important loss-of-coherence problem generated by phase noise. The phase noise term $\theta(\bullet)$ is reduced substantially to $\theta(t_d)$ by the DAMD detection process; $\theta(t_d)$ itself is a zero-mean gaussian random variable with a variance of $\theta(t_d)$ that is given by [see also Eqs. (8.8b) and (8.13)]

$$\text{var}[\theta(t_d)] = (2\pi)^2 N_0 t_d = 2\pi B_L t_d = \frac{(2k-1)\pi B_L}{2f_i} << 1 \quad\quad (8.22a)$$

where we assume that $B_L << f_i$. Some typical numbers that demonstrate this are $k = 1$, $B_L = 15$ MHz (see SLDs used for data in Fig. 4.27), and $f_i \approx 4.6$ GHz is used in a system that will be described in the next section. To see how DAMD maintains coherence, we note that the normalized autocorrelation function, evaluated at $\tau = t_d$, for phase noise is

$$\gamma(t_d) = \exp\left[-\frac{t_d}{\tau_{coh}}\right] = \exp[-\pi B_L t_d] = \exp\left[-\frac{(2k-1)\pi B_L}{4f_i}\right] \approx 1 \quad (8.22b)$$

For the numbers quoted above, $\gamma(t_d) = 0.9975$, which shows that signal remains strongly correlated. Let us now contrast this with the need to maintain coherence over the entire baud as would be needed for a non-DAMD-based detection scheme such as the FSK detector shown in Fig. 8.9.

$$\gamma(T) = \gamma\left(\frac{1}{R}\right) = \exp[-\pi B_L T] = \exp\left[-\frac{\pi B_L}{R}\right] \qquad (8.22c)$$

In most optical heterodyne systems $R \ll 4f_i$, so the loss of correlation is much larger for the non-DAMD detection case. As a numerical example the system covered in the next subsection transmits 2 Gbits/s, which results in a normalized correlation loss of 0.977. If we consider $1 - \gamma$ as a measure of loss of coherence, the performance of the DAMD-based system is ~10 times better. Note also that the situation described by Eq. (8.22c) is dependent on the data rate, which is not the case for DAMD detection. At lower data rates the improvement in coherence loss will be even larger for DAMD detection. The large improvement in coherence retention for DAMD detection can be traced directly to the truncation of the random-phase walk at the very delay short time t_d. As far as maintaining coherence is concerned, the performance of the DAMD improves as the IF-stage frequency increases. This particular aspect of DAMD detection is an important asset for the system described in Chap. 10.

In the remainder of this subsection we shall examine how DAMD detection affects signal-to noise ratio (SNR). We first consider the detection of analog FM and note that such a system can transmit digital information in the form of subcarriers. We begin by rewriting Eq. (8.21h) in the following form:

$$z(t) \approx \frac{A^2}{2}\left\{\frac{\pi f_\Delta}{2f_i}\, x(t) + \theta(t_d)\right\} \qquad (8.22d)$$

where we have selected $k = 1$. We can see that from this equation that phase noise is now an additive gaussian process, and we no longer need to worry about the nonlinear aspects of phase noise. It is well known that analog FM detection process is particularly suited to reducing the effects of additive noise. Such a system is described in Chap. 10.

We shall now consider FSK detection using a DAMD. We begin this task by combining Eqs. (8.21i) and (8.21k) to obtain

$$Y = (-1)^k\frac{A^2}{2}\int_0^T\left\{\pm\frac{(2k-1)\pi}{4}\,\frac{f_d}{f_i} + \theta(t_d)\right\}dt \qquad (8.23a)$$

We now assume that $B_L \ll 1/T$, so we may treat the very small quantity $\theta(t_d)$ ~ constant over the duration of a baud which leads to

$$Y = (-1)^k\frac{A^2 T}{2}\left\{\pm\frac{(2k-1)\pi}{4}\,\frac{f_d}{f_i} + \theta(t_d)\right\} = CZ \qquad (8.23b)$$

where $\qquad C = (-1)^k\dfrac{A^2 T}{2}$

and where
$$Z = \pm \frac{(2k - 1)\pi}{4} \frac{f_d}{f_i} + \theta(t_d) \tag{8.23c}$$

We now employ Z as the detection variable for the FSK system. However, we shall ignore the constant multiplier C since it does not affect either the SNR or the resulting P_e, but we will definitely use the polarity of C to select the appropriate decision rule in Eq. (8.21j). We note that Z is a gaussian random variable, and since the phase noise term $\theta(t_d)$ is a zero-mean gaussian random variable, we find that the mean and variance of Z are given by

$$E[Z] \doteq \bar{Z} = \frac{\pm (2k - 1)\pi f_d}{4f_i} \tag{8.23d}$$

and
$$\sigma_z^2 = \text{var}[\theta(t_d)] = \frac{(2k - 1)\pi B_L}{4f_i} \tag{8.23e}$$

Furthermore, since the detector response as represented by Z has the same variance for both signal states, and the means of the two states are equidistant from the origin, we have a binary symmetric channel with a detection threshold set at zero. The probability of error for this arrangement is given by

$$P_e = \frac{1}{\sqrt{2\pi\sigma_z^2}} = \int_0^\infty \exp\left\{-\frac{(Z - \bar{Z})^2}{2\sigma_z^2}\right\} dZ \tag{8.23f}$$

now let
$$X \doteq \frac{Z - \bar{Z}}{\sigma_z}$$

then
$$dZ = \sigma_z dX$$

which leads to

$$P_e = \int_{Z/\sigma_z}^\infty \frac{\exp\left[-\dfrac{X^2}{2}\right]}{\sqrt{2\pi}} dX = Q\left(\frac{\bar{Z}}{\sigma_z}\right) \tag{8.23g}$$

where we used Eq. (7.10a). Next we use Eqs. (8.23c) and (8.23d) to obtain the following two equations:

$$\frac{\bar{Z}}{\sigma_z} = \frac{\sqrt{(2k - 1)\pi}}{2} \frac{f_d}{\sqrt{B_L f_i}} \tag{8.23h}$$

$$P_e = Q\left\{\frac{\sqrt{(2k - 1)\pi}}{2} \frac{f_d}{B_L f_i}\right\} \tag{8.23i}$$

TABLE 8.1   Some Numerical Values

Case	B_L, MHz	f_d, MHz	f_i, MHz	k	$\overline{Z}/\sigma_z$	P_e
1	24	1400	5,000	1	3.62	1.6×10^{-4}
2	24	1400	5,000	2	6.26	1.9×10^{-10}
3	24	1400	10,000	1	2.56	5.9×10^{-3}
4	24	1400	10,000	2	4.43	4.9×10^{-6}
5	24	1400	10,000	3	5.72	5.5×10^{-9}
6	24	1400	10,000	4	6.77	6.7×10^{-12}
7	2	1400	500	1	12.5	2.5×10^{-36}
8	2	1400	20,000	1	6.27	1.9×10^{-10}

We shall now use the approximation for $Q(\bullet)$ represented by Eq. (7.14) to obtain the following trends in P_e as a function of the parameters:

1. $Q(u)$ decreases monotonically as the argument u increases.

2. P_e decreases as f_d increases, but it increases as the phase noise bandwidth B_L increases.

3. A somewhat surprising[6] result is that P_e increases as f_i increases or equivalently, t_d decreases. We shall discuss this phenomenon below.

The trends listed above follow from the previous two equations and Eq. (7.14). Some numerical values are presented in Table 8.1. Since we are considering OHD here, the phase noise bandwidth parameter B_L that appears in Eqs. (8.23h) and (8.23i) is the sum of the individual phase noise bandwidths of the signal and local lasers. We assume that both the local and signal source SLDs have equal bandwidths for each case listed in Table 8.1 and equal to half of the B_L listed in the B_L column. Thus, for example, case 1 lists B_L as 24 MHz, which means that both SLDs have 12-MHz bandwidths, which happens to be the actual bandwidth of the SLDs used in the system described in the next subsection.

In Table 8.1, except for parameter k, all the parameters used in case 1 are based on values employed in the system that we shall study in Subsec. 8.4.3. The parameter values employed for the other cases listed in the table were selected to demonstrate some of the trends described above and demonstrate some combinations the satisfy the $P_e < 10^{-9}$ criterion often employed in FOCS. The cases that satisfy this criterion are set off by additional space above and below in Table 8.1. Cases 1 and 2, as well as cases 3 to 6, show that P_e decreases as k increases. This is a result of the $\sqrt{2k-1}$ factor in the numerator of Eq. (8.23h). However, by increasing the parameter k we also

---

[6]It is surprising because as f_i increases, the delay time t_d decreases, and at first glance one would expect this to reduce the effects of phase noise.

increase t_d, which may cause loss in coherence during DAMD detection. Comparing cases 1 and 7, we see that, as expected, reducing B_L improves the error rate. In fact, all the cases in the table would satisfy the 10^{-9} criterion (usually by a wide margin) if we used lasers with 1-MHz phase noise bandwidths. Even case 3 (the worst case) would yield $P_e = 4.\times 10^{-19}$.

One possible application for the type of data listed in Table 8.1 is to find an optimum set of parameters to achieve minimum error rate. This brings up the question of why we used the same f_d for all the cases listed in Table 8.1. There is no special importance to this particular value of f_d, but we note that of all the parameters employed in Table 8.1, f_d is the only one that significantly affects transmission bandwidth. Although B_L has a small effect on channel bandwidth, it is too small to be significant, while f_i, t_d, and k are all properties of the receiver. Were it not for bandwidth considerations, one would simply select as large a value for f_d as is feasible. Any determination of an optimum value for f_d must include consideration of its effect on channel bandwidth and the consequences, both good and bad, of a large channel bandwidth in addition to considering the error rate.

Probably the most surprising parameter behavior is that as f_i increases (or, equivalently, as t_d decreases), the error rate increases. Although both the mean [Eq. (8.23d)] and the variance [Eq. (8.23e)] are directly proportional to the delay time t_d [see also Eq. (8.21d)], the error rate is determined by the ratio $\underline{Z}/\sigma_z$. However, this ratio is proportional to $\sqrt{t_d}$, and as t_d decreases, so does $\underline{Z}/\sigma_z$, and as a consequence P_e increases. The point here is that although the variance is small for small t_d, so is the mean. Thus it would seem that DAMD detection deteriorates system performance of an OHD FSK system. However, when a DAMD is used exclusively for FSK signals, it is not necessary to actually reproduce the waveform $x(t)$ by use of the $\sin X \approx X$ (for small X) approximation. Then we need *not* select the delay time t_d according to Eq. (8.21d) and the DAMD response will be the following modified form of Eq. (8.21c):

$$z(t) = \frac{A^2}{2} \cos\{\omega_i t_d + [2\pi f_\Delta t_d] x(t) + \theta(t_d)\} \qquad (8.23\text{j})$$

Now for FSK, let $x(t) = \pm 1$ according to the digital data, which leads to

$$z(t) = \frac{A^2}{2} \begin{cases} \cos\{\omega_i t_d + [2\pi f_\Delta t_d] + \theta(t_d)\} & \text{if } 1 \\ \cos\{\omega_i t_d - [2\pi f_\Delta t_d] + \theta(t_d)\} & \text{if } 0 \end{cases} \qquad (8.23\text{k})$$

We note the following aspects about the signals in Eq. (8.23k):

1. If we ignore phase noise, then it is possible to achieve better performance by increasing t_d beyond a quarter period of ω_{if}. This process can increase the detection distance between the two signal states of Eq. (8.23k)

2. Now as t_d increases so does the variance of $\theta(t_d)$. Furthermore Eq. (8.21d) is no longer satisfied nor is Eq. (8.21h) Phase noise is no longer additive and appear inside a cosine function which greatly complicates the analysis.

3. The selection of t_d is a composite of the considerations described in parts 1 and 2 above, and overall t_d is selected to minimize P_e.

4. We also note that the delay time must still be reasonably short to maintain coherence over the baud duration T.

When all these aspects are taken into account, the use of a DAMD receiver for FSK detection greatly improves system performance, both with and without phase noise.

Before we move on to the next subsection, it is very important to note that all the equations and numerical results described in this subsection were obtained with only phase noise present. Other noise sources such as shot noise, thermal noise, and excess noise were omitted from the model. Therefore, we haven't discussed system sensitivity parameters such as photons per bit. Although the signal-to-noise ratio has been mentioned in this section, it refers to the mathematically defined ratio $(\text{mean})^2/\sigma^2$, not physical powers—only the presence of phase-noise-limited system performance. Clearly, when we include the other types of noise, the error rates will increase significantly. In particular, all the numerical data listed in Table 8.1 represent minimums on achievable error rates imposed by phase noise, not actual error rates.

8.4.3 Example of a 2-Gbit/s FSK system

We next consider an actual system that was constructed and tested by Gimlett et al. at Bellcore [12]. A block diagram of this system is depicted in Fig. 8.16. This is an example of an OHD FSK system, and it incorporates many of the topics described in this text. Some of these are

1. The signal source is a DC-modulated DFB single-frequency SLD operating at 1.520 μm. Several interesting features of this source that are either shown or implied by Fig. 8.16 are

    a. A bias-T device is used for coupling the digital data signal onto the SLD drive current. The transmitter also includes some impedance-matching circuitry as well as some electronics for matching the specific response characteristics of the SLD.

    b. Even though there is no external modulator, optical isolation is introduced to the system to avoid reflections back into the laser cavity. These could come from other potential external reflectors in the system such as optical surfaces in splices.

    c. Approximately 10 percent of the transmitter SLD output is coupled into a feedback loop to stabilize the SLD center frequency. A confocal etalon (a stabilized form of a Fabry–Peort cavity) serves as a frequency reference for the transmitter.

2. The signal is then transmitted over a 101kKm length of single mode fiber made up of 14 separate sections. The total fiber loss was 27.5 dB. This includes loss incurred from 15 *biconical connector* splices. If we assume an overall average loss coefficient of 0.2 dB/km for the 101-km cable, this

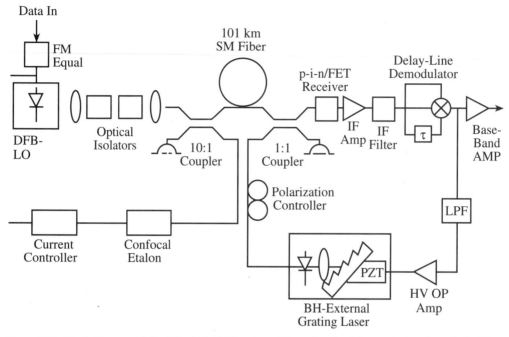

Figure 8.16 Block diagram of Gimlett et al. FSK system. (*Reproduced, with permission, from Ref. 12.*)

leaves ~7.5 dB of the loss to share among the 15 splices. Therefore, the average loss per splice was about 0.5 dB.

3. The local SLD employed a grating mounted externally to the laser chip but carefully aligned so as to be part of the laser cavity. This grating was mounted on a piezoelectric mount and used to adjust the SLD output frequency. The drive signal to the piezoelectric mount was obtained from the discriminator output, and the entire loop acted as an automatic frequency control for the receiver.

4. The polarization of the local laser output was adjusted probably manually[7] to match the polarization of the received signal. This nondynamic polarization correction works here because this laboratory version of the system is not subject to the random pressure fluctuations experienced by an underground cable.

5. Optical detection is achieved with a *p–i–n* photodiode connected to an FET mounted on the same chip. This is followed by an IF-stage amplifier. Besides amplification, another task for this group of devices is the need to impedance-match the IF stage output to the discriminator.

6. Finally we note the presence of a delay-and-multiply discriminator for FSK detection.

---

[7]I cannot state for certain that polarization was adjusted manually, but the two loops shown in Fig. 8.16 are very reminiscent of such devices that I have observed in other laboratories.

The IF stage was designed for an $f_{if} = 4.6$ GHz. As noted in item 3 above, the local laser was dynamically adjusted to maintain a difference frequency between local and signal lasers in the vicinity of f_{if}. A modulation index for FSK may be defined as

$$m \doteq \frac{f_1 - f_0}{R} \doteq \frac{f_d}{R} \qquad (8.24)$$

Although data were obtained for other values of m, the main results reported were obtained for $m = 0.7$. This coincides with $f_d = 1.4$ GHz at a rate of 2 Gbits/s. We note that it would be impossible to detect this FSK signal with the dual-filter arrangement displayed in Fig. 8.9.

Now, if we compare the Bellcore system with a system with modulation index large enough to use the dual-filter detection system, it is easy to see that the discriminator system requires considerably less channel bandwidth. The authors of Ref. 12 do state, however, that when m becomes large enough to enable use of dual-filter detection, it would probably yield better performance than DAMD detection for the same bandwidth.

Gimlett et al. also investigated the effects of phase noise on their system. The results of their theoretically based investigations are presented in Fig. 8.17, where the term dBm stands for dB above one milliwatt. These data

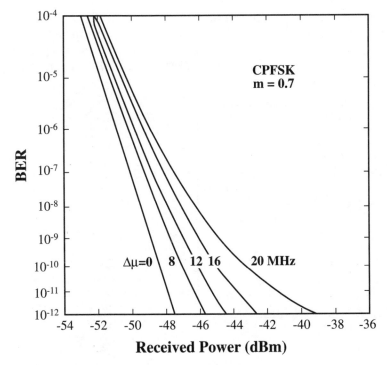

Figure 8.17 Theoretical P_e for several IF linewidths Δv. (*Reproduced, with permission, from Ref. 12.*)

were obtained under the assumptions of shot-noise-limited reception and photodetectors with 100 percent quantum efficiency. Since an OHD receiver is used here, the phase noise bandwidth parameter $\Delta\nu$ is the sum of both transmitter laser and local laser linewidths. However, in this particular system the external grating structure of the local laser should result in a much smaller linewidth than the DFB chip laser used in the transmitter. Therefore, it would seem that the transmitter laser will supply almost all of the phase noise, that is, most of $\Delta\nu$.

An examination of Fig. 8.17 shows that the spacing between curves for different phase noise bandwidths increases as the error rate decreases. For example, it is possible to perceive a floor on error rate forming in the vicinity of $P_e = 10^{-12}$ for $\Delta\nu = 20$ MHz. This phenomenon is not yet apparent for the $\Delta\nu = 8$-MHz case. Therefore, the effects of phase noise on system performance increase as either P_e decreases and/or $\Delta\nu$ increase. We can also use the data plotted in Fig. 8.17 to estimate the decibel penalty caused by phase noise for this system by noting the difference in dBm required to achieve a specific error rate and the same quantity for the $\Delta\nu = 0$ curve. Thus, for example, at $P_e = 10^{-9}$ there is ~4-dB penalty for the phase noise with $\Delta\nu = 20$ MHz while at $P_e = 10^{-12}$ the penalty is ~9 dB. It is not possible to compare these results directly with those of Salz [8] in Fig. 8.12 because the detection regimes are entirely different.

Since the detection system in the Gimlett et al. paper [12] is designed exclusively for FSK detection, it is possible to improve system performance by selecting the delay to optimize the overall signal-to-noise ratio. They take advantage of the fact that only two signals need be considered. The optimum

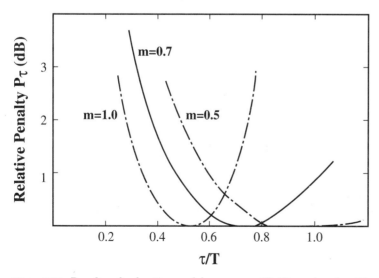

Figure 8.18 Penalty of suboptimum delay versus τ/T. (*Reproduced, with permission, from Ref. 12.*)

delay results in a delay that is larger than a quarter-period, and hence there is more phase noise (*about a 1.8-dB penalty*) than is obtainable with a quarter-period delay. Actually this is a moot point here because the deviation frequency f_d is larger than one-fifth of the IF carrier frequency and analog detection will be distorted by this relatively small carrier-to-bandwidth ratio. The effect of varying the time delay to determine an optimum delay for the system described in the Gimlett paper is displayed in Fig. 8.18. Note that the quarter-period delay of the IF carrier frequency at 4.6 Hz is 54.3 ps. Figure 8.18 shows that the optimum delay at $m = 0.7$ occurs at $\tau/T \sim 0.7$, which at 2 Gbits/s (hence $T = 500$ ps) translates into a delay of $\tau = 350$ ps. This is nearly 7 times larger than the delay for the analog discriminator described earlier in this section. Hence the resulting phase noise will be considerably larger. Nevertheless, phase noise is only one consideration out of many, and other detection parameters such as output signal strength must also be taken into account. The overall system performance is optimized for this particular system by the 350-ps delay.

Finally we examine the error rate performance of this FSK system. The error rate is plotted against the received power in Fig. 8.19. One interesting aspect of

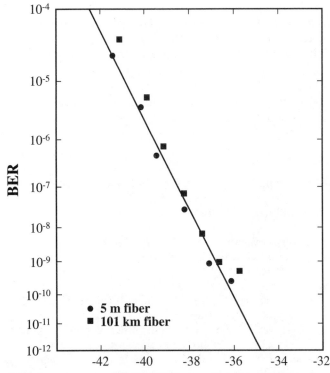

Figure 8.19 System error rate. (*Reproduced, with permission, from Ref. 12.*)

Fig. 8.19 is that the plots for the 5-m and the 101-km fibers are almost identical. Of course, in order to achieve a specific output power for each fiber, the input power for the 5-m fiber is much smaller than that for the 101-km fiber (27.5 dB, as quoted earlier). The fact that these two drastically different fiber lengths produce almost identical error rate performance curves demonstrates that the effects of dispersion, which occur only in the longer fiber, are negligible.

The received power level P required to achieve $P_e = 10^{-9}$ can be obtained from Fig. 8.19. This value turns out to be -36.7 dBm, which translates to ~ 213.8 nW. This, in turn, corresponds to

$$\frac{213.8 \times 10^{-9}}{hc/\lambda} = \frac{213.8 \times 10^{-9}}{130.8 \times 10^{-21}} = 1.772 \times 10^{12} \text{ photons}$$

At 2 Gbits/s this corresponds to a photons per bit rate (ppb) of

$$\text{ppb} = \frac{1.772 \times 10^{12}}{2 \times 10^9} = 886 \text{ photons/bit} \tag{8.25}$$

Gimlett et al. also quote $\eta \bar{P} = -39.2$ dBm as the power that actually enters the photodiode. This is supposed to take into account the quantum efficiency of the detector. However, this latter power level, which would seem to yield a lower value for ppb, does not apply in heterodyne detection as long as the local laser contribution to the photodiode input is large enough to negate the effects of receiver electronics noise. The quantum efficiency equally affects the local laser portion of the input as well as the signal portion, and η would cancel out from the overall signal-to-noise ratio, as does the local laser amplitude B. Of course, if η is too small, then the codicil on the local laser masking the receiver noise is no longer valid, but even in that case the -39.2-dBm power level would not yield the desired error rate because of the additional degradation in performance due to thermal noise.

Problems

1   Repeat the calculations for Example 8.1, in which the criterion is 3 dB.

2   Use Fig. 8.11 to determine the minimum data rate for a FSK system where both lasers have 4-MHz bandwidths and we are willing to accept a 5-dB penalty in performance.

3   Let the power of a $\lambda_0 = 1.55$-μm laser be denoted by P_L. Find the constraint on $\eta_d P_L$ whereby a receiver with termination R ohms is *relatively* independent of η_d. [See Eqs. (6.24d) and (6.25).]

4   In the Gimlett et al. paper discussed at the end of the chapter, a received power level of -39.2 dBm was quoted. How many photons per bit does this correspond to?

5   Although the gain–bandwidth profile of an excited He-Ne mixture is on the order 1.5 GHz due to Doppler broadening, laser bandwidths on the order of a few hertz have been observed (at least on a transient basis). Such narrow linewidths, hence low phase noise levels, have *not* been observed in semiconductor lasers (even with open structures such as shown in Fig. 8.7). Explain this phenomenon in terms of our discussion of phase noise in this chapter.

Appendix 8A. Derivation of Equation (8.7b)

Recall from the material in Sec. 8.2 and Fig. 8.6 that there are two contributions to the phase fluctuations:

1. The *orthogonal component,* which is proportional to $\sin\theta_i$ (our use of θ_i here differs from Salz's usage).

2. The *in-phase component of the spontaneous photon* ([pro] to $\cos\theta_i$), which has an effect that is more difficult to see but turns out to have the greatest effect on phase noise. It initially produces an amplitude change which causes a slight change in the index of refraction of the active region of the SLD. This, in turn, affects the phase of the lasing field traveling along the entire length of the cavity.

The total phase change is shown in Henry's paper [2] to be

$$\Delta\phi = \frac{\alpha}{2I} + \frac{1}{\sqrt{I}}\{\sin\theta_i - \alpha\cos\theta_i\} \tag{8A.1}$$

where $\alpha = \Delta n'/\Delta n''$, and where $\Delta n'$ and $\Delta n''$ are the real and imaginary parts of n, respectively. The parameter α can be estimated from experiment data. It greatly increases the SLD bandwidth over that predicted by theory based on fixed n. The derivation of the preceding value for α requires details of the mathematical model for laser dynamics. This is described in Henry's paper [2], and we won't consider it further here. The main task of this appendix is to examine the statistical underpinning for the result represented by Eq. (8.7b).

Now recall that θ_i is uniformly distributed over 2π. Then from Eq. (8A.1), we obtain

$$<\Delta\phi> \text{ per photon} = \frac{\alpha}{2I} \quad \text{and} \quad <\Delta\phi> = \frac{\alpha Rt}{2I}$$

where Rt is the average number of spontaneous photons in t seconds (Henry's notation).

Now we first compute the phase fluctuation contributed by a single spontaneous event. For a single event, let

$$X = (\Delta\Phi - <\Delta\Phi>) = \frac{1}{I^{1/2}}[\sin\theta_1 - \alpha\cos\theta_1]$$

Then

$$q\sigma_x^2 = \frac{1}{I}\frac{1}{2\pi}\int_0^{2\pi}\{\sin^2\theta + \alpha^2\cos^2\theta - 2\alpha\sin\theta\cos\theta\}d\theta$$

Thus

$$\sigma_x^2 = \frac{1+\alpha^2}{2I}$$

Now, to obtain the total fluctuations, including the effects of random numbers and timing of spontaneous events $<\Delta\Phi^2>$, we shall employ a technique known as *random sums* [4]. In this technique the number of events n and the contributions of each event are random. Let

$$s = \sum_{i}^{n} = \Delta\Phi_i$$

Then
$$\sigma_s^2 = <\Delta\Phi^2> = \eta^2\sigma_n^2 + \sigma_x^2\lambda<n>$$

where for a single event $\eta = <\Delta\Phi_i> = \alpha/2I$ and σ_x^2 is as given above and the Poisson nature of spontaneous emission $\sigma_n^2 = <n> = Rt$:

$$\sigma_s^2 = Rt\left[\alpha^2\left\{\frac{1}{(2I)^2} + \frac{1}{2I}\right\} + \frac{1}{2I}\right]$$

Now, if $2I \gg 1$ (typically $I \approx 40,000$), then we obtain the same result as Henry:

$$<(\Delta\phi)^2> = \frac{Rt(1 + \alpha^2)}{2I} \tag{8A.2}$$

This also happens to equal Eq. (8.7b).

Appendix 8B. Derivation of Equation (8.19j)

8B.1   Introduction

We assume that frequency ω_0 is transmitted so that the upper leg of Fig. 8.9 contains signal plus noise while the lower leg contains only noise. In terms of the in-phase and quadrature components, we have

$$X = \int_0^T \{x^2(t) + y^2(t)dt\} \tag{8B.1}$$

$$Y = \int_0^T \{v_1^2(t) + v_2^2(t)dt\} \tag{8B.2}$$

where X represents the output of the ω_0 branch and Y represents the output of the ω_1 branch whose in-phase and quadrature components are labeled by $v_1(t)$ and $v_2(t)$, respectively. Furthermore, since there is no phase lock involved in the detection process for FSK, the reference for in-phase and quadrature components is arbitrary. As was the case in FSK without phase noise, an error occurs when $Y > X$. The probability of error for this binary symmetric channel is

$$P_e = P[X - Y < 0] = P_e[q < 0] \qquad \text{where} \qquad q = X - Y$$

8B.2 Connection between P_e and $C_q(\xi)$

We are now ready to compute an approximate error rate for an optical-heterodyne-detected FSK system. Now, in terms of the decision variable q, we obtain

$$P_e = P[q < 0] = \int_{-\infty}^0 f(q)dq \tag{8B.3}$$

The probability density for q is related to the characteristic function by

$$f(q) = \frac{1}{2\pi} \int_{-\infty}^{\infty} C_q(\xi) \exp[j\xi q] d\xi \tag{8B.4}$$

Then, from Eqs. (8B.3) and (8B.4), we have our connection between P_e and $C_q(\xi)$:

$$P_e = \lim_{\epsilon \to 0} \frac{1}{2\pi} \int_{-\infty}^{\infty} [C_q(\xi)\{\int_{-\infty}^{0} \exp[(j\xi - \epsilon)q]dq\}d\xi] \tag{8B.5}$$

where ϵ is required to form a contour that avoids passing directly through a pole that forms at zero in the next equation, and where we have exchanged the order of integration. We shall first carry out the inner integration over q. After some algebra, we obtain the form that we will be use to estimate P_e:

$$P_e = \lim_{\epsilon \to 0} \frac{1}{2\pi j} \int \frac{C_q(\xi)}{(\xi + j\epsilon)} \, d\xi \tag{8B.6}$$

Now it is well known [9,10] that for large values of $2WT$, the eigenfunctions of this particular KLE are approximately given by

$$\Psi_k(t) \approx \frac{\sin[(\pi(2Wt - k)]}{\pi(2Wt - k)} \tag{8B.7}$$

while the eigenvalues of the KLE can be approximated by a constant inside the passband and zero outside the passband. This approximation gets better as $2WT$ gets larger. In this particular example the eigenvalues are approximately

$$\lambda_k \approx \begin{cases} 2 & \text{for } k \le n = 2WT \\ 0 & \text{for } k > n = 2WT \end{cases} \tag{8B.8}$$

Now, using Eqs. (8B.7) and (8B.8), we derive Eq. (8.19g), which we repeat here for convenience:

$$C_q(\xi) = E_{\Delta(t)} \left\{ \frac{\exp\{j\xi\sum_k (\bar{x}_k^2 + \bar{y}_k^2)/(1 - 2j\xi\lambda_k)\}}{\prod_k (1 - 2j\xi\lambda_k)\prod_k (1 + 2j\xi\lambda_k)} \right\} \tag{8B.9}$$

Note that the averaging in Eq. (8B.9) is carried out with respect to the phase noise represented by the subscript on E. However, it turns out that, for reasons described below, the direct effects of phase noise vanish, although there is an effect due to the increased bandwidth that phase noise imposes on OHD.

Now, from Eqs. (8.19g), (8.19c), and (8.19d) along with a change of variable of integration, one can arrive at the following form for P_e (see Salz [8] for details):

$$P_e = -\frac{1}{2\pi j} \int_{-j\infty}^{j\infty} \frac{dz}{z} \frac{\exp\{Pz/(1 - z)\}}{(1 - z)^n(1 + z)^n} \, dz \tag{8B.10}$$

This is Eq. (8.19j).

References

1. M. W. Fleming and A. Mooradian, "Fundamental line broadening of single mode (GaAl)As diode lasers," *Appl. Phys. Lett.* **38,** 511 (1981).
2. C. H. Henry, "Theory of the Linewidth of Semiconductor Lasers," *IEEE J. Quant. Electron.* **18,** 259–264 (1982).
3. G. P. Agrawal, *Fiber-Optic Transmission Systems,* Wiley, New York, 1992, p. 119.
4. A. Papoulis, *Probability, Random Variables, and Stochastic Processes,* 2d ed., McGraw-Hill, New York, 1984.
5. G. J. Foscini and G. Vannucci, "Characterizing filtered light waves corrupted by phase noise," *IEEE Trans. Inform. Theory* **34,** 1437–1448 (1988).
6. L. G. Greenstein, G. Vannucci, and G. J. Foscini, "Optical power requirements for detection OOK and FSK signals corrupted by phase noise," *IEEE Trans. Commun.* **37,** 405–407 (1989).
7. G. J. Foscini, G. Vannucci, and L. G. Greenstein, "Envelope statistics for filtered optical signals corrupted by phase noise, *IEEE Trans. Commun.* **37,** 1293–1302 (1989).
8. J. Salz, "Coherent lightwave communications," *AT&T Tech J.* **64,** 2153–2209 (1985).
9. D. Slepian and H. Pollack, "Prolate spheroidal functions—Fourier analysis and uncertainty—I," *Bell Syst. Tech. J.* **40,** 43–63 (1961).
10. M. Schwartz, W. R. Bennet, and S. Stein, *Communication Systems and Techniques,* McGraw-Hill, New York, 1966, Sec. 1.2.
11. A. D. Whalen, *Detection of Signals in Noise,* Academic Press, New York, 1971, p. 114.
12. J. L. Gimlett, R. S. Vodhanel, M. M. Choy, A. F. Elrefaie, N. K. Cheung, and R. E. Wagner, "A 2-Gbit/s optical FSK heterodyne transmission experiment using a 1520-nm DFB laser transmitter," *J. Lightwave Technol.* **LT-5,** 1315–1334 (1987).
13. I. S. Gradshteyn and I. M. Rhyzik, *Table of Integrals, Series and Products,* Academic Press, New York, 1965.

Discussion of Systems

9.1 Some Comparison between Detection Systems

9.1.1 Introduction

In this section we shall consider some comparisons between direct detection and coherent detection systems. This discussion will be based on two recent papers, each strongly preferring one type of detection over the other. The subject matter in these two papers is presented in a clear, direct manner and should be comprehensible to both undergraduate and graduate students. These papers are

R. E. Wagner and R. A. Linke, "Heterodyne lightwave systems: Moving towards commercial use," *IEEE Mag. Lightwave Commun. Syst.,* pp. 28–35 (Nov. 1990) (Ref. 1).

P. E. Green and R. Ramaswami, "Direct detection lightwave systems: Why pay more?" *IEEE Mag. Lightwave Commun. Syst.,* pp. 36–49 (Nov. 1990) (Ref. 2).

As the titles of the articles clearly indicate, the Wagner–Linke paper [1] is pro–heterodyne detection while the Green–Ramaswami paper [2] favors direct detection. In the following two subsections we shall highlight some of the main arguments presented in these papers in outline format. (*Note:* In the following discussion *some* of the editorial-type comments by the author of this text are expressed in italics or in some cases presented in parentheses or footnotes.)

9.1.2 Discussion of Wagner–Linke [1] paper

We begin this discussion of the Wagner–Linke paper [1] with a direct quote from that paper:

At every bit rate where a heterodyne system has been compared against a state-of-the-art conventional system in the laboratory, sensitivity, and transmission span records were taken by the heterodyne system. Commercialization, however, is a different matter.

In spite of the impressive performance of heterodyne detection, most fiber-optic communication systems (FOCS) being deployed today employ direct detection and OOK modulation. Predetection gain is obtained for direct-detection receivers by either an APD or an optical preamplifier; the latter are favored in WDM FOCS. Wagner and Linke suggest that the future deployment of long-range undersea optical cables is a potential application for heterodyne technology. *Optical heterodyne systems may have been a candidate for this task at the time Ref. 1 was written, but today there seems to be more interest in the direct detection of solitons transmitted over a cable containing a large number of optical amplifiers. These systems employ OOK modulation as well as an optical preamplifier. Solitons are used in long range undersea cables to avoid the considerable dispersion that would result if ordinary digital pulses were employed in the absence of regenerative repeaters.* One particular version designed by AT&T is a 9000-km undersea cable system that uses a very large number of optical amplifiers spaced at 30-km intervals to maintain sufficient signal levels so that solitons retain their shape. These ~300 amplifiers must be powered from the ends of the undersea cable (a formidable task, indeed!). *It is very doubtful that this awkward system represents the ultimate design for such systems. Therefore, it is very possible that future undersea cables may turn to heterodyne detection wherein the improved detector sensitivity can provide much larger interrepeater spacings. We also note that regenerative repeaters limit dispersion effects to those incurred in individual interrepeater stretches; a fresh signal is generated at each repeater stage that is devoid of the dispersion effects.*

To compare direct detection (DD) and coherent detection (CD), we consider the two systems shown in Fig. 9.1. Each system has been designed for sharp channel selectivity, and all systems have approximately the same degree of complexity. For example, a laser is required by both detection systems; the local laser is required for optical heterodyne detection (OHD), and a laser is required to pump the optical preamplifier in the direct-detection (DD) receiver. Typically, a considerably stronger laser is employed to pump the optical amplifier, while the local laser used in OHD has much sharper constraints on output wavelength. The local laser can be stabilized using an automatic frequency control (AFC) loop such as we described at the end of Chap. 8 when we discussed an experimental FSK system designed by Bellcore. One important aspect of these systems is the ease and speed at which they can be tuned from one channel on a network to another. The channel selection in the DD system displayed in Fig. 9.1*a* is essentially a Mach–Zender interferometer in which the tuning is accomplished by adjusting the length of one of the interferometer paths. On the other hand, OHD can be tuned far more simply and more rapidly by adjusting the bias current of the local oscillator.[1] The interferometer is typically much larger than the microwave electronics required for the combination of the IF

---

[1]The local laser can also be tuned by adjusting the ambient temperature of the local SLD, but this is a relatively slow process compared to current control. Nevertheless, temperature stability must be maintained for stable operation.

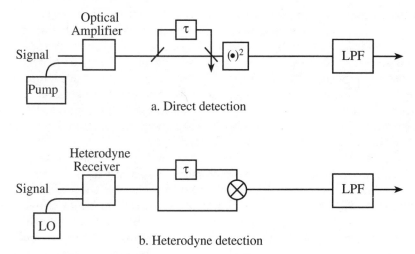

a. Direct detection

b. Heterodyne detection

Figure 9.1   Direct (a) and heterodyne (b) detection systems. (*Reproduced with permission, from Ref. 1.*)

stage and AFC loop. However, there are much smaller electronically controlled optical filters that can be tuned either very rapidly or with sharp tuning capability, but not with both of these desirable characteristics in the same device. We shall consider this point again in Sec. 9.2.

Several advantages of OHD systems over conventional DD systems mentioned in the Wagner–Linke paper [1] are

1. By using IF-stage components in the microwave frequency range, OHD systems can borrow freely from the vast repertoire of techniques developed over the years for microwave frequency systems. On the other hand, multigigabit data rates at IF frequencies in the range of 2 to 5 times the data rate (such as the system we studied in Sec. 8.4.3) can severely strain existing microwave technology. *One possible way to simplify this problem is to use higher IF frequencies.*[2]

2. One major advantage of OHD over DD is that practical filters can be made with much sharper rolloff characteristics at microwave frequencies than at optical frequencies. This is particularly important in densely packed WDM systems, and it is a direct consequence of item 1 above.

3. We now consider some further advantages of the OHD tuning system. With the technology available today, it is possible to construct multiple section lasers (such as those discussed in Sec. 4.5, on tunable lasers), that are continuously tunable over nearly a terahertz range. Furthermore, these devices are such that channel switching can be accomplished in ~1 μs, which

---

[2]In Chap. 10 we shall describe a system that does exactly this.

is faster than is possible using optical filters. In Sec. 9.2 we shall discuss a very large wide area network being designed for western Europe. One of the prime design considerations for this system is the ability to reconfigure major terminals called *nodes* very rapidly. A strong selling point in favor of OHD here is the ease and speed with which this reconfiguration can be implemented by returning local lasers. The fact that heterodyne receivers in general can be tuned over a broad range of frequencies without altering the receiver bandwidth is one of the prime reasons why heterodyne detection is used for virtually all rf and microwave receivers.

4. We noted in earlier sections of this book that OOK signals generated by direct modulation of an SLD causes severe chirping, which, in turn, results in much wider transmission bandwidths than are required by the data rate. Thus conventional OOK systems have bandwidths 10 to 100 times greater than required by the data rate. One consequence of this broader bandwidth is that it generates much greater material dispersion. However, we also note that this chirping could be avoided in OOK systems by employing external modulators. There has been considerable interest recently in using direct-modulation absorption-based modulators in a new class of SLDS which avoid much of the chirping [12].

Some possible system applications for OHD system structures are displayed in Fig. 9.2. (*Note:* Both Figs. 9.1 and 9.2 were taken directly from the Wagner–Linke paper [1].)

The uppermost diagram in Fig. 9.2 represents long-distance point-to-point communications in which the OHD system can take advantage of the greater receiver sensitivity to extend the distance between repeaters. In some cases of intermediate distance it may be possible to have completely passive fiber cable links. In the next chapter we shall consider one possible OHD system which on paper shows some promise for achieving this. If very long lengths of cable are used, it may be more economical to replace regenerative repeaters with simpler optical amplifiers, provided dispersion can be kept to tolerable levels. There are good reasons, however, for the preference for regenerative repeaters over amplifiers in wire-conductor-based cable systems where both types of repeater systems have been readily available for many years. It is reasonable that the same level of technology and economics will be available for FOCS. For direct modulation of the SLD source, the narrower bandwidth of OHD minimizes the overall dispersion. Regardless of whether external or direct modulation is used, the best amplifier noise rejection occurs when the signal is bandwidth-limited.

The middle diagram in Fig. 9.2 represents a star-coupled OHD network. In this arrangement any terminal on the network can communicate with any other terminal on the network. Furthermore, each terminal is assigned its own transmission wavelength, and each user can access the transmission wavelengths of the other terminals by tuning the local oscillator. One problem with star networks is that by passing through the N-branch star coupler, the signal reaching each terminal is reduced by a factor of N (in practice, this

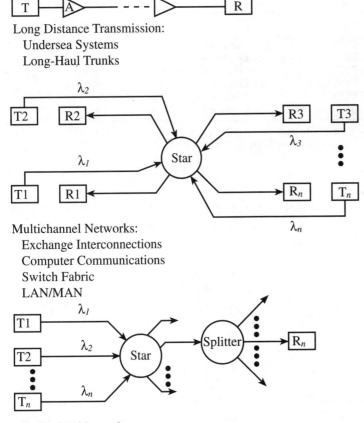

Long Distance Transmission:
  Undersea Systems
  Long-Haul Trunks

Multichannel Networks:
  Exchange Interconnections
  Computer Communications
  Switch Fabric
  LAN/MAN

Distribution Networks:
  CATV Distribution

Figure 9.2 Three possible OHD systems. (*Reproduced with permission, from Ref. 1.*)

attenuation may be considerably higher). It is also apparent that unless this setup is used over long distances, it is not suited to optical amplifiers (where would you insert them?). Thus the improved sensitivity and flexibility (rapid tuning) of OHD can be very useful for such LAN applications.

The setup depicted in the lowest diagram of Fig. 9.2 has been proposed for use in CATV (cable television) systems. In particular, GTE [3,4] has studied such systems, some of which even included broadcasting the local laser reference for heterodyne detection. Future CATV systems will most likely carry a very large number of channels, each having moderate[3] bandwidth. Optical heterodyne detection should be a strong candidate for this task because of its

---

[3]For high-resolution television, the bandwidth of each channel may be as much as 100 MHz, and channel packing consideration could be of prime importance here.

TABLE 9.1   Some OHD Field Tests

Laboratory*	Data rate	Modulation	Route	Year
KDD	565 Mbits/s	FSK	Submarine cable	1988
BTRL	565 Mbits/s	DPSK	Cambridge–Bedford	1988
AT&T	1.7 Gbits/s	FSK	Roaring Creek–Sunbury	1989
BTRL	622 Mbits/s	FSK, DPSK	Edinburgh–Newcastle	1989
BTI	620 Mbits/s	DPSK	U.K.–Guemsey	1989
NTT	2.5 Gbits/s	FSK	Matuyama–Ohita–Kure	1990

*KOD—Kokusai-Denshin Denwa Co. Ltd.; BTRL—British Telecom Research Laboratories; AT&T—AT&T Bell Laboratories; BTI—British Telegraph Laboratories; NTT—Nippon Telegraph and Telephone.

capability to facilitate channel separation in high-channel-packing-density environments. In practice, the channel packing limitations are not generated by the usable fiber bandwidth (which is enormous), but they are determined by either the non-mode-hopping range of the local lasers for OHD, or in the case of DD, the tuning range of the tunable optical filters employed for channel selection. The steepness of the filter rolloff skirts will also play an important role in designing this particular type of system. This consideration is a point in favor of OHD over DD here.

Table 9.1 lists OHD field experiments given in Table III of the Wagner–Linke paper [1].

It is readily apparent from the companies listed in Table 9.1 that there has been a great deal of interest overseas in OHD for a number of years.

In Table IV of their paper, Wagner and Linke present [1] some problems in OHD that have been solved as well as some that are only partially solved. Since a number of the problems and their solutions have already been discussed here, we shall not reproduce this table. However, we do recommend that the reader peruse the entire paper as well as Table IV in the Wagner–Link article [1]. One important aspect that is mentioned in that table that we have alluded to previously, but not yet dealt with, is the problem of polarization alignment of the local laser with the received optical signal.

As we noted in Chap. 2, the polarization of the light propagating through an optical fiber can be altered by a pressure-sensitive birefringence effect. Although this effect is small in a local sense, it can produce major polarization fluctuations over a long optical fiber. One way to combat this problem is to measure the polarization of the optical input signal and match the polarization of the local laser to it. This turns out to be both cumbersome and slow.

Another way to solve the polarization problem is to make the receiver polarization independent. Figure 9.3 presents an example of a polarization diversity OHD receiver that basically consists of two OHD receivers in parallel. The received signal is decomposed into two orthogonal polarizations, and because of the polarization-dependent beam splitter, each photodiode

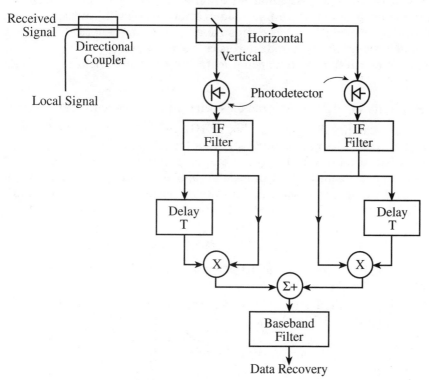

Figure 9.3 Polarization-independent coherent receiver. (*Reproduced with permission, from Ref. 5.*)

responds to only one of those polarizations. A prime feature of this particular receiver setup that has been proposed by Glance [5], is that only a single local laser is required for detection. The local laser output is circularly polarized, and the polarization-sensitive beam splitter sends the two equal-amplitude linear polarizations down separate paths. The input signal will have arbitrary polarization, which will also decompose into two linear polarizations. The sum of the outputs from each facet of the beam splitter is separately detected by the photodiodes indicated in Fig. 9.3, and the outputs are summed before a binary decision is made. This particular receiver is designed to receive DPSK modulation where the phase reference is the preceding pulse. Therefore the 90° phase difference between the two orthogonal components of linear polarization does not affect the detection process. It does, however, necessitate the use of two separate DPSK detectors.

The receiver shown in Fig. 9.3 could be converted to an FSK receiver by making the time delays one-fourth of the period of the IF frequency (thereby creating a separate DAMD for each polarization leg of the receiver). The detector then becomes a discriminator. Finally, another possible solution to the polarization problem is to use polarization preserving optical fibers. These fibers are considerably more expensive than ordinary optical fiber, but large-volume production could reduce this cost considerably.

Some other requirements that complicate the commercialization of OHD are

1. Laser stability in OHD is required for both signal source and local lasers.

2. While laser stability is needed for both OHD and DD systems, the requirements for OHD are much more stringent and hence more expensive. Commercialization of OHD may depend on the development of complete receivers and transmitters on integrated optoelectronic chips (OEICs). In that regard we note that the isolators that reduce reflections are not suited to incorporation in microchips.

3. OHD AFC is required at both transmitter and receiver, while within reasonable bounds it is not required at all for OOK DD systems.

4. The solutions to the random polarization problem are either cumbersome. and/or expensive.

5. A high degree of frequency stability for laser outputs is required for LAN applications using OHD technology. There has been considerable progress in solving this problem by using stable optical transitions as standards. Nevertheless, it may be necessary to broadcast these standards in network configurations.

6. Lasers employed in OHD require careful temperature control and current control to a much higher degree than do the pump lasers in order to drive optical preamplifiers in DD systems.

9.1.3 Discussion of Green–Ramaswami [2] paper

We begin our discussion of this paper with the following direct quote:

> The objectives of this article are to explore how far one can get with optical networks of various kinds that use only direct detection (DD), and to mention some noncoherent solutions that have been achieved. We conclude that there is almost nothing that coherent systems can do that properly designed noncoherent systems cannot do more cheaply and simply.

One of the major advantages attributed to OHD in the Wagner–Linke paper (W&L) [1] was the relative ease of rapid tuning the receiver. However, Green and Ramaswami (G&R) [2] present a number of commercially available electrically tunable filters. In particular, a tunable optical filter employing the electrooptic effect that is capable of tuning in nanoseconds is described in the G&R paper. However, the tuning range of this filter is very limited. In any case, there are much more practical alternatives to the large interferometer-type tuner displayed in the W&L paper. We note here that OHD is still a strong candidate in a very large-scale FOCS network being planned for Europe (see Sec. 9.2). The simplicity and speed with which OHD systems can be tuned is one of the main reasons for considering OHD here.

Green and Ramaswami clearly state that their main background is in networks as opposed to long-range point-to-point communications, and most of their discussion concerns large-scale (in terms of the number of nodes) networks. The following is a quote from that paper: "Experience has shown that networks with less than 100 nodes or so are not very useful, while networks larger than several thousand nodes are difficult to manage."

It is this type of network with distances under about 100 km that occupies much of the discussion in the G&R paper. A typical network of this type in displayed in Fig. 9.4. We note the following aspects of this figure in contradistinction to Fig. 9.2, which was taken from Wagner and Linke:

1. Most of the receivers in the DD network are *passive,* meaning that no local lasers are used for reception. Although not explicitly shown in Fig. 9.4, optical preamplifiers may be required to boost signals at some of the more distant terminals in order to compensate for losses incurred at the star coupler as well as transmission losses. Receiving terminals that require an optical preamplifier are certainly not passive because of the presence of the pump laser.

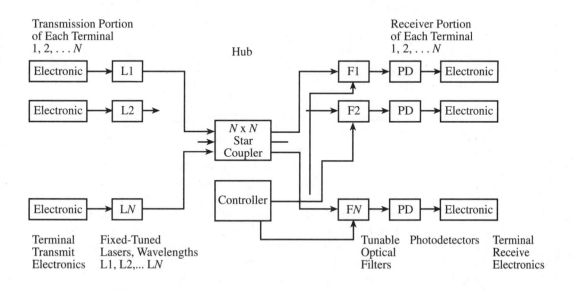

Figure 9.4 Direct-detection star network. (*Reproduced with permission, from Ref. 2.*)

2. Electronically tunable filters are used to select the desired WDM channel in DD networks. These filters, like all tunable filters have Q factors (hence the filter bandwidth and frequency rolloff characteristics) that change with filter tuning. Therefore, tunable filters do not have the sharp rolloff characteristics across the tuning band that can be obtained using heterodyne detection. One of the prime advantages of heterodyne detection in general is that the frequency response characteristics of the IF stage are not affected by the local oscillator setting. As a consequence of these heterodyne detection features, the spacing between the various channels is considerably larger for DD WDM networks than for OHD networks. However, with 12 THz of useful frequency range available in the neighborhood of 1.55 μm, this channel spacing advantage for OHD may not be an important design consideration until large-scale networks with gigabit/channel data rates are needed. It is important to note here that the channel spacing required to attain 10^{-9} error rates is considerably larger than indicated by the 3-dB filter bandwidth [6] for both OHD and DD systems.

3. Figure 9.4 shows separate transmitter (LHS of figure) and receiver portions (RHS of figure), but it is certainly feasible, if not absolutely, necessary to have a transmitter and a receiver at each node. A very important block in Fig. 9.4 is the centrally located controller, which prevents collisions on the network and maintains order.

Direct detection in the G&R [2] paper is limited to two signaling formats. In addition to OOK, they generate FSK and FSK/ASK (small amplitude) by direct modulation of the transmitter laser. The prime reason for FSK/ASK here is to generate OOK by direct modulation without the chirping problem. The FSK signal is converted into an OOK signal by blocking one of the two FSK tones with a bandpass filter. Thus the signal that passes through the filter will have the same structure as an OOK signal. However, this OOK signal will have been generated with much smaller switching transients and a greatly reduced amount of chirping. The detection of the ordinary FSK channel version is performed with the optical filter discriminators as indicated in Fig. 9.5. The detection of FSK/OOK does not require the filter-discriminator detection system.

In Fig. 9.5b and c the filtering and the wavelength demultiplexing (separation) is accomplished by optical filters. For example, in the FSK detection in Fig. 9.5b the 0 and 1 signals are sent down different paths. The receiver then bases its decision on which path had the most energy during a baud in the same way as in the dual-filter detection displayed in Fig. 8.9. In the FSK/ASK setup of Fig. 9.5c, a single filter is used and the small-scale ASK modulation is such that when a 1 is sent, the optical frequency of the transmitted signal will coincide with the passband of the optical filter and produce a significant filter output.

On the other hand, when a 0 is transmitted, the signal frequency is such that it is blocked by the filter and the only response is due primarily to noise. Thus the detection of this signal amounts to DD OOK. However at no time is the transmitter laser turned entirely off. The transmitter SLD can be directly modulated by actual signals with finite rise times and spectral bandwidths

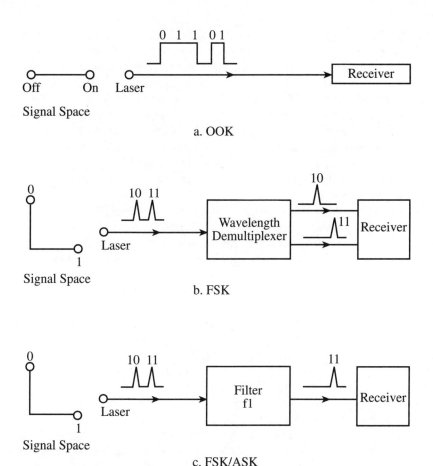

Figure 9.5 Modulation and detection forms. (*Reproduced with permission, from Ref. 2.*)

using a bias-T network This yields a continuous transition from one digital state to the other. The primary advantage of this arrangement is that one can employ direct modulation and at the same time avoid most of the chirp difficulties encountered in true OOK transmission.

Green and Ramaswami surmise that an optical preamplifier on a DD system using the modulation schemes described above can perform nearly as well as OHD systems. There is some mention of 38 photons/bit (ppb) for DD versus 36 ppb for OHD. A basic theme throughout their paper is that this can be achieved at much lower cost using DD. On the other hand, they do allow that if PSK is used, then OHD would experience an additional advantage of 3 dB for heterodyne detection and 6 dB for homodyne detection. However, both of these complicated CD-based systems would be subject to severe data rate limitations imposed by phase noise. As noted on the previous chapter, these limitations become even more severe when DPSK is used.

There seems to be no rebuttal to this from Wagner and Linke, so I shall take up the gauntlet here. There is no disagreement with the statements made in G&R concerning the equivalent performance of OHD and the direct detection–optical preamplifier combination (DDOPC) applied to OOK transmission. In Chap. 8 we saw that there is a 3-dB penalty when optical heterodyne detection is used and in Chap. 6, that there is also at least a 3-dB noise figure encountered when a large-gain optical preamplifier (OP) detection system is used. This agreement can be extended to the comparison of OOK and FSK when an average power constraints applies to transmitter power. The reason for this is that the 3-dB gain obtained from the increased detection distance for FSK over OOK is offset by a 3-dB advantage for OOK because no energy is transmitted when a 0 is transmitted. However, when maximum power limitations exist at the transmitter, FSK will yield a performance of 3 dB better OOK. Due to nonlinear distribution, a peak power limitation will affect system design. The maximum transmitted signal power is limited equally for both systems by nonlinear effects in the optical fiber. In a similar vein, the increased detection distance yields a 6-dB advantage for PSK.

Both sets of authors (W&L and G&R) as well as the rest of the FOCS community seem to have overlooked one very important aspect of heterodyne detection. That is the Bohr correspondence principle as applied to FOCS reception, which states that when the number of photons in a mode is large enough, the field in that mode behaves very much like a continuous classic electromagnetic field. The larger the number of photons in the mode, the better the approximation. The local oscillator generates a very large number of photons per baud even at gigabit / second data rates. Thus the total field entering the photodiode in typical OHD can be treated conventionally, and furthermore such noise reduction techniques as FM feedback can be utilized to significantly improve system performance. We shall go into this point at greater length in Chap. 10, where we describe one potential OHD system that has a theoretical ppb of ~6 without resorting to either optical phase lock or photon-counting detection. As mentioned in Sec. 9.1.1, the OHD receiver can be stabilized by employing AFC techniques, a particularly useful characteristic when reconfiguring a node or returning a receiver. The presence of an optical preamplifer in a DD receiver does not generate the type of signal that can drive an AFC control; simply stated, optical amplification of a weak input signal along with amplified spontaneous emission yields an output that is too irregular to operate an AFC.

9.2   Wide Area WDM Network

9.2.1   Introduction

In this section we shall consider the following paper:

Giok-Djan Khoe, "Coherent multicarrier lightwave technology for flexible capacity," *IEEE Commun. Mag.*, pp. 22–33 (March 1994) (Ref. 7).

This paper, like the two papers we considered in the previous section, is written in a clear and simple (almost tutorial) style that contains few, if any, equations. It reports on the progress and plans for a large-scale information network to be deployed in Europe. One very strong candidate for that system is WDM with optical heterodyne detection, and the flavor of this article seems to strongly favor coherent detection. The paper also contains some important updates on the status of available tunable lasers and electronically controllable optical filters.

Before we discuss the paper, it would be useful to list several of the acronyms frequently used there:

RACE—Research and technology development in Advanced Communication technologies in Europe

COBRA—Coherent Optical systems for Business traffic Routing Access

RENO—REconfigurable NOde

BCPN—Business Customer Premises Network

The source of the letter appearing in the acronym is capitalized in these descriptions.

9.2.2 Some general system requirements

We begin our discussion by noting the multiplexing hierarchy for this substantial system depicted in Fig. 9.6. Each optical carrier in the WDM portion of this hierarchy will consist of a 2.5-Gbit/s channel. The information in this channel is then subdivided into smaller packages by using TDM demultiplexing techniques. One of the goals for this network is to be able to reconfigure each principal node (RENO) so as to enable the node to process several such channels simultaneously. This flexibility permits the system to reroute the data so that it can continue to process information even when individual nodes are overloaded, possibly even reroute signals when a complete outage occurs on one of the trunk lines. The overall system may experience extreme increased needs at an individual node created by some special event such as the Olympics. It is to satisfy the RENO requirement that the COBRA group wishes to use heterodyne detection techniques to reconfigure the nodes, in particular, the rapid tunability of heterodyne detection systems that are of paramount importance here.

The vast dimensions of this network make star networks impractical here, although star couplers could be useful for local regional networks with transmission limited to the vicinity of a major network node. If a star coupler were used for the overall system, there would no doubt be some political ramifications associated with selecting the location of that single central star coupler through which all transmissions would need to pass. To avoid such political problems, a tap-in/tap-out type of configuration such as that shown in Fig. 9.7 has drawn much interest.

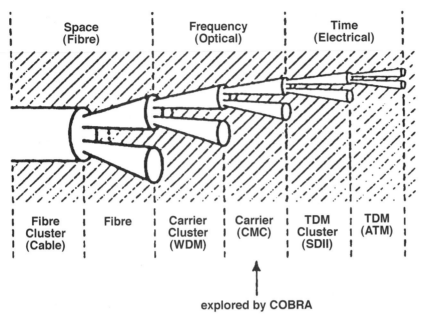

Figure 9.6 Multiplexing hierarchy for large scale network. (*Reproduced with permission, from Ref 7.*)

As shown in Fig. 9.7, each terminal is connected to the network by separate input and output taps. In a network of this type there is attenuation to all the channels at each tap (input and output) attached to the network, not just on the specific channel the receiver is tuned to. The net attenuation from many taps may be significant, and it is likely that broadband optical amplifiers will be required to maintain suitable signal levels. Each terminal employs OHD, and therefore specific channels are selected by adjusting the local SLD. Since the system is designed to convey many channels, there is some potential for image channel problems here. We shall put off any discussion of this phenomenon until the next section. It is interesting to note that Khoe never mentions this effect, so perhaps it doesn't occur at all in this system.

A number of trials have already been performed to test various major components of this broadband network. Other trials have been scheduled for the near future to be carried out before the final determination is made as to whether DD or CD will be used. One test that was performed successfully was to use components made by three separate manufacturers interchangeably in a modest-sized system employing optical heterodyne detection.

The proposed BCPN subsystem will provide a star-coupled system for approximately 100 business users, with each user having a 155-Mbit/s capability. Thus the total system capability will be 15.5 Gbit/s. This will be a WDM-based system, and there should be no collisions between multiple users

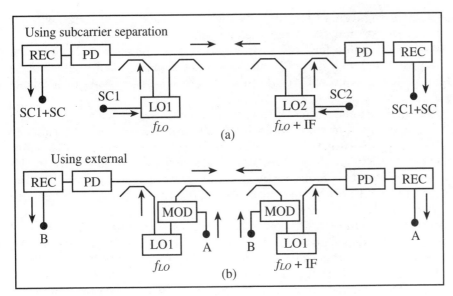

Figure 9.7   Heterodyne receivers that can also be operated as transmitters using just one laser. (*Reproduced with permission, from Ref. 7.*)

provided the system is managed properly.

9.2.3   Some updates on status of OHD components

The Khoe paper [7] includes an update on the status of some of the basic system components that were described in Sec. 9.1. The overall network described in the Khoe paper requires the capability to retune the receivers and transmitters rapidly enough to satisfy the reconfigurable node specifications of the proposed system. Khoe seems to favor the adoption of OHD to meet these requirements. Many of the reasons for his preference are based on the present-day status of some of the key components, in particular those that affect the rapidity of tuning and frequency rolloff characteristics of the primary tuning mechanism. We now present some specific observations regarding this subject:

1. Khoe notes that lasers that are continuously tunable over 7 nm of wavelength operating at 1550 nm are commercially available at costs comparable to those of an optical amplifier system. To minimize mode hopping, the local laser is tuned by varying the laser diode current while maintaining the SLD at a constant temperature as much as possible. Devices of this type have

---

[4]This relatively small drift implies that no mode hopping has occurred.

demonstrated long-term ($\sim$1-month) frequency stability ($\sim$200 MHz[4]) without resorting to an external wavelength reference.

2. Lasers well suited for FSK with only 0.49 MHz phase noise bandwidth have been reported by Fujitsu [7], and as we noted in Sec. 8.4,[5] a phase noise bandwidth this narrow will result in a system virtually immune to phase noise effects.

3. Electrically tuned optical filters can be constructed with sharp steep rolloff characteristics, or they can be built to readjust the resonant frequency at speeds comparable to those obtainable in heterodyne detection. Unfortunately, both of these desirable characteristics do not occur in the same device. This turns out to be a major drawback of optical filtering employed in DD of FSK systems as opposed to filtering at microwave frequencies as used in OHD receivers.

4. Another problem with the optical preamplifier–direct detection combination using erbium-doped optical amplifiers (EDFA) is that the user is limited to operation in the neighborhood of 1550 nm. These traveling-wave amplifiers can amplify over a wide range of wavelengths (on the order of 30 nm); however, they do not as yet operate in the 1300-nm range. There are semiconductor-based optical amplifiers that can operate at shorter wavelengths, but as we noted in Chap. 4, these amplifiers are subject to severe crosstalk when used to amplify a large number of channels. Because of the long upper-state lifetime of the EDFA, these devices do not experience much crosstalk between channels. On the other hand, OHD is not subject to these problems.

We conclude this section by including a pair of graphs (Fig. 9.8) that depict recent developments in laser diodes and two tables from the Khoe paper. The references mentioned at the bottom of the graphs are listed at the end of this chapter. We also include several tables listing experiments that have been carried out already (Table 9.2) and experiments that are projected for the future (Table 9.3).

9.3   Image Channel Rejection

9.3.1   Introduction

The material presented in this section is based on the following paper:

G. Lachs, R. E. Henneing, Y. K. Choi, and S. M. Zaidi, "Multichannel image-rejection coherent detection system," *J. Lightwave Technol.* **LT-8,** pp. 1874–1881 (1990) (Ref. 8).

Unlike the papers reviewed earlier in this chapter, this paper [8] is not written in tutorial style. Most of the calculations are contained in the appendix of

---

[5]In particular, note the discussion concerning Table 8.1.

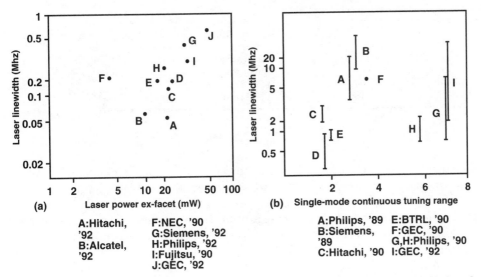

A:Hitachi, '92   F:NEC, '90
B:Alcatel, '92   G:Siemens, '92
                      H:Philips, '92
                      I:Fujitsu, '90
                      J:GEC, '92

A:Philips, '89   E:BTRL, '90
B:Siemens, '89   F:GEC, '90
C:Hitachi, '90   G,H:Philips, '90
                 I:GEC, '92

Figure 9.8 Some recently reported diode lasers with small noise bandwidths and with broad continuous ranges. (*Reproduced with permission, from Ref. 7.*)

the paper, and we shall make no attempt to reproduce that material here. Instead, we will concentrate on the basic concepts and employ as few equations as possible.

One means of utilizing the enormous bandwidth capability of optical fibers is to employ wavelength-division multiplexing (WDM). Early versions of this type of system used wavelengths sufficiently spaced so that the channels could be separated at the receiver by optical filtering techniques. Recently, however, WDM fiber-optic systems employing coherent detection (CD) have been described in the literature [9,10] which permit a much denser packing of communication channels than could be separated into individual channels by employing optical filtering. In addition to the obvious telecommunication uses

TABLE 9.2   Some Recent (1994) Laboratory and Field Tests

Year	Bit rate	Features
1990, BT	622 Mbits/s; DPSK, FSK	Two-channel test, via field percent undersea cable; booster and in-line optical amplifier tested
1991, AT&T	1.7, 2.5 Gbits/s; FSK	1 month 1.7-Gbits/s test via 35-km concatenated field fiber; laboratory test on 1.7 and 2.6 Gbits/s with OAs
1991/92, NTT	2.5 Gbits/s; CPFSK	1000 km installed fibers, 8 coherent repeaters, 3 booster OAs, 1-year long-term tests
1991/92, RACE	10 × Mbits/s; FSK	10-channel TV distribution for more than a year continuously, field test in COBRA project
1992, TFL	636 Mbits/s; SF-FSK	Real telecommunications traffic via 157-m submarine cable bidirectional operation for 90 days

TABLE 9.3   Some Planned Laboratory and Field Tests

Type	Functions	Goal 1994	Goal 1995
RENO	2.5 Gbits/s, RENO-H	Laboratory trial	Field trial
BCPN	2.5 Gbits/s, TV studio	Laboratory trial	Field trial
BCPN	155 Mbits/s with ATM	Field trial	—
BCPN	140 Mbits/s, videoconference	Field trial	—

for such fiber-optic communications there exists another class of potential applications for this technology in which densely packed WDM channels may well be a basic requirement, namely, internal communications within next-generation supercomputers. For example, the integrity of the parallel processing structure likely to be employed in future supercomputers can be maintained by employing WDM fiber-optic communication links for internal board-to-board, chip-to-chip, as well as console-to-console communications. The parallel data structure can be transmitted within a single optical fiber interconnect by having each data channel operate at its own individual wavelength. One possible scenario for these systems is to utilize integrated monolithic optoelectronic interfaces for coherent detection of densely packed WDM channels.

Optical frequency heterodyning of these WDM channels will require IF frequencies that lie in the microwave range, and the interchannel frequency spacings will also lie in the microwave range. The problem of rejecting the image frequency channel, however, is not as easily solved in optical frequency systems as in lower-frequency systems. A number of schemes for accomplishing the channel separation of densely packed WDM channels that at the same time reject the image channel have been described in the literature, and the paper cited above [8] presents one such technique.

Consider the set of densely packed spectra carried by a single optical fiber depicted in Fig. 9.9. The strong signal at point L represents the local laser. It is important to note that these lines all represent optical frequency signals (note the break in the f axis to indicate that these are nowhere near baseband). Although optical detection is not a truly quadratic process (the sum frequency is not generated in the detection process), there is optical interference between all WDM carriers whose difference frequencies from one another lie within the IF-stage passband. To avoid total chaos here, it is especially necessary for the local laser to be much stronger than the incoming WDM channels. It is not an accident that the vertical scale (ordinate) in the figure is plotted in decibels. Therefore, if such a system is to be employed for short-range internal communications within a computer, all signal sources will need to have very low-power outputs. Quantum-well lasers may fit the bill here very well for they are inherently low-power sources and eminently suited for OEIC technology.

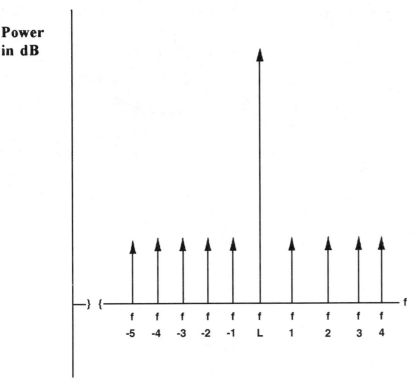

Figure 9.9 Densely packed WDM spectrum.

The manner in which the image channel can distort the signals received by an optical heterodyne receiver may be seen in Fig. 9.10. In rf heterodyne detection systems the image channel is eliminated by bandpass filter characteristics of the rf-stage amplifier before it gets to the mixing stage as depicted in Fig. 8.1. In the optical heterodyne detection of densely packed WDM signals, however, both the desired channel and the image channel arrive at the mixer stage, which in OHD happens to be the receiver photodiode. Thus a composite signal consisting of the desired channel and the image channel will appear at the IF-stage output with overlapping spectra (see lower spectrum depicted in Fig. 9.10).

The method employed here [8] to solve the image channel problem is based on the "phase shift" method for the detection of single-sideband-modulated channels. Figure 9.11 presents a block diagram for such a system adapted for application to the OHD of densely packed channels. It is important to note at this point that we are not proposing that SSB modulation itself be used, and the primary function of this system is to separate the channels whose carrier frequencies are higher than the local laser frequency from those channels whose carrier frequencies are lower. In this manner we avoid the image chan-

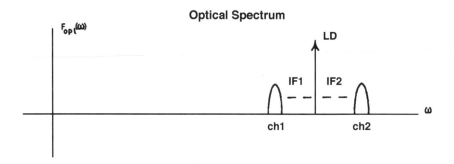

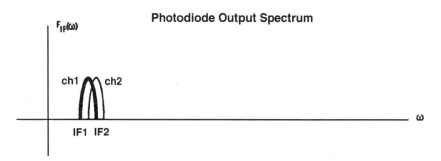

Figure 9.10 Example of image channel problem.

nel problem (no overlapping spectra)[6]. This system will perform this subdivision of channels regardless of the type of modulation employed in the individual channels. Once the channels have been separated into groups, the two system outputs can then be further separated into individual channels by conventional microwave frequency filtering techniques.

The principal technical problem encountered in this system is the requirement of accurately maintaining the appropriate phase delays at optical frequencies, which, in turn, requires that the path lengths be matched extremely accurately. In particular, the two paths emanating from the local laser (labeled A and B in Fig. 9.11) must be matched to within a fraction of an optical wavelength in order to maintain the proper phase relationships for the signals incident to the two photodiodes. It turns out the path lengths for C and D, and for E and F, need be matched only to within a fraction of a wavelength of the difference frequencies which are in the microwave range. Thus accuracy requirements for these path lengths are on the order of 0.5 mm.

To correct the phase error which occurs when path lengths A and B are not perfectly matched, we employ a voltage-controlled phase-shift device in an

---

[6]If the channels on either side of the local laser are sufficiently spaced, there will be no interchannel crosstalk.

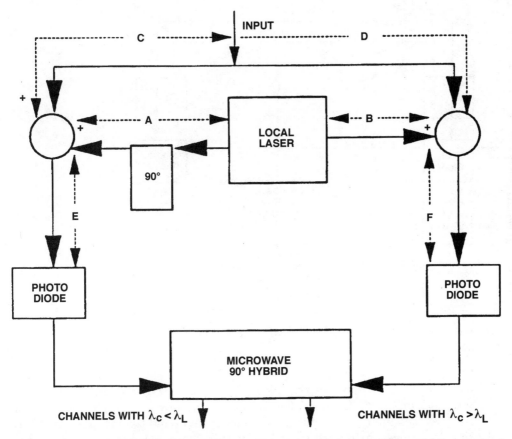

INPUT

Figure 9.11   SOB-based technique for separating channels with frequencies above and below the local oscillator.

arrangement such as that shown in Fig. 9.12. A number of such devices have been reported in the literature primarily for other applications. The control signal for the phase shifter can be derived by appropriate processing of the outputs of the photodiodes shown in Fig. 9.12. We now present a quantitative description of the phase error, and we describe one method for extracting a suitable feedback signal to use for phase control.

The system depicted in Fig. 9.11 is analyzed in App. A of the basic paper [8] where the derivations of the principal results described below may be found. We note here, however, that these results are based on the assumption of matched polarizations for all the communication channels as well as the local laser. The received optical signal is assumed to have the form

$$x(t) = \sum_{k=1}^{M} A_k \cos[\omega_k t + \phi_k] \tag{9.1}$$

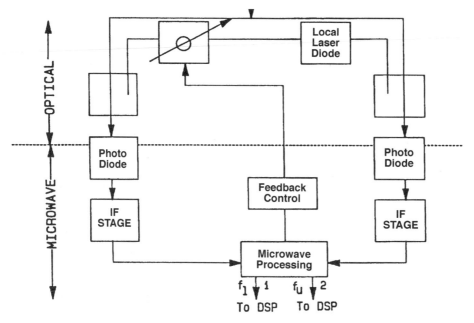

Figure 9.12 Image channel rejection system showing voltage-controlled phase shifter and feedback path.

where M = the number of channels whose post-OHD frequencies lie within the passband of the IF stage. The amplitude, frequency, and phase are denoted by A_k, ω_k, and ϕ_k, respectively; the kth channel and the modulation of each channel are assumed to be embedded within these parameters. Then, provided that the path length pairs C and D, and E and F, are matched to within a few percent of the *difference frequency wavelengths*, the amplitudes of all the image channels at both outputs of the system shown in Fig. 9.11 are proportional to $\sin(\phi_e)$, where

$$\phi_e = \frac{\phi_D - \Delta}{2} \tag{9.2}$$

and where ϕ_D is the phase error between paths C and D and where Δ is the phase error for paths A and B. The amplitudes of the desired channels turn out to be proportional to $\cos(\phi_e)$.

The feedback control signal $V_{\phi,e}$ is obtained by passing the product of the two photodiode outputs through a low-pass filter with a cutoff frequency several orders of magnitude lower than the modulation rates. Assuming that the individual channels are sufficiently separated in wavelength so that crosstalk is negligible, one obtains

$$V_{\phi,e} = \{A_L^2 \sum_{k=1}^{M} A_k^2\} \sin(\phi_e) + N_s(t) \tag{9.3}$$

where A_L represents the local laser amplitude. The term $N_s(t)$ arises as a result of cross-product terms between the channels whose carriers are lower than the local laser and the channels whose carriers are higher. It is shown in App. A of the paper [8] that, as a result of the averaging process induced by the low pass-filter, $N_s(t)$ will be negligible for FSK and PSK signaling systems. Furthermore, by suitable adjustment of the local laser frequency, $N_s(t)$ can be made negligible for all modulation schemes. Now, since all frequency and phase information has been eliminated from the first term on the right-hand side of Eq. (9.3) and the premultiplier of $\sin(2\phi_e)$ is nonnegative for all modulation schemes, this image channel rejection system is immune to phase noise. With $N_s(t)$ removed from the right-hand side of Eq. (9.3), the remaining term is proportional to $\sin(2\phi_e)$. Thus $V_{\phi,e}$ can be employed as a suitable feedback control signal. This feedback signal will not correct the path-length-induced phase errors Δ and ϕ_D on an individual basis, but it will correct the combination represented by ϕ_e. However, by making $\phi_e \approx 0$, we effectively eliminate the image channels from the system outputs, and at the same time maximize the amplitudes of the desired channels.

9.3.2 Experimental setup

Many of the topics that were discussed in the paper cited above [8] are also discussed in the experimental verification of the concept described earlier in this book. Thus it should be instructive to look at this setup. Block diagrams for this are given in Figs. 9.13 and 9.14.

For the *transmitter portion,* Fig. 9.13 displays the following devices discussed earlier in the text:

1. The transmitter employed two laser sources, and the output frequency of the two laser outputs were set ~6 GHz apart. Eventually the local laser employed in the receiver portion of the experiment was tuned to be centered directly between the transmitter lasers. This resulted in an IF frequency of ~3 GHz for each channel.
2. Note that each transmitter laser has the following equipment associated with it:
   a. Current controlled driver and temperature controller. The temperature was maintained as stable as possible, and the tuning of the lasers was accomplished by a fine adjustment on the current controller.
   b. Each transmitter laser also has a bias-T for coupling in a modulation signal if needed (and it was). Although we modulated only one of these lasers, this capability turned out to be very useful for the experimental verification of the concept.
   c. Each laser output was routed through two isolators to provide 60-dB attenuation of reflections from system surfaces. The block denoted by LDS (laser delivery system) was a commercial device for coupling the laser beam into an optical fiber.

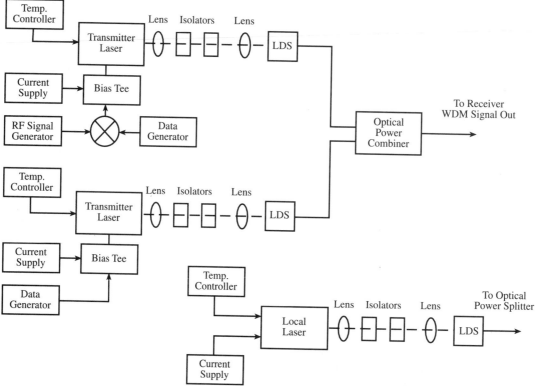

Figure 9.13 Block diagram for transmitter portion. (*Reproduced with permission, from Ref. 8.*)

    d. The signals which have been inserted into polarization preserving optical fibers are then optically coupled (added), and the combination is sent to the receiver.

3. Recall that this is an experimental setup and the primary object was to demonstrate image channel rejection. This will help explain why the local laser appears in the transmitter portion. Physically all three lasers were placed next to each other on the same end of the optical bench, while the receiver portion was pretty much spread over the remainder of the bench. It was very convenient to have all the power supplies required for the lasers as well as the temperature controllers in one location. This equipment, as well as other power supplies and so on, were mounted on a tabletoplike shelf suspended from the ceiling and hanging directly over the optical bench.

The *receiver portion* of the prototype of the system is shown in Fig. 9.14. Note the two parallel paths, which for the most part mimic the structure shown in Figs. 9.11 and 9.12 as well as the two feedback loops. One of these

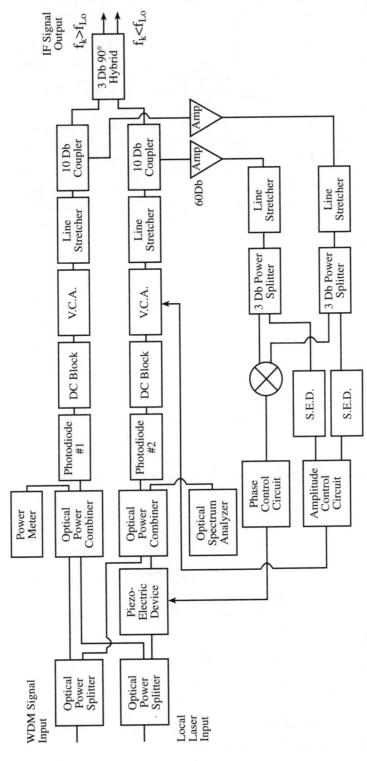

Figure 9.14 Block diagram for receiver portion (*Reproduced with permission, from Ref. 8.*)

loops provides the control signal for the phase error caused by the path-length difference between legs A and B in Fig. 9.11. In this prototype model it is accomplished by adjusting the actual length by an amount on the order of a fraction of a wavelength with a piezoelectric fiber stretcher (not the line stretchers).

Both legs of the receiver operate entirely within the microwave frequency range after the two photodiodes. The signals in each path are tapped off at the 10-dB couplers, and then the control signal is obtained by processing as described earlier (i.e., multiplied and low-pass-filtered). The control signal is depicted in Fig. 9.15. The upper figure shows the initial acquisition process where the phase may settle at the correct phase within a modulo 2π error. The initial start-up phase could settle at 180° error, but this would be an

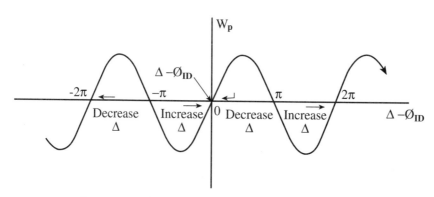

Convergence Algorithm for Phase Control

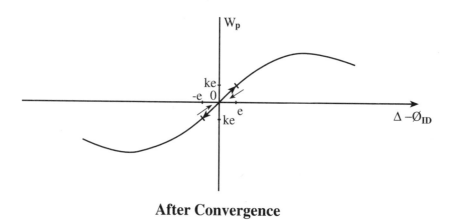

After Convergence

Figure 9.15 Correction voltage as a function of phase error.

unstable equilibrium point. The lower figure shows how the voltage adjusts itself to correct phase error to either side of the nominal value. The other feedback loop in Fig. 9.15 serves to balance the signal strengths in the two main paths. The blocks denoted by the acronym S.E.D. represent square-law envelope detectors. We note the following about the receiver in this detection system:

1. There is considerable amplification throughout the receiver that is not shown explicitly in Fig. 9.15.

2. Both the phase control and amplitude control circuits are subsystems with considerable circuitry.

3. Optical couplers were used as signal splitters by having one input and two outputs. Optical couplers were also used to add the signals at the transmitter and to add the local laser to the input signal at the receiver.

Although the receiver electronics seem imposing in the prototype, it is not that complex. A proposal was presented at the 199 SPIE meeting [11] for a chip version of the entire system. One OEIC chip would contain the optical portion of the receiver (sans local laser) up to and including the photodiodes. This would incorporate a voltage-controlled phase-shift portion. The other chip would contain all the microwave circuitry, including the hybrid and the outputs containing the separated channels.

9.3.3 Experimental results

The first test simply used unmodulated carriers as signal sources. The two outputs from the prototype image channel rejection system were sent to two separate microwave frequency spectrum analyzers. A typical set of outputs is displayed in Fig. 9.16. These particular Polaroid photographs of the spectra were recorded simultaneously. The test signals in this case are two unmodulated carriers where one test signal is tuned to be 3.2 GHz higher and the other, is 2.5 GHz lower, than the local laser frequency. These two signals yield different IF-stage output frequencies in order to make the individual spectral lines visible in this figure. The spectrum on the left side of Fig. 9.16 was recorded with an Anritsu Model MS710F spectrum analyzer, and the spectrum on the right side was recorded with an Hewlett-Packard Model 71210A spectrum analyzer, which accounts for the difference in scale between the two spectra. These figures display better than 25-dB image channel output rejection.

We note, however, that in the setup shown in Fig. 9.16 the two channels do not truly overlap at the IF-stage output. To test the system with overlapping spectra after the IF stage that are nevertheless clearly identifiable, we generated the following signal. The output of one of the signal source laser diodes was tuned 3.0 GHz higher than the local laser frequency. This laser was then FM-modulated with a steady-state sinusoid at 200 MHz connected to the

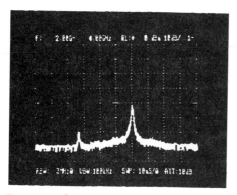 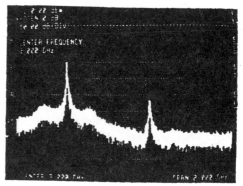

Figure 9.16 Output spectra with two unmodulated channels. (*Reproduced with permission, from Ref. 8.*)

bias-T associated with that laser. An FM carrier modulated by a pure sinusoid can be expressed as

$$y_{\text{fm}}(t) = A \sum_{k=-\infty}^{\infty} J_k(\beta) \cos\left[(\omega_c + k\omega_m)t\right] \tag{9.4}$$

where ω_c is the optical carrier frequency and ω_m is the frequency of the modulating signal. The parameter β, called the *modulation index,* is defined by

$$\beta = \frac{A_m f_\Delta}{f_m} \tag{9.5}$$

where A_m is the amplitude of the rf sinusoidal component and for the situation here f_Δ is a constant that represents the FM modulation sensitivity of the diode laser. We next adjust the amplitude of the rf signal until $\beta \approx$ 2.405, at which point $J_0(\beta) \approx 0$ and the central spectral line of the FM spectrum vanishes. The spectrum of this signal is displayed as the upper plot of Fig. 9.17, while the lower plot is the observed spectrum of an unmodulated laser.

Thus the IF spectrum of this channel after the heterodyne detection will overlap with the center of the IF spectrum of the other FM-modulated channel. The output spectra for these channels were recorded with the same equipment as used to obtain the data shown in Fig. 9.16.

Figure 9.18 shows the image signal rejection capability of the detection system more clearly. Both image channel outputs are suppressed by more than 20 dB, while the principal channel in each case is clearly visible. These data clearly demonstrate that the image signal rejection system functions even for the channels whose IF spectra are overlapped. These figures also demonstrate the channel separation (thus channel doubling) capability of the proposed image signal rejection system.

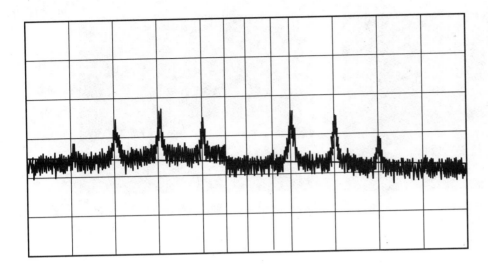

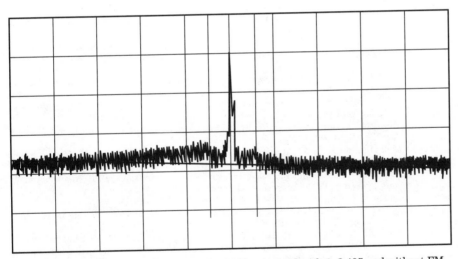

Figure 9.17 Signal laser output spectra with FM by sinusoid with β≈2.405 and without FM.

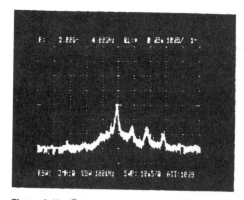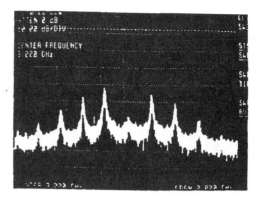

Figure 9.18 Output spectra with only one channel frequency-modulated. (*Reproduced with permission, from Ref. 8.*)

References

1. R. E. Wagner and R. A. Linke, "Heterodyne lightwave systems: Moving towards commercial use," *IEEE Mag. Lightwave Commun. Syst.* 28–35 (Nov. 1990).
2. P. E. Green and R. Ramaswami, "Direct detection lightwave systems: Why pay more?" *IEEE Mag. Lightwave Commun. Syst.* 36–49 (Nov. 1990).
3. R. Olshansky and V. A. Ianzisera, "Subcarrier multiplexed lightwave systems for broad-band distribution," *IEEE J. Lightwave Technol.* **7,** 524–529 (1989).
4. R. W. Gross, W. Rideout, R. Olshansky, and G. R. Joyce, "Heterodyne video distribution systems sharing transmitter and local oscillator lasers," *IEEE J. Lightwave Technol.* **9,** 1329–1342 (1991).
5. B. S. Glance, "Polarixation independent coherent optical receiver," *IEEE J. Lightwave Technol.* **5,** 274–276 (1987).
6. L. G. Kazovsky, "Muktichannel coherent optical communications," *IEEE J. Lightwave Technol.* **5,** 1095–1102 (1987).
7. Giok-Djan Khoe, "Coherent multicarrier lightwave technology for flexible capacity," *IEEE Commun. Mag.* 22–33 (March 1994).
8. G. Lachs, R. E. Henneing, Y. K. Choi, and S. M. Zaidi, "Multichannel image-rejection coherent detection system," *IEEE J. Lightwave Technol.* **LT-8,** 1874–1881 (1990).
9. B. S. Glance, J. Stone, K. J. Pollock, P. J. Fitzgerald, C. A. Burrus, Jr., B. L. Kasper, and W. Stutz, "Densely spaced FDM coherent star network with optical signals confined to equally spaced frequencies," *IEEE J. Lightwave Technol.* **6,** 1770–1781 (1988).
10. R. A. Linke and A. H. Gnauck, "High-capacity coherent lightwave systems," *IEEE J. Lightwave Technol.* **6,** 1750–1769 (1988).
11. G. Lachs, S. Zaidi, A. Singh, R. E. Henning, D. Trascitti, H. Kim, P. K. Bhattacarya, J. Pamulpati, P. J. McCleer, G. I. Haddad, and S. Peng, "Integration of a coherent optical receiver with adaptive image rejection capability," *SPIE Proc.* **1474,** 248–259 (April 1991).
12. H. Q. Hou and T. Y. Chang, "Nearly chirp-free electroabsorption modulation using InGaAs-InGaAlAs-InAlAs coupled quantum wells," *IEEE Photonics Lett.* **7,** 167–169 (1995).

New Horizons for Coherent Detection

10.1 Bohr Correspondence Principle

10.1.1 Introduction

At the beginning of Chap. 4 we introduced the photon-optics model. This description included a short discussion of how electromagnetic fields are quantized. We began by noting that a field in a cavity may be expressed in the modal expansion

$$\tilde{\mathbf{E}}(\mathbf{r},t) = \sum_q \sum_k \tilde{A}_q \, U_q(\mathbf{r}) \exp\{j2\pi f_q t\} \, \hat{\mathbf{p}}_k \qquad (10.1a)$$

where the index q represents the spatial portion of the mode designation and the index k represents two orthogonal polarization states. Taking the real part of this equation, we arrive at

$$\vec{\mathbf{E}}(\mathbf{r},t) = \sum_q \sum_k |\tilde{A}_q| \, U_q(\mathbf{r}) \cos\{j2\pi f_q t + \phi_q\} \, \hat{\mathbf{p}}_k \qquad (10.1b)$$

where ϕ_q is the phase part of the complex amplitude $\tilde{A}_q$. Thus we clearly see that the net field in the cavity may be considered as a superposition of simple harmonic oscillators. In the classic representation of electromagnetic fields, these are classic oscillators, while the quantum description of electromagnetic fields employs quantum-mechanical oscillators for the modal expansion.

As a model of a classic (one-dimensional) oscillator, we consider a particle of mass m attached to a spring in a frictionless environment. Then, if the mass is displaced to, and released from, its equilibrium position at $x = 0$ (*defined to be at the site of minimum potential energy*), it will oscillate back and forth with an instantaneous position that is given by

$$x = A \cos(\omega t + \phi) \qquad (10.2a)$$

We assume the usual spring force versus displacement equation where $F = -kx$. Then it is well known from classic mechanics that the resonant frequency of the spring is $\omega = \sqrt{k/m}$ (radians/s). Furthermore, when the mass is located at position x, the potential energy stored in the spring is $\frac{1}{2} kx^2$.

The characteristics of the quantum-mechanical version of the simple harmonic oscillator can be obtained by solving the Schroedinger equation, in which the energy stored in the spring is used for the potential-energy term. This leads to the following differential equation for the wavefunction of a quantum-mechanical oscillator:

$$\frac{d^2 \Psi_n(x)}{dx^2} + \frac{2m}{(\hbar)^2}\left[E_n - \frac{1}{2} kx^2\right]\Psi_n(x) = 0 \qquad (10.2b)$$

where $\hbar = h/2\pi$. It turns out (as noted in Chap. 4) that the solutions of Eq. (10.2b) result in the discrete set of energy levels given by $E_n = (n + \frac{1}{2}) hf$, where $hf = \hbar\omega$ is the photon energy.

The primary goal in this section is to demonstrate that the behavior of quantum-mechanical oscillators in quantum states with high quantum numbers,[1] and the solutions for classic harmonic oscillators with the same average potential energy, are very much alike. The primary basis for this comparison will be the similarity of both the probability densities and the probability distributions for the location of the particle mass between the quantum and classic models. It is very important to note here that the formal quantum-mechanical description of an electric field is directly related to the quantum-mechanical position operator [10,14]. This forms the principal basis for the equivalence of the electromagnetic fields in both oscillator models.

10.1.2 Statistical comparison between classic and quantum oscillators

In quantum mechanics the probability density for the location of a particle $p(x)$ in terms of the wavefunction is given by

$$p_n(x) = |\Psi_n(x)|^2 \qquad (10.2c)$$

where Ψ_n is the wavefunction for the particle in a quantum-mechanical oscillator in an energy state characterized by the energy level (eigenvalue) $E_n = (n+\frac{1}{2}) hf$. This wavefunction turns out to be [1,2]

$$\Psi_n(x) = \frac{\exp[-\alpha^2 x^2/2]\, H_n(\alpha x)}{[n! 2^n \sqrt{\pi}]^{1/2}}\, \exp\left\{\frac{-jE_n t}{\hbar}\right\} \qquad (10.3a)$$

---

[1]For the simple harmonic oscillator considered here, a large quantum number means a large number of photons in the cavity mode being modeled by the oscillator.

$$p_n(x) = |\Psi_n(x)|^2 = \frac{\exp[-\alpha^2 x^2]\,|H_n(\alpha x)|^2}{[n!2^n\sqrt{\pi}]} \qquad (10.3b)$$

where $\alpha = \sqrt{m\omega/\hbar}$ and where the $H_n(\bullet)$ is the nth Hermite polynomial [3,4] defined by

$$H_n(y) = (2y)^n - 2\left(\frac{n}{2}\right)(2y)^{n-2} + 2^2 3\left(\frac{n}{4}\right)(2y)^{n-4} - 2^3 3 \cdot 5\left(\frac{n}{6}\right)(2y)^{n-6}\cdots \qquad (10.3c)$$

As can be seen from Eq. (10.3c), the Hermite polynomials are fairly complicated and many texts, including Refs. 1 and 2, describe them in terms of generating functions.[2] Our primary interest here will be Ψ_n, hence Hermite polynomials, for large values of n.

The position of the mass in a frictionless classic oscillator is given by

$$x = A\cos(\omega t + \phi)$$

If we consider the position of the particle at a sampling instant t selected completely at random,[3] then the probability density random variable x (the location of the point mass attached to the spring) can be determined from using the following statistical model. Let

$$x = A\cos(\theta) \qquad \text{where the density of } \theta \qquad p(\theta) = \begin{cases} \dfrac{1}{2\pi} & \text{for } -\pi \le \theta \le \pi \\ 0 & \text{otherwise} \end{cases}$$

This yields the following probability density [5,6]:

$$p_{cl}(x) = \begin{cases} \dfrac{1}{\pi\sqrt{A^2 - x^2}} & \text{for } -A \le x \le A \\ 0 & \text{otherwise} \end{cases} \qquad (10.3d)$$

where $p_{cl}(x)$ is the probability density of x for the classic oscillator. The value of the amplitude A_n for the classic oscillator that matches the quantum-mechanical oscillator in the state with energy $E = (n+\frac{1}{2})\,hf$ is given by [1,2]

$$A = A_n = \sqrt{2(n + \tfrac{1}{2})}$$

Substituting A_n for A in Eq. (10.3) yields the probability density of the classic oscillator $p_{cl}(x)$ and is plotted as a dashed line in all four cases shown in

---

[2]It also may be of interest to electrical engineers that Ref. 1 contains an appendix that solves the Schroedinger equation for the simple harmonic oscillator [Eq. (10.2c)] using Laplace transform techniques.

[3]The term $p_{cl}(x)$ can also be derived using phase as the random variable.

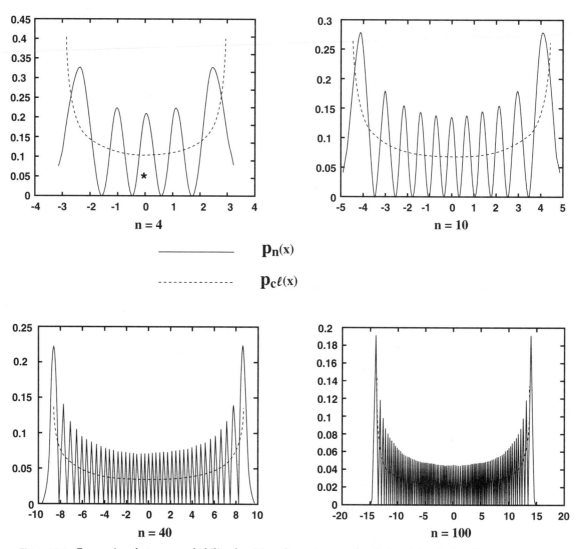

Figure 10.1 Comparison between probability densities of quantum-mechanical and classic densities.

Fig. 10.1. It is quite logical that within the range $-A_n \leq x \leq A_n$ this density has a minimum at $x = 0$ because the particle velocity is maximum here; that is, t spends the least time at $x = 0$. Similarly, the particle spends more time in the vicinity of the zero-kinetic-energy points $r \pm A_n$. In fact, the density at the $\pm A_n$ extremes $\to \infty$. Nevertheless, the area under the density is finite and properly normalized, thanks to the presence of the square root in the denominator. Recall that the density itself is not the probability of an event and that probabilities of events are obtained by integrating the probability density over some region of interest.

The solid curves in Fig. 10.1 are plots of $|\Psi_n|^2$ for all four examples depicted in the figure. The similarity between the classic and quantum-mechanical versions can be examined by comparing the results of integrating these densities over the same regions. It can be seen that if we select a small interval in the neighborhood of the asterisk displayed in the $n = 4$ portion of Fig. 10.1, then $p_n(x) > p_{cl}(x)$ over this entire region. This clearly is poor fit even though there is some similarity between the densities. However, it can be seen also that as n increases, the oscillations of $p_n(x)$ around $p_{cl}(x)$ become regular and very closely packed and it is clear that integrating over comparable regions (as in the $n = 4$ case) will likely provide a good fit. This property, called the *correspondence principle,* is a subset of a more general version known as the *Bohr correspondence principle.* This type of density comparison is often found in quantum-mechanics textbooks.

One aspect that differs from classic and quantum-mechanical oscillators is the fact that the probability density for the classic oscillator is zero for $|x| > A_n$ while the probability density quantum model is not limited to this region. As a consequence, the classic density is fully normalized within the $\pm A_n$ range while the quantum-mechanical density is not normalized over this region. However, the probability that $|x| > A_n$ is very small and decreases as n increases. This may be seen from Fig. 10.2, where the densities have been plotted over an extended scale.

Another way to demonstrate the correspondence principle is to employ the probability distribution functions $\mathcal{P}_{cl}(x)$ for the classic oscillator, and $\mathcal{P}_n(x)$ for the quantum-mechanical oscillator. These are defined by

$$\mathcal{P}_{cl}(x) = \int_{-\infty}^{x} p_{cl}(y)dy \quad \text{and} \quad \mathcal{P}_n(x) = \int_{-\infty}^{x} p_n(y)dy$$

These are plotted in Fig. 10.3. Note that we have displaced the classic model by 0.2 in order to highlight how the similarity between the distributions

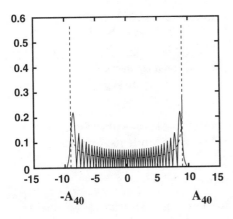
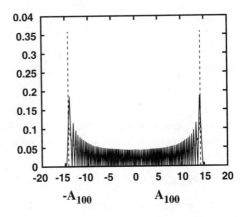

Figure 10.2 Extended plots of densities for $n = 40$ and $b = 100$.

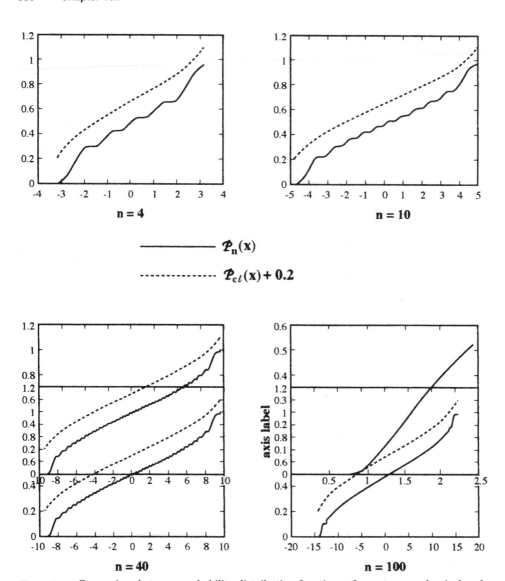

Figure 10.3 Comparison between probability distribution functions of quantum-mechanical and classic densities.

improves as it increases. In particular, we note how these curves get smoother as n increases. This is the essence of the Bohr correspondence principle.

10.2 Correspondence Principle and Coherent Detection

The primary consequence to optical detection of the similarity between classic and quantum-mechanical descriptions of the simple harmonic oscillator, described above, is that due to modal expansion [Eq. (10.1a)]; this correspondence will also hold for electromagnetic fields. This will be the case *provided the net received field is sufficiently strong.* When these conditions are satisfied, the reception of optical signals may be treated as a classic detection process and the effects of quantization may be ignored. In addition to simplifying the analysis of OHD FOCS, it opens the door to many powerful signal-processing techniques. We shall soon describe this aspect, but first we note that for coherent detection to be useful, the local laser must be strong enough that the receiver-generated electronic noise (primarily thermal noise) is overwhelmed by the shot noise generated by the OHD process. This requirement on the local laser strength results in a surplus of photons present in the depletion region[4] of the photodiode to achieve Bohr correspondence status, regardless of the strength of the received signal. We now examine this statement more carefully.

The number of modes in a specified volume V and bandwidth Δf is given by Eq. (4.18c), repeated here for convenience:

$$p = \frac{8\pi V f_{21}^2 (\Delta f)_{21}}{(c/n)^3} = \frac{8\pi V (\Delta f) n^3}{c \lambda_0^2} \tag{10.3e}$$

We first encountered this parameter in our discussion of lasers in Chap. 4. The last term on the right-hand side presents p in terms of wavelength and is the form that we shall employ here. The other important quantity is the number of photons per second at a particular wavelength $\phi(\lambda_0)$ for a given power P_0:

$$\phi(\lambda_0) = \frac{P_0}{hf} = \frac{P_0 \lambda_0}{hc} \tag{10.3f}$$

From these two equations we obtain the following formula for the average number of photons per mode, which we denote by $\bar{\rho}_m$:

$$\bar{\rho}_m = \frac{\phi(\lambda_0)}{p} = \frac{P_0 \lambda_0^3}{8\pi h V (\Delta f) n^3} \tag{10.3g}$$

Now, with the aid of these equations, we present Example 10.1, which will demonstrate that the average number of photons per mode is indeed a very large number.

Example 10.1 In this example we shall select very conservative values for the parameters which will yield much larger values for the number of modes parameter

---

[4]The small volume of the depletion region in a photodiode results in the number of modes parameter p being $< 10^7$ [see Eq. (4.18c)]. Even when the received power is as low as 1 nW (a very small input power level), the number of photons/s will be on the order of 10^{10}, which is $\gg p$. Thus we have an average of $\sim 10^3$ per mode and the correspondence principle applies. See Example 10.1.

and much smaller average photons per mode than those encountered in actual systems. Let

$V = 1 \text{ cm}^3 = 10^{-6} \text{ cm}^{-6}$ (*this is very large for a high-speed photodiode*)

$\Delta f = 10 \text{ GHz} = 10^{10} \text{ Hz}$ (*high-speed* data link)

$P_0 = 1 \text{ mW} = 10^{-3}$ (very modest power level for a local oscillator)

Using these parameters, we obtain the following plot of the number of modes parameter p and $\bar{\rho}_m$ versus λ_0 over a range from the blue end of the visible spectrum up to upper end of wavelengths useful in contemporary FOCS.

Figure 10.4 clearly shows that the number of modes is extremely large here, much larger than the $p \sim 10^7$ found for a GaAs semiconductor laser discussed in Chap. 4. Although the depletion region of a p–i–n photodiode is usually much larger than the active region of an SLD, the cubic centimeter volume used in this example is much larger than would be found in high-speed optical receivers.

In spite of the conservative parameters described above, the average number of photons per mode depicted in Fig. 10.5 is larger than 800 over the entire range of wavelengths plotted in these figures. Under normal operating conditions encountered in OHD-based FOCS, this number would be orders of magnitude larger, and the conditions for the Bohr correspondence principle to apply and the detection process to behave classically would be satisfied many times over. Note that we employed $n = 10$ photons to demonstrate what is meant by this equivalence in connection with our discussion of Fig. 10.1. It would be a much more accurate approximation for $n \sim 800$.

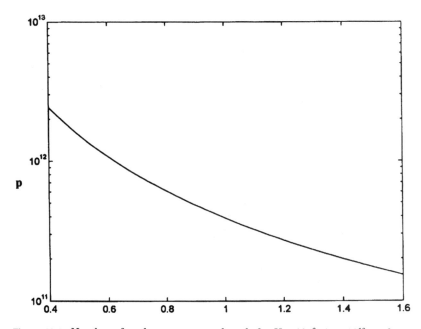

Figure 10.4 Number of modes versus wavelength for $V = 10^{-6}$, $\Delta_f = 10^{10}$, and $n = 3.6$.

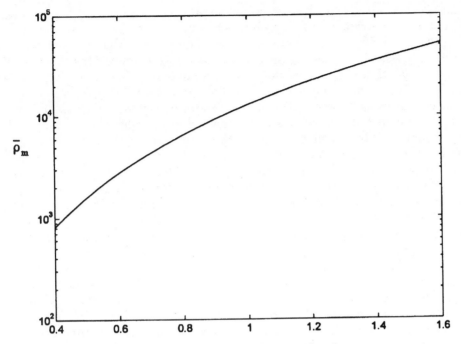

Figure 10.5 Average number of photons per mode versus wavelength.

This process opens up the prospect of employing a number of powerful signal processing and detection techniques that are not available to direct detection even when direct detection is preceded by an optical amplifier (OA). One reason for this is that many of the most powerful SNR improvement techniques employ phase and/or frequency modulation of the optical carrier. We note here that optical carrier phase modulation cannot be processed at all by direct detection, while direct-detection systems employing crude FM detection using optical filters before the photodiode provide insufficient frequency resolution. This aspect of the comparison between direct and coherent detection was not considered at all by either side in the first two papers discussed in Chap. 9 (Refs. 1 and 2 in Chap. 9). We now present a detection system that utilizes the classic nature of the optical heterodyne detection process to enhance receiver sensitivity.

10.3 A Coherent Detection System for Enhanced Sensitivity

10.3.1 Introduction

The material in this section is based primarily on Ref. 7:

G. Lachs, S. M. Zaidi, and A. K. Singh, "Sensitivity enhancement using coherent heterodyne detection," *IEEE J. Lightwave Technol.* **12,** 1036–1041 (1994).

Recent device developments such as optical amplifiers and electronically tunable optical filters constructed on microchips have accelerated the deployment of direct-detection (DD) fiber-optic systems. By placing an optical amplifier directly before a DD system, the rationale goes, the same effect is achieved as the conversion gain present in a CD system. This viewpoint assumes that there is, at best, a 3- to 6-dB advantage for CD (using PSK) over DD (which cannot detect the phase of an optical carrier) [8,9].

The comparisons between CD and DD that were discussed in Chap. 9, however, did not include the effects of the Bohr correspondence principle. One consequence of this principle is that the coherently detected signal will behave as a classic signal disturbed by additive gaussian noise regardless of how weak the signal emerging from the fiber-optic channel at the receiver is. This is because the total photon rate incident on the photodiode is very large because of the strength of the local oscillator. As a result of the interaction generated by the superposition of a relatively strong local laser and a weak received signal, the weak signal affects a much larger number of photons than is represented by the photon rate in the received signal by itself. *This permits application of powerful processing techniques that have been developed for such signals. On the other hand, when a very weak optical signal (a few photons per pulse) is processed by optical amplification followed by DD, the detection process cannot be treated classically.* The lower limits on suitable received signal strength for DD will be determined by quantum-mechanical considerations. The difference between the two detection schemes arises from the fact that in CD the electromagnetic fields are added linearly before detection because Maxwell's equations are themselves linear in differential equations in free space. This superposition is almost a classic process in the quantized field versions except for negligible "zero-point field" contributions, which appear only once in the formal quantum-mechanical description of the superposition process[5] [10]. Optical amplification, on the other hand, is a quantum-mechanical process involving stimulated emission driven by the intensity (i.e., $\alpha |E|^2$).

The often quoted minimum requirement of 10 photons per bit (ppb) to achieve an error rate of $P_e = 10^{-8}$ for intensity-modulated direct detection (IMDD) is based on an ideal photon-counting detection [11]. This includes the unrealistic assumption that no errors occur for OFF-state transmissions. The corresponding theoretical lower limits for the photons per bit parameter of detection systems employing integrated current as the primary detection variable lie in the 18 to 36 range, depending on digital modulation scheme [12]. A number of authors also state that these results could be improved by using coding, but as yet no system is even close to achieving the preceding

---

[5]This comment refers to the fact that variance of an electromagnetic field component contains a zero-point field contribution for each separate field being superposed. However, as discussed in Ref. 10, the variance of the corresponding component of the sum field contains only one zero-point field contribution. This is another example of how quantum statistics varies from classic statistics because the latter requires that the variance of the sum equal the sum of the variances.

theoretical values to test this hypothesis. In this section we shall describe a particular CD detection system that computer simulation studies have demonstrated to be capable of ~6.0-ppb receiver sensitivity. This result was obtained employing well-known analog FM detection techniques. Although it represents our best results to date using computer simulations, it does not connote a lower bound on the photons per bit. Also note here that the results quoted in this section are based entirely on computer simulations and as yet no such system has actually been constructed.

10.3.2 Some basics of coherent detection

In this subsection we elaborate on the concept that in coherent detection (CD) systems the heterodyned signal may be treated as a classic (non-quantum-mechanical) signal no matter how weak the incoming signal. Coherent detection of an optical signal utilizes the superposition of the received optical signal with a local laser to generate a signal at the difference frequency (usually in the microwave range). The probability that an electron from a particular atom in the detector is excited to the conduction state in the time interval $0-T$ is given by [13]

$$p_a = K_1 \int_0^T |E_s(t) + E_L(t)|^2 \, dt \qquad (10.4a)$$

where the received signal $E_s(t)$ and the local laser signal $E_L(t)$ are given by

$$E_s(t) = A \cos(\omega_s t + \phi_s) \qquad (10.4b)$$

In Eq. (10.4a) K_1 is a wavelength-dependent constant which depends on the species of atom in

$$E_L(t) = B \cos(\omega_L t) \qquad (10.4c)$$

the detector and the system of units being employed. The detector output current $i_d(t)$ is then given by

$$i_d(t) = R[P_s + P_{Lo} + 2\sqrt{P_s P_{Lo}} \cos(\omega_i t + \phi_s)] \qquad (10.4d)$$

where R = responsivity of the photodiode receiver
  P_s = received optical carrier power
  P_{Lo} = local laser power
  $\omega_i = |\omega_s - \omega_L|$.

All sum frequency terms are omitted since they are not generated to begin with because of the normal ordered structure of the creation and annihilation operators in photon detection [14].

The $P_s + P_{Lo}$ portion of the bracketed term in Eq. (10.4d) generates a DC bias current in the photodiode output, and this bias current, in turn, generates shot noise which can be modeled as white gaussian noise. The cosine term at the difference frequency ω_i represents the signal component of the output. In a typical CD system $P_{Lo} \gg P_s$, and one can assume that shot noise (generated primarily by P_{Lo}) overwhelms the thermal noise generated in the post-optical-detection portion of the receiver. The signal-to-noise ratio (SNR) at this stage is then given by

$$(\text{SNR})_R = \frac{2P_s P_{Lo}}{P_s + P_{Lo}} \approx 2P_s \qquad (10.4e)$$

The interference between the superposed field constituents enables the weak received signal to affect a much larger number of photons than are contained in the received signal by itself. This may be observed by noting that while the number of received signal photons detected during the interval 0–T is proportional to P_s, the total number of photons that contribute to the IF signal is proportional to the product $P_s P_{Lo}$. In the typical CD system this is much larger than P_s, and it is the source of the conversion gain commonly associated with CD.

When the number of photons represented by $P_s P_{Lo}$ is sufficiently large, the Bohr correspondence principle [3] applies and the optical heterodyne output can be treated as a classic signal perturbed by *white gaussian noise* whose spectral density is proportional to P_{Lo}. This property gives CD the potential to utilize frequency and phase modulation of lasers along with the powerful SNR-improving detection procedures that are available with these modulation schemes. We next describe how FM carrier modulation combined with CD can be utilized to enhance receiver sensitivity to a degree not achievable by DD.

We next consider the coherent detection of frequency-modulated optical carriers. For rf frequencies, it is well known that frequency modulation (FM) is one of the most effective systems for extracting analog signals disturbed by white gaussian noise. These systems trade bandwidth for SNR improvement. The usual formula for the SNR after FM detection is given by [15]

$$(\text{SNR})_D = 3\left(\frac{f_\Delta}{W}\right)^2 \left(\frac{S_R}{\eta W}\right) S_x \qquad (10.5)$$

where W = bandwidth of modulating signal, Hz
  f_Δ = maximum instantaneous frequency deviation from the carrier, Hz
  η = white gaussian noise power density, W/Hz
  S_R = average received signal power at the input to the FM receiver, W
  S_x = average signal power in the modulation signal

Since the FM detection is performed on the microwave-range IF frequency, Eq. (10.5) also holds for coherent detection of FM-modulated optical carriers. This equation indicates a quadratic improvement in SNR as f_Δ is increased, but the increase in f_Δ will be accompanied by an increase in transmission bandwidth. The receiver front-end bandwidth is correspondingly increased, with a proportionate increase in the total noise entering the FM detection system. When the SNR at the input to the FM detector is too small (typically < 10), threshold effects are encountered which seriously deteriorate system performance and render Eq. (10.5) invalid. Furthermore, the performance deteriorates sharply as the bandwidth increases beyond the threshold point. Threshold effects place a lower limit on the usable received signal strength and hence a lower bound on the number of received photons per second that still satisfy the threshold condition, which eventually sets a minimum on the photons/bit parameter. Not all modulation schemes experience threshold effects, and other systems may achieve better results than we describe here.

One technique that has been shown to be effective for combating the threshold effect is *FM feedback* (FMFB) [16–18], which is essentially an analog detection technique; however, digital information can be incorporated into the system by employing subcarriers. It is specifically the classic nature of CD reception of optical frequency signals that makes it possible to utilize FMFB to enhance receiver sensitivity. While a frequency-modulated optical carrier using very large values of f_Δ could be demodulated in principle using optical filters and DD, FMFB cannot be employed with DD of weak optical inputs. It simply would not operate because of the impulsive nature of the detected current at low photon rates. The presence of an OA preceding DD does not improve matters in this regard; if anything, it worsens FMFB, due to the noise added by the OA process. We have incorporated FMFB into our detection system, and it is responsible for a large portion of the SNR improvement that we have observed. FMFB is discussed further in Sec. 10.3.4.

10.3.3 Modulation scheme

A block diagram of the overall communication system that we investigated using computer simulation is shown in Fig. 10.6. Digital information is embedded in four QPSK-modulated subcarriers which are summed in a frequency-division structure that is then used to frequency-modulate an optical carrier. The subcarrier frequencies are harmonics of $1/T$ and thus are orthogonal to each other over a common T-second pulse interval. This type of signal structure is known as a *biorthogonal* digital modulation scheme [19], which incorporates 8 bits per pulse. Clearly many other schemes are possible.

To clarify the specification of the maximum frequency deviation f_Δ, we normalize the maximum signal input to the FM modulator. This is achieved by dividing the combination of the subcarriers by the sum of four subcarrier amplitudes prior to modulation. The carrier frequency of the modulator output was selected to be 40 GHz to match the receiver portion of the simulated system, while the amplitude of the transmitted signal was also set to unity. The 40-GHz value was selected because photodiodes are now commercially

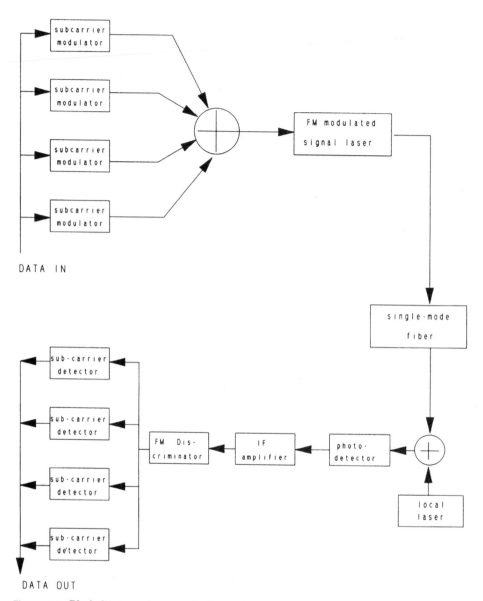

Figure 10.6 Block diagram of communication system.

available in this frequency range and because this particular system works better with high IF frequencies. Thus, for example, if a 4-GHz IF frequency is used, the transmission rate would turn out to be smaller by a factor of 10 in order to maintain the same f_{if}/B_T ratio as in the 40-GHz case. When f_{if}/B_T gets too small, the subcarriers become distorted and are no longer orthogonal.

10.3.4 Detection system

The post-optical-detection portion of the detection system for the QPSK-sub-carrier FM-carrier modulation scheme is composed of an FM receiver portion followed by a set of QPSK detectors.

The FM detection system section consists of a hard limiter, delay-and-multiply discriminator, filters, and a frequency compression feedback circuit as displayed in Fig. 10.7.

It can be seen from Fig. 10.7 that the FMFB detection system employs a second mixing stage where a voltage-controlled oscillator (VCO) is mixed with the IF-stage output. Both of these frequencies will be in the microwave frequency range, and the VCO output will be controlled by the discriminator output. By tracking the instantaneous frequency of the output from the first IF stage, we are able to use a much narrower predetection bandwidth for the feedback filter than the gross receiver bandwidth represented by B_{in}. This results in a narrower effective bandwidth for the overall FMFB system, and it will present the discriminator with a larger SNR than would be the case without FMFB. This, in turn, means that we can accept a smaller received power and still remain above the FM detection threshold. Finally the reduced input power requirement translates into a better (lower) photons/bit (ppb) receiver sensitivity.

The delay-and-multiply discriminator (DAMD) has been described in some detail in Chap. 8. In particular, such FM detectors are often employed for detection in ultra-high-bit-rate FSK systems, but in our application here we are using a DAMD primarily for analog detection of an FM-modulated carrier. For the detection of analog signals, the DAMD is composed of a mixer and a delay line whose delay is one-quarter of the period of the intermediate frequency (see Fig. 10.8, which is essentially Fig. 8.15 repeated here for convenience).

A distinct advantage of the discriminator (with its extremely short delay at 40 GHz) shown in Fig. 10.8 is that it makes the detection process virtually impervious to phase noise effects. To begin with, the phase noise term in Eq.

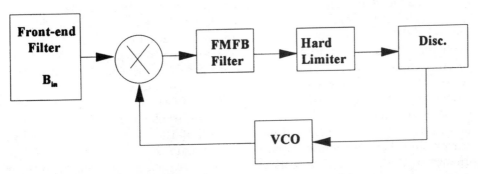

Figure 10.7 FM detector with FMFB.

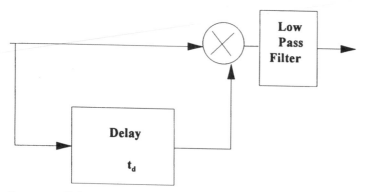

Figure 10.8 Delay-and-multiplier discriminator.

(8.22h) is so small that it does not affect the FMFB process at all. In addition, the front-end bandwidth of the receiver is wide enough so that even before the signal reaches the discriminator, the phase noise will be a slowly varying random constant phase term (see discussion in Sec. 8.2.4) that does not affect the receiver SNR at all. Thus, under normal operating conditions (above threshold), Eq. (10.7) is valid irrespective of the phase noise.

Now, the detector for each QPSK subcarrier channel shown in Fig. 10.6 contains two additional mixers in the detection portion (one for cosine and one for sine) within the boxes labeled "subcarrier detector" followed by an integrate-and-dump block such as shown in Fig. 10.9. Both the Y_i and Y_q outputs can be considered a binary channel. Thus each subcarrier conveys 2 bits per pulse, and the overall collection of four subcarriers contains 8 bits per pulse interval; this subcarrier digital modulation scheme is the biorthogonal communication system mentioned in Sec. 10.3.3.

10.3.5 Computer simulation

Gaussian noise is incorporated into the simulation by adding computer-generated, independent identically distributed (iid), zero-mean, gaussian random variables to each sample in the DSP simulation. This process takes place at the front end of the receiver just prior to the receiver input filter B_{in}. In this arrangement the only parameter that was required to specify the SNR at the receiver input was the variance of the iid gaussian random variables. The resultant signal plus noise is then passed on to the detection system.

The simulation was performed using COMDISCO SPW software. This is a digital signal processing (DSP)-based software in which the sampling rate f_s is one of the principal parameters of the simulation. It must be selected to ensure that there is a sufficient number of samples in a period of highest frequency in the system (f_{if} in this case) and simultaneously yield an integer number of samples delay in the delay-and-multiply discriminator. The choice

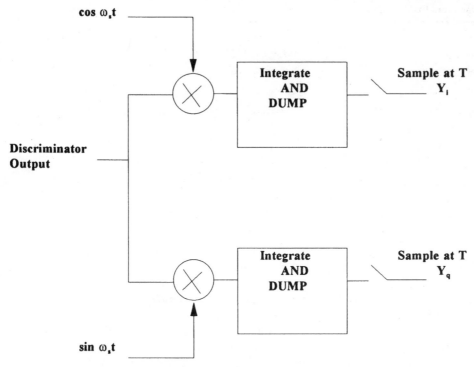

Figure 10.9 Subcarrier detector system.

of f_s = 480 GHz resulted in 12 samples per period at f_{if} and at the same time permits 3-sample delay in the delay-and-multiply discriminator without FMFB and 4-sample delay with FMFB. As shown in Fig. 10.7, the FMFB demodulator contains two bandpass filters (front-end and feedback loop filter), a multiplier (mixer), a hard limiter, and a discriminator in its forward transmission path. The output of the discriminator is delayed using the bulk delay block. This delay is essential for the software implementation of the feedback loop. The delayed output of the discriminator is used to drive a voltage-controlled oscillator (VCO), and the frequency-modulated output is fed back to the multiplier. The front-end filter center frequency is 40 GHz, and its passband bandwidth is 3.9 GHz; the feedback filter center frequency is 30 GHz, and its passband bandwidth is 2 GHz. This accounts for the 4-sample delay for the setup with feedback. The center frequency of the VCO in the feedback path is 10 GHz.

Unfortunately, the high value for f_s (480 GHz) also resulted in 9600 samples per data symbol, which slowed down the simulation process considerably. Typical running times were on the order of 10 h for 5 million samples on a dedicated Sun SPARC-1 workstation. The set of parameters which produced our best results to date are listed in Table 10.1.

TABLE 10.1   Parameters for Best-Case Simulation

Sampling frequency	480 GHz
QPSK subcarriers; $f_{s1}, f_{s2}, f_{s3}, f_{s4}$	0.1, 0.15, 0.2, 0.25 GHz
QPSK subcarrier amplitudes; $a_{s1}, a_{s2}, a_{s3}, a_{s4}$	0.9, 0.95, 1.05, 1.10 V
Pulse duration	20 ns
Maximum frequency deviation, f_{Δ}	1.0 GHz/V
f_{if}	40.0 GHz
Feedback loop filter center frequency	30.0 GHz
B_{in}	4.0 GHz
Noise variance per sample, σ_m^2	27.0

It should be noted that the simulation configuration does not include optical signals and that only the DSP equivalent of rf and microwave frequencies are used. The basic time unit represented 1 ns, so a frequency of 40 Hz represented 40 GHz. Similarly, the subcarrier frequencies of 100, 150, 200, and 250 MHz were scaled to 0.1, 0.15, 0.2, and 0.25 Hz, respectively, while the symbol duration $T = 20$ corresponds to 50 million symbols per second. Thus the data rate in this simulation corresponds to 400 Mbits/s.

We found that another often used FM signal processing technique known as *preemphasis-deemphasis* did not work for the particular digital modulation scheme of the system represented by Fig. 10.6. For example, our simulations showed that preemphasis led to a very large transmission bandwidth, which, in turn, resulted in serious distortion at the receiver. Some system performance enhancement was achieved, however, by small variations of the subcarrier amplitudes from their original uniform settings. The largest amplitude was applied to the highest-frequency subcarrier because this subchannel suffered the greatest distortion.

The error rates were then estimated by applying gaussian statistics to the outputs of the integrate-and-dump blocks displayed in Fig. 10.9. We note here that as long as the input signal is above threshold, the discriminator output (even in the presence of a limiter) is well approximated by a superposition of signal with gaussian noise [20–22]. The length of the simulation was increased until we noticed that the error rate no longer changed significantly as the total number of samples increased. Then we applied several different seeds to both our random noise generator and bit pattern generator to verify our results.

10.3.6 Calculation of photons per bit

To determine the number of photons per bit (ppb) represented by our simulation parameters, we first find the number of photons per pulse r_s generated by the received signal alone. We do this by comparing the signal-to-noise ratio at the output of B_{in} obtained in terms of system parameters with the same quantity obtained in terms of simulation parameters.

We begin by noting that the shot noise power density (η) at the input to the receiver generated by the local laser is given by

$$\eta = 2q\bar{I} \approx 2q^2K_dB^2 = 2q^2\frac{r_L}{T} \tag{10.6a}$$

where r_L is the photon-per-pulse rate due to the local laser alone, q is the charge of an electron, and $\bar{I}$ is the average photodiode current. The total noise power entering the receiver (N_{in}) is given by

$$N_{in} = \eta B_{in} = 2q^2\frac{r_L}{T}B_{in} \tag{10.6b}$$

The signal portion of the current generated by the heterodyne component is proportional to $2AB$. This leads to

$$S_{in} = 2q^2K_d^2A^2B^2 = 2q^2\frac{r_Lr_s}{T^2} \tag{10.6c}$$

Now, dividing Eq. (10.6c) by Eq. (10.6b), we find that the signal-to-noise ratio passing through B_{in} is

$$(\text{SNR})_{in} = \frac{S_{in}}{N_{in}} = \frac{r_s}{TB_{in}} \tag{10.6d}$$

It should not come as a surprise that this result is independent of r_L.

To compute the $(\text{SNR})_{in}$ in terms of the simulation parameters, we note that the addition of iid gaussian random variables to each sample at the receiver input is equivalent to adding white gaussian noise with the power density

$$\eta = \frac{\sigma_m^2}{f_s/2} \tag{10.7}$$

where σ_m^2 is the variance of the gaussian iid variables added to each sample at the input to the receiver (i.e., before B_{in}). Furthermore, since we have the freedom to choose receiver amplitude A arbitrarily, we set it to unity and adjust σ_m^2 until the error rate criterion is met. Thus we have $S_{in} = \frac{1}{2}$. In terms of the simulation parameters, we get

$$N_{in} = \frac{2\sigma_m^2B_{in}}{f_s} \tag{10.8a}$$

$$(\text{SNR})_{in} = \frac{f_s}{4\sigma_m^2B_{in}} \tag{10.8b}$$

Then, from Eqs. (10.6d) and (10.8b), we obtain the number of photons per symbol:

$$r_s = \frac{f_s T}{4\sigma_m^s} \qquad (10.9)$$

Finally we obtain the number of photons per bit (ppb) by dividing r_s by the number of bits per pulse. As an example, we consider the parameters listed in Table 10.1. Specifically, $f_s = 480$ ns^{-1}, T = 20 ns, and $\sigma_m^2 = 27$. Now, with 8 bits per pulse, this leads to 11.1 photons per bit.

It is interesting to note at this point that receiver sensitivity, including the quantum efficiency of the photodiode which is embedded within K_d, does not appear in the principal result at all, namely, Eq. (10.9). Obviously we require that the detection sensitivity parameter K_d be large enough so that the local-laser-generated shot noise masks the receiver electronic noises: thermal, shot, and dark current noises. We note that the local laser is usually super-posed with the incoming signals by using optical couplers made of the same glass-based optical fibers used to carry the signal. To avoid signal-distorting nonlinear effects, the local laser level must be kept below ~8 mW (depending on core radius), which sets a lower bound on useful K_d. This constraint still leaves plenty of room for a broad range for acceptable basic detector sensitivities where K_d neither affects the overall receiver sensitivity of CD receivers nor results in nonlinear distortion. In direct-detection receivers the effects of the K_d parameter can be negated by employing predetection gain by using either an OA or an APD, but all these systems add excess noise to reception process beyond and above the usual shot noise limitations.

10.3.7 Discussion of results

The primary goal of the particular modulation scheme of this fiber-optic com-munication system is to improve receiver sensitivity. In this case it does so at the expense of bandwidth (a quantity that is abundantly available in optical fibers). Specifically, we utilized 4 GHz of channel bandwidth to transmit 400 Mbit/s. On the other hand, gigabit-per-second-range transmission could be attained with this type of system by employing wavelength-division multiplex-ing (WDM). Alexander et al. [23] attained ~37-ppb sensitivity using 4-FSK. Although M-ary FSK is a form of FM, the signal-to-noise improvement for that system is not determined by Eq. (10.5). In particular, increasing the spacing between reference tones does not improve FSK performance. By using narrow deviation, 4-ary, FSK, Gross et al. [24] obtained a spectral efficiency of 0.53 bit Hz^{-1} s^1. This was, however, accomplished with a sensitivity of ~1600 ppb.

One way to increase the number of photons per pulse is to increase the size of the signaling alphabet. Two general schemes that accomplish this without increasing bandwidth (in this case for each of the subcarriers) are M-PSK and QAM. Two specific examples are QPSK and 8-PSK whose signal spaces are shown in Fig. 10.10; an example of 16-QAM is shown in Fig. 10.11. The detec-tion of these signals is accomplished by appropriate processing of the Y_i and Y_q outputs of the detection system depicted in Fig. 10.11.

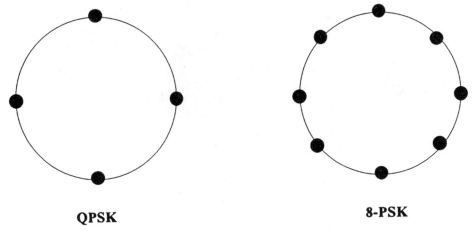

QPSK   **8-PSK**

Figure 10.10 Some *M*-PSK signal spaces.

Note that each QPSK pulse carries 2 bits while each 8-PSK pulse carries 3 bits of information. In the case of 8-PSK, the detection must be performed on the whole pulse as opposed to QPSK which can be treated as individual binary channels. The 16-QAM signaling constellation represents 4 bits per pulse. We note here that the combined data rates of two QPSK channels is equivalent to the data rate of one 16-QAM channel. Nevertheless, it turns out that the two QPSK subcarriers work better for this system.

Each of these coding schemes spread the signal points through a plane whose axes at the receiving end are the Y_i and Y_q of Fig. 10.7. The signal points displayed in these figures represent output coordinates when the receiver is perfectly synchronized and in the absence of noise. In general, the received point does not fall directly on an ideal signal point and most detection schemes for uniform a priori input probabilities perform detection by selecting the signal point which is closest to the Y_i, Y_j coordinates of the received point. It can be shown that for large SNR, the error rate is, to a good approximation, dominated by the error rate for a decision involving the signals associated with the minimum distance between ideal signal points. In other words, it is desirable to design the signal structure so as to maximize the minimum distance between signals subject to the condition of a prespecified average signal energy per pulse.

Now, the amplitude of the subcarrier for any of these signals is proportional to the euclidian distance of the signal point from the origin. The examples in Figs. 10.10 and 10.11 have very regular structures, and we note that for the *M*-PSK signal structure each signal has the same amplitude and hence energy. This is not the case, however, for 16-QAM, where there are actually three different amplitudes involved, namely, one amplitude for each group consisting of the inner four points, the four corners, and the remaining eight points.

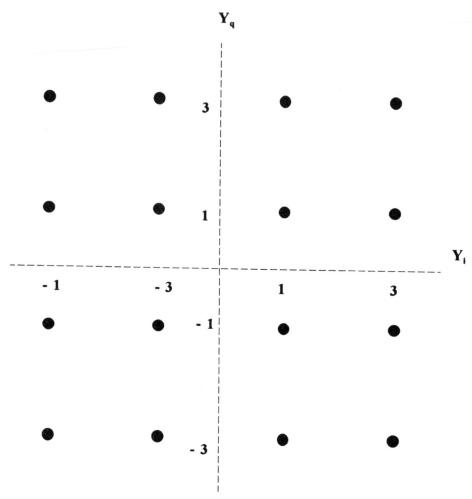

Figure 10.11 16-QAM signal space.

In our simulations we found that for the particular system considered in this study two QPSK channels operated better than one 16-QAM channel. The primary reason for this, we believe, is that the inner radius members will generate smaller instantaneous frequency variations and hence poorer SNR performance from the FM detector. If the sum of a set of QAM carriers is used to drive the FM laser modulator, one must design both the transmitter and receiver to handle the worst bandwidth case where all subcarrier channels have maximum amplitude and yet at the same time have a system that will satisfy the $P_e \leq 10^{-9}$ error criterion. This simply did not work well in the particular system of Fig. 10.4 with one 16-QAM in place of two QPSK subcarrier channels, but this does not rule out the possible use of QAM signal structures in other CD system arrangements.

It can be readily observed from Eq. (10.9) that the number of photons per bit can be reduced provided one can find a way to increase σ_m^2 without deteriorating the overall system performance. However, in this particular system we find that increasing σ_m^2 beyond a certain value results in total chaos because of the disastrous FM threshold effect. While FMFB helps a great deal in mitigating this severe degradation, and thus considerably enhances the receiver sensitivity, there are limits to what it can accomplish. One way of improving the performance of the FMFB system is to increase f_Δ, but this requires a corresponding increase of the front-end bandwidth B_{in}. The bandwidth B_{in} alone does not determine system performance since the FMFB filter reduces the effective noise bandwidth of the overall detection system. However, increasing B_{in} causes a more subtle effect on system performance because the IF carrier frequency-to-bandwidth ratio (IF/B_{in}) is simultaneously reduced in this process. The usual FM performance as represented by Eq. (10.7) is predicated on $f_{if}/B_{in} \gg 1$. The results reported here for the simulation have been achieved with $f_{if}/B_{in} = 10$. We have found that when the f_{if}/B_{in} ratio becomes small, considerable distortion results. Some symptoms of this distortion are

1. The subcarriers were no longer orthogonal; hence there was considerable crosstalk between binary components of the biorthogonal signal structure.

2. We also found that the FMFB system would not operate properly.

As an example of these problems, we found that we were able to achieve a 2-Gbit/s rate using a single QPSK-modulated subcarrier with the parameters f_s = 320 GHz, T = 1 ns, 2 bits/s, and B_{in} = 12 GHz to attain an error rate better than 10^{-9}. This resulted in $f_{if}/B_{in} = 3.333$. We were, however, unable to employ FMFB under these conditions, and we found $\sigma_m^2 = 1.5$ to be the best we could do. This results in 26.7 ppb. When we attempted to use multiple subcarriers we also found that the distortion caused by this small value of f_{if}/B_{in} entirely destroyed the orthogonality of the subcarriers.

One conceivable option to improve the performance of the existing system would be to increase the IF frequency in order to maintain the desired IF/B_{in} = 10. Unfortunately, the choice of IF is restricted by the speed of present-day photodetector electronics. The value we used, 40 GHz, represents the state of the art. However, for academic purposes and to reinforce our theory that the performance of the proposed system is not quantum-limited, we conducted some preliminary experimentation with IF = 80 GHz. This leads to 9.8 ppb, which is not considered to be a very reliable number because increasing the IF mandates a corresponding increase in the sampling frequency. Therefore, in order to obtain the desired number of data points, excessively long computer runs are required. Nevertheless, it does indicate the potential of surpassing the quantum barrier. This study, however, did provide us with clues on how to proceed to achieve a ppb rate of ~6. The results of this study are described in Sec. 10.4.

10.3.8 Conclusion

The material contained in this section is based on the results of computer simulation techniques used to study the proposed subcarrier modulated system. A receiver sensitivity of 11.1 ppb at 400 Mbits/s was obtained. The study indicated that FM carrier modulation combined with coherent detection can indeed enhance the receiver sensitivity and make it comparable to the so-called quantum limit. It is important to note that our study indicates that the 11.1-ppb figure does not represent the peak of performance for coherent detection systems. The major contribution of this study is that it points to a direction that holds substantial promise for enhancing the ultimate receiver sensitivity of CD systems (lower number of photons per bit) and hence render their performance devoid of quantum-mechanical considerations.

The system performance can be improved by employing coding. In particular, by simply incorporating 4–8 trellis coding, the error rate can be reduced to $\sim 10^{-19}$ (obtained by extrapolating known upper-bound curves for this code [25]). It is not possible, however, to take full advantage of this by simply reducing the input signal (hence the number of photons per bit) until P_e $\sim 10^{-9}$ is achieved, because then the threshold condition will be violated. Once the signal strength falls below threshold, the statistics of the integrate-and-dump outputs will no longer be gaussian.

The key to obtaining further improvements in receiver sensitivity is to incorporate more bits per pulse without violating threshold. Conceivably this could be accomplished by employing 8–16 trellis coding for each subcarrier, or perhaps adding more subcarriers. This will require some adjustment of parameters to achieve P_e $\sim 10^{-9}$. We shall describe some results obtained using these techniques in the next section.

10.4 Further Developments on System

10.4.1 Introduction

The system described in Sec. 10.2 has been described in Ref. 7. This section will describe some of the progress made on this system with performance that improves on the system reported in Ref. [7]. Specifically, we shall eventually describe a system that projects a receiver sensitivity of ~ 6 photons per bit. This value is clearly below the intensity-modulated direct-detection (IMDD) limit of 10 photons per bit, which assumes photon counting as the detection mechanism. The 6-ppb result, however, was obtained without either photon-counting detection or optical phase lock. The next generation of the system was basically the same as that described in Sec. 10.2, but it employed larger signal constellations than those described in Sec. 10.2 and used trellis coding to improve receiver sensitivity. Trellis coding is eminently suited to the modulation scheme of this system. Coherent detection (CD) makes high-resolution FM detection possible, while the QPSK subcarrier modulation is perfectly suited to trellis coding. Trellis coding is discussed later in this section, but first let us examine some computer running-time problems encountered during simulation.

10.4.2 Simulation running-time problem

The results noted in the previous section demonstrated that the utilization of complex signaling alphabets was a viable means for improving the ppb sensitivity criterion. This trend was then used as a guide in designing the second generation of that system. However, this added complexity resulted in very long computer runs, which, in turn, limited the number of variations in both the parameters and system structures that we could investigate. The first hurdle we had to overcome in order to analyze the performance of the more complicated system was to find a way to speed up the simulation process. We next demonstrate the computer running-time problem by considering the numerical values for some of the parameters used for the simulation described in the previous section.[6] Specifically, we recall 480 Gsa/s (billion samples per second), pulse duration $T = 0.02$ μs (50 million pulses/s), and $P_e \leq 10^{-9}$. This results in 9600 samples that need to be processed per pulse. A typical computer simulation running time (on a dedicated Sun SPARC-1 workstation) was 13.5 h to process 650 pulses which generated the same number of samples at the output of each of the eight integrate-and-dump blocks, one for each of the 8 bits per pulse. By finding the mean and variance of the 650 samples for each integrate-and-dump output, we estimated the error rate using gaussian statistics. If we were to estimate the error rate by simply counting the errors that occur, we would require at least 10^{10} pulses, which would require approximately 10^{8} h. Even a computer that is 1000 times faster would still require 1,000,000 h per run.

To obtain optimum performance from such a complicated system (see Fig. 10.4), many critical parameters (sometimes in groups) need to be adjusted to obtain a trial-and-error type of optimization. Furthermore, after most parameter changes, it is also necessary to make specialized simulation runs in order to resynchronize the reference signals for the digital detection circuits. This results in a large number of computer runs for each system realization. Thus it is obvious that any reduction in simulation times is greatly appreciated. The systems that we examined for the second-generation system used either 8 or 10 subcarriers as a means to increase the number of bits per pulse. These subcarriers were spaced at 25-MHz intervals instead of 50 MHz as in the original system, and this led to a simulation employing twice as many samples (19,200) per pulse. The combined effect of all these factors would have resulted in an increase in running time by a factor of approximately 4 were it not for the running-time reductions that we achieved.

---

[6]The material presented in this subsection is not required for the remainder of this chapter. It has been included primarily for students interested in computer simulations.

To reduce computer running time, we took advantage of the fact that the frequencies of the signals that precede the FM modulator and those that follow the discriminator are much lower than the IF frequency signals. The programs were run in three stages where we used the simulation equivalent of 12 Gsa/s to generate the input to the FM modulator (stage 1) and to process the discriminator system output (stage 3) while still using the simulation equivalent of 480 Gsa/s for the portion in between these extremes (stage 2). This process has resulted in a six-fold reduction in running time, most of which came from the application of the reduced sampling rate to the third stage. This is not surprising since stage 3 is clearly the most computationally intensive of the three stages. These three stages were run sequentially in separate programs and coupled to one another by using finite impulse responses smoothing filters and repeating the sampling to bridge from one stage to the subsequent stage.

10.4.3 Some comments on trellis coding

In general, error control coding trades off data rate for improved system performance in terms of either error detection and/or error correction capability. In a typical block coding setup, for example, the signaling alphabet may have only k information bits in a transmitted block containing n bits. The remaining $n - k$ bits are determined by the k information bits in a way that makes error correction possible. Since this set of dependent bits carries no new information, they are known as *redundant bits*. Thus, in a communication channel that can handle a fixed number of bits per second, the information data rate is reduced to $k/n \times 100$ percent of the transmitted bit rate. An alternative way to implement coding is to increase the transmitted bit rate while maintaining the same information rate. In this case the performance improvement is achieved at the expense of increased channel bandwidth.

Now, one can view each of the 2^n possible bit patterns of a block of n bits as a point in an n-dimensional space. Since only 2^k of these points represent signals, there will be greater spacing between these signal points than there would be if only $n - k$ dimensions were available to spread these points in. The process of error correction in these block coding cases is more complicated than the simple distance concepts discussed in connection with Figs. 10.10 and 10.11 because each bit is detected separately followed by decoding to remove the errors. The binary detection of each bit in the block results in one point in n-dimensional space representing one of 2^n possible binary sequences. The decoding process selects the ideal signal point (one of the 2^k possibilities) that differs from the detected sequence in the least number of positions. For example, if all the 2^k ideal signal points (n-bit sequences) differed from each other in at least $2\ell + 1$ binary positions and no more than ℓ errors occurred during transmission, then the second decoder would select a completely correct sequence for the final output. Because of the exponential manner in which the number of points increases with the number of bits, it turns out that this coding process is very efficient and the price paid in terms of bandwidth or data rate are relatively small. However, the decoding process for higher-level error correction is, in practice, far more complicated than indicated by the preceding simplified description.

A variation of the error correction coding procedure known as *convolutional coding* has become very popular in recent years. In this system each binary output from the transmitter is a linear combination of the present bit and a number of preceding bits. The decoding algorithm makes each decision on the basis of a sequence of inputs, which usually is simpler to implement than block-coding-based error correction detection. One particularly useful decoding scheme, known as the *Viterbi algorithm* [26], employs minimum distance decision within the decoding process for convolutional coded signals. Convolutional coding employing the Viterbi algorithm for detection plays an important role in trellis code modulation, which we discuss next.

In 1982 Ungerboeck [27] introduced an ingenious coding scheme called *trellis code modulation* (TCM), which blended convolutional coding directly with digital modulation concepts. His system improved the error rate performance of a digital system without increasing signal bandwidth. This system employed higher signaling alphabets. The signal constellation for a 4–8 TCM system is shown in Fig. 10.12.

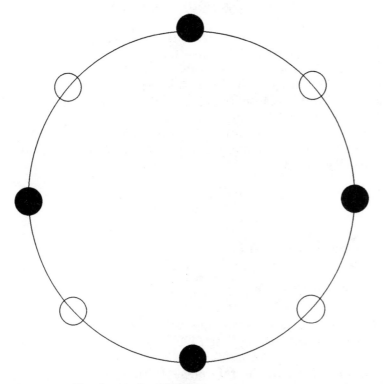

Figure 10.12   Signal space for 4–8 TCM.

The signal space depicted in Fig. 10.12 indicates that the actual information represented by the solid dots is an alphabet of size 4 and represents 2 bits of information per pulse. The actual transmitted signal is convolutionally encoded in an eight-fold alphabet as represented by both the solid and open dots shown in Fig. 10.12. It is then decoded using the Viterbi algorithm, and the process results in a gain of 3 dB or better, depending on conditions, in signal-to-noise performance of the receiver. It is important to note that this improvement is achieved without an increase in bandwidth and with only a modest increase in system complexity. The utilization of an eight-fold alphabet to transmit a four-fold information alphabet is the source of the redundancy.

Since we are already using QPSK for our subcarriers, it will be straightforward to adapt trellis coding to this system. We shall also investigate the performance of an even higher-order trellis code in which an eight-fold information alphabet is transmitted using 16-PSK modulation. Before we examine the consequences of trellis coding, we note that published analyses and data relating to the improvements generated by TCM are based on the assumption of white gaussian noise (WGN), while the noise spectrum encountered in our system is quadratic as a result of the FM detection process. Computer simulations were made separately to verify that we did indeed still obtain at a 3-dB advantage for quadratic spectra. Figure 10.13 shows one set of results obtained from these simulations [28].

Figure 10.13 clearly shows that there is at least a 3-dB improvement obtained by using TCM even when the noise has a quadratic spectrum. Similar graphs were obtained for the other subcarrier frequencies. In these simulations it was possible to generate enough sample pulses to estimate error rates in the 10^{-6} vicinity because we didn't need to simulate the IF part of the system. This, in turn, permitted us to use fewer samples per pulse and process more pulses than in the simulations for the whole system. The quadratic spectrum was generated by numerically differentiating simulated white gaussian noise. Recall that if

$$f(t) \longleftrightarrow] F(j\omega)$$

then

$$\frac{df(t)}{dt} \longleftrightarrow j\omega \, F(j\omega)$$

Hence the output spectrum is quadratic.

10.4.4 Sensitivity improvements obtained using TCM (~6 ppb)

In this section we shall summarize some ppb rates obtained by employing TCM. The specific results obtained by Singh [29] are summarized in Table 10.2.

In the following discussion all the references to item number refer to entries in Table 10.2. All the results listed in Table 10.2 were obtained by

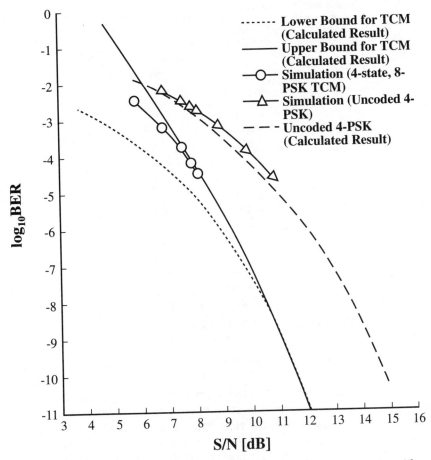

Figure 10.13   Error rate for a 4–8 TCM subcarrier at $F_{sc} = 0.15$ GHz in noise with a quadratic spectrum.

trial-and-error optimization of system parameters which included varying subcarrier amplitude and synchronization. Then the input noise power density parameter σ_m^2 was adjusted to the maximum value whereby all channels simultaneously satisfied the desired error rate criterion. For those cases where TCM was not employed (items 1, 2, and 4 in Table 10.2), this was $P_e < 10^{-9}$. For the remaining cases where we assumed that TCM was to be used, we first determined what error rate without the coding would translate into and error rate (for each subcarier) that did satisfy the error rate criterion when TCM was included. Then we adjusted σ_m^2 to the minimum value that allowed all the subcarrier channels to have $P_e < 10^{-9}$.

Item 1 (in Table 10.2) transmits 16 bits per pulse at a rate of 25×10^6 pulses per second, which is equivalent to 400 Mbits/s. This is exactly the same

TABLE 10.2   Photons/bit for Various Systems and Modulation Schemes

Item	Number of subcarriers	Modulation	Trellis coding ?	Photons/bit
1	8	QPSK	No	9.84
2	10	QPSK	No	8.70
3	8	QPSK	Yes, 4–8 state	7.89
4	10	8-PSK	No	7.61
5	10	QPSK	Yes, 4–8 state	6.97
6	10	8-PSK	Yes, 8–16 state	5.81

data rate as in the system described in Sec. 10.2, so the question arises as to why we get ~10 ppb here and ~11 ppb in Sec. 10.2. There are three reasons for this:

1. Since we transmit more subcarriers at half the pulse rate for the system in item 1 (Table 10.2), there are twice as many samples per pulse here. This results in a more accurate simulation in general.

2. The spectrum of the input to the FM modulator is slightly smaller, and this results in a larger FM SNR improvement [see Eq. (10.5)].

3. In Sec. 6.2.4 we noted that orthogonality doesn't guarantee statistical independence of the components. Thus there is a degree of interdependence between channels in the biorthogonal modulation scheme being utilized here. These fluctuations appear in the variance computed for each subchannel and increase the error rate. Now one can show [30] that the correlation between orthogonal components for a Fourier-type expansion of the type used in the biorthogonal system tend to vanish as the time–bandwidth product increases. Thus the subcarrier channels should be more independent of each other in item 1 than in Sec. 10.2 and the corresponding should variances decrease.

Items 2 and 4 also represent improvements in the ppb parameters that were obtained without employing coding; however, in these cases the improvement was achieved by increasing the number of bits per pulse. This increased the data rate to 500 Mbits/s. Improvements were achieved here solely by adding complexity to the signal structure. TCM was employed to improve system performance for items 3 (400 Mbits/s), 5 (500 Mbits/s), and 6 (500 Mbits/s); the best result was ~6 ppb. The overall improvement for item 6 over item 5 was obtained by increasing the coding complexity from trellis 4–8 to trellis 8–16. It is important to note here that this improvement was not obtained by merely changing the code. It was necessary to decrease the input noise parameter σ_m^2 for the item 6 calculations in order to permit detection with more densely packed signal points. As a consequence, the improvement of ppb in item 6 is not the $\frac{2}{3}$ factor you would expect from pure coding. There is no doubt that a diminishing-return type of scenario is beginning to come into play here.

Let us now determine the maximum distance between repeater for a system using the system listed as item 6 in Table 10.2. Let $\lambda_0 = 1.55$ μm. Then

$$\text{Photons s}^{-1}\text{ mW}^{-1} = \frac{10^{-3}\lambda_0}{1.9865\times10^{-25}} = 7.8 \times 10^{15}$$

and

$$\text{Input ppb} = \frac{7.8 \times 10^{15}}{7.5 \times 10^{8}} = 1.04 \times 10^{7}$$

where for item 6 we have 30 bits/pulse $\times$ 25 $\times$ 10^6 pulses/s = 750 Mbits/s. Then the allowable attenuation is

$$10 \log_{10}\left\{\frac{1.04 \times 10^{7}}{6}\right\} = 62.3 \text{ dB}$$

Then assuming the loss factor to be 0.12 dB/km, we obtain 62.3/0.12 = 519 km ~311 mi between repeaters.

The input power was deliberately kept down to 1 mW to allow for several WDM channels without causing any nonlinear crosstalk effects. Suppose, for example, that eight WDM channels are used which corresponds to a 6-Gbit/s rate over the cable. It would seem also that this could be used as a two-way cable with the reverse direction also being at 6 Gbits/s. The received signals at both ends will be too weak to cause nonlinear distortion. Of course, the wavelength and spectra of the reverse-direction WDM channels should not overlap with the wavelengths of the forward-direction channels.

References

1. L. I. Schiff, *Quantum Mechanics,* McGraw-Hill, New York, 1955, Chap. 4, Sec. 13.
2. A. Messiah, *Quantum Mechanics,* Vol. I, Wiley, New York, 1961, Chap. XII.
3. I. S. Gradshteyn and I. M. Ryzhik, *Tables of Integrals, Series, and Products,* Academic Press, New York, 1965, p. 1033.
4. M. Abramowitz and I. A. Stegun. *Handbook of Mathematical Functions with Formulas, Graphs, and Mathematical Tables,* National Bureau of Standards, 1964, Chap. 22.
5. C. W. Helstrom, *Probability and Stochastic Processes for Engineers,* 2d ed., Macmillan, New York, 1991, p. 86.
6. A. Papoulis, *Probability, Random Variables, and Stochastic Processes,* 2d ed., McGraw-Hill, New York, 1984, pp. 99–100.
7. G. Lachs, S. M. Zaidi, and A. K. Singh, "Sensitivity enhancement using coherent heterodyne detection," *IEEE J. Lightwave Technol.* **12,** 1036–1041 (1994).
8. R. E. Wagner and R. A. Linke, "Heterodyne lightwave systems: Moving towards commercial use," *IEEE Mag. Lightwave Commun. Syst.* 28–35 (Nov. 1990).
9. P. E. Green and R. Ramaswami, "Direct detection lightwave systems: Why pay more?" *IEEE Mag. Lightwave Commun. Syst.* 36–49 (Nov. 1990).
10. G. Lachs, "Theoretical aspects of mixtures of thermal and coherent radiation," *Phys. Rev.* **38**(4B), B1012–B1016 (1965).

11. J. Salz, "Coherent lightwave communications," *ATT Tech. J.* **64,** 2153–2209 (1985).
12. B. E. A. Saleh and M. C. Teich, *Fundamentals of Photonics,* Wiley, New York, 1991, p. 912.
13. W. Louisell, *Radiation and Noise in Quantum Electronics,* McGraw-Hill, New York, 1964, Sec. 5.4.
14. R. J. Glauber, "Optical coherence and photon statistics," in C. DeWitt, A. Blandin, and C. Cohen Tannoudi, Eds., *Quantum Optics and Electronics Les Houches 1964,* Gordon & Breach, New York, 1965.
15. A. B. Carlson, *Communication Systems,* McGraw-Hill, New York, 1986, p. 333.
16. L. H. Enloe, "Decreasing the threshold in FM by frequency feedback," Proc. IRE 18–30 (Jan. 1962).
17. J. G. Chaffee, "The application of negative feedback to frequency-modulation systems," *Bell Syst. Tech. J.* **18,** 404–437 (1939).
18. Elie J. Baghdady, "The theory of FM demodulation with frequency-compressive feedback," *Proc. IRE* 226–245 (Sept. 1962).
19. J. M. Wozencraft and I. M. Jacobs, *Principles of Communication Engineering,* Wiley, New York, 1965, pp. 261–263.
20. P. F. Panter, *Modulation, Noise, and Spectral Analysis,* McGraw-Hill, New York, 1965, Chap. 14, p. 428.
21. R. F. Pawula, "On the theory of error rates for narrow-band digital FM," *IEEE Trans. Commun.* **Com-29,** 1634–1643 (Nov. 1981).
22. J. G. Proakis and M. Salehi, *Communication Systems Engineering,* Prentice-Hall, Englewood Cliffs, N.J., 1994, p. 414.
23. S. B. Alexander et al., "4-ary FSK coherent optical communication system," *Electron. Lett.* **26**(17), 1346–1348 (Aug. 16, 1990).
24. R. Gross, B. Enning, and R. Olshansky, "Narrow deviation 4-ary FSK heterodyne video system," *Proc. IOOC/ECOC 1991,* Paris, paper WeC8-3, pp. 565–568.
25. E. Biglieri, D. Divsalar, P. J. McLane, and M. K. Simon, *Introduction to Trellis-Coded Modulation with Applications,* Macmillan, New York, 1991, Sec. 4.3, Fig. 1 (block diagram of system, Sec. 4.3).
26. A. J. Viterbi and K. K. Omura, *Principles of Digital Communication and Coding,* McGraw-Hill, New York, 1979.
27. G. Ungerboeck, "Channel coding with multilevel/phase signal," *IEEE Trans. Inform. Theory* **IT-28** (Jan. 1982).
28. H. Kim, "Subcarrier modulated optical coherent communications using QPSK and trellis-coded modulation schemes," Ph.D. thesis, University of South Florida, 1994, unpublished.
29. A. K. Singh, "Performance analysis of a coherent communication system employing quadrature PSK and FM feedback using multirate simulation," Ph.D. thesis, University of South Florida, 1994, unpublished.
30. A. Papoulis, *Probability, Random Variables, and Stochastic Processes,* McGraw-Hill, New York, 1965, Chap. 13.

Index

ABOUT THE AUTHOR

Gerard Lachs, Ph.D., is Professor of Electrical Engineering at the University of South Florida in Tampa. His research interests include communications theory, fiber optic digital communications, signal detection and processing in human sensory perception, quantum electronics, and laser applications.